Applied Micromechanics of Complex Microstructures

Applied Micromechanics of Complex Microstructures

Computational Modeling and Numerical Characterization

Majid Baniassadi
School of Mechanical Engineering, College of Engineering, University of Tehran, Tehran, Iran

Mostafa Baghani
School of Mechanical Engineering, College of Engineering, University of Tehran, Tehran, Iran

Yves Remond
ICube/CNRS, University of Strasbourg, Strasbourg, France

Elsevier
Radarweg 29, PO Box 211, 1000 AE Amsterdam, Netherlands
The Boulevard, Langford Lane, Kidlington, Oxford OX5 1GB, United Kingdom
50 Hampshire Street, 5th Floor, Cambridge, MA 02139, United States

Copyright © 2023 Elsevier Inc. All rights reserved.

No part of this publication may be reproduced or transmitted in any form or by any means, electronic or mechanical, including photocopying, recording, or any information storage and retrieval system, without permission in writing from the publisher. Details on how to seek permission, further information about the Publisher's permissions policies and our arrangements with organizations such as the Copyright Clearance Center and the Copyright Licensing Agency, can be found at our website: www.elsevier.com/permissions.

This book and the individual contributions contained in it are protected under copyright by the Publisher (other than as may be noted herein).

MATLAB® is a trademark of The MathWorks, Inc. and is used with permission. The MathWorks does not warrant the accuracy of the text or exercises in this book. This book's use or discussion of MATLAB® software or related products does not constitute endorsement or sponsorship by The MathWorks of a particular pedagogical approach or particular use of the MATLAB® software.

Notices

Knowledge and best practice in this field are constantly changing. As new research and experience broaden our understanding, changes in research methods, professional practices, or medical treatment may become necessary.

Practitioners and researchers must always rely on their own experience and knowledge in evaluating and using any information, methods, compounds, or experiments described herein. In using such information or methods they should be mindful of their own safety and the safety of others, including parties for whom they have a professional responsibility.

To the fullest extent of the law, neither the Publisher nor the authors, contributors, or editors, assume any liability for any injury and/or damage to persons or property as a matter of products liability, negligence or otherwise, or from any use or operation of any methods, products, instructions, or ideas contained in the material herein.

ISBN: 978-0-443-18991-3

For Information on all Elsevier publications
visit our website at https://www.elsevier.com/books-and-journals

Publisher: Matthew Deans
Acquisitions Editor: Dennis McGonagle
Editorial Project Manager: Mason Malloy
Production Project Manager: Fizza Fathima
Cover Designer: Vicky Pearson Esser

Typeset by MPS Limited, Chennai, India

Dedication

This book is dedicated to our wives,
Maryam, Roshanak, and Marie-Claude,
for their love,
endless support,
and
encouragement.

Contents

About the authors	**xi**
Preface	**xiii**
Acknowledgments	**xv**

1. Introduction to computational modeling of complex microstructures — **1**

1.1 Statistical descriptor	1
1.1.1 Correlation functions	1
1.1.2 Two-point correlation functions	2
1.1.3 N-point correlation functions	4
1.1.4 Other important statistical descriptors	5
1.2 Representative volume element	5
1.3 Numerical reconstruction of heterogeneous materials	7
1.4 Effective properties	12
1.4.1 Calculation of effective properties	13
1.5 Asymptotic homogenization	20
1.6 Hull space and materials design	21
1.7 Conclusion remarks	23
References	23

2. Numerical characterization of micro- and nanocomposites — **29**

2.1 Introduction	29
2.2 Realization of composites with different types of inclusions	32
2.2.1 Composites loaded with ellipsoidal inclusions	32
2.2.2 Composites loaded with helical inclusions	34
2.3 Experimental reconstruction of inclusionary composites	36
2.3.1 Small-angle neutron scattering	36
2.3.2 Reconstruction using SEM-FIB images	44
2.4 Numerical homogenization of thermomechanical properties	47
2.4.1 Finite element homogenization of polymer nanocomposites with the interphase zone	47
2.4.2 Homogenization of nanocomposites with interfacial debonding using FEMs	50
2.4.3 Homogenization of nanocomposites with nonlinear matrix properties using finite element methods	61
2.4.4 Finite element homogenization of coupling thermomechanical properties	72
2.5 Conclusion and remarks	85
References	85

viii   Contents

3. Numerical realization and characterization of random heterogeneous materials   **95**
   3.1 Introduction   95
   3.2 Realization of multiphase random heterogeneous materials using a Monte Carlo approach   97
      3.2.1 Realization of the SOFC anode   101
   3.3 Multiphase reconstruction of heterogeneous materials using a full set of TPCFs   108
      3.3.1 Approximation of TPCFs for three-phase microstructures   109
      3.3.2 Number of independent sets of TPCFs   110
      3.3.3 Approximation formulation   110
      3.3.4 Modification of the phase recovery algorithm for three-phase reconstruction   112
      3.3.5 Isotropic reconstruction   114
      3.3.6 Microstructural properties   118
   3.4 Numerical characterization of thermomechanical properties: finite element method   122
      3.4.1 Results and discussion   125
      3.4.2 Elastic properties   127
      3.4.3 Thermal conductivity and thermal expansion   129
   3.5 FFT approach   131
      3.5.1 Linear elastic properties   132
      3.5.2 Thermal properties   134
   3.6 Conclusion and remarks   137
   References   138

4. Numerical characterization of tissues   **145**
   4.1 Introduction   145
   4.2 Finite element investigation of effective mechanical behavior of the cerebral cortex tissue using 3D homogenized representative volume elements   146
      4.2.1 Viscoelastic homogenization of the brain tissue   149
      4.2.2 Homogenization of the heterogeneous brain tissue under quasistatic loading: a viscohyperelastic model of a 3D representative volume element   164
   4.3 Characterization of the liver tissue   181
      4.3.1 Materials and method   183
      4.3.2 Homogenization with the combined FE−optimization method   184
      4.3.3 A heterogeneous model including vessels and the surrounding tissue   186
      4.3.4 Homogenized model   188
      4.3.5 Generalization of limited homogenized samples with artificial neural networks   189
      4.3.6 Results and discussion   189

4.4	Statistical reconstruction and mechanical characterization of the bone microstructure	194
	4.4.1 Specimen preparation	196
	4.4.2 Statistical reconstruction of the bone microstructure	196
	4.4.3 Finite element homogenization	199
	4.4.4 Results and discussion	201
4.5	Conclusion and remarks	210
	References	210

5. Mechanical characterization of Voronoi-based microstructures   **221**

5.1	Introduction	221
5.2	Computational elucidation of isotropic and anisotropic microstructures with Voronoi tessellation	223
	5.2.1 Homogenization and elastic percolation of composite structures with regular tessellations of the microstructure	224
	5.2.2 Homogenization and elastic percolation threshold in isotropic and anisotropic microstructures with random Voronoi tessellation	232
5.3	Conclusion remarks	249
	References	249

6. Numerical modeling of degraded microstructures   **253**

6.1	Introduction	253
	6.1.1 Degradation of PLLA/magnesium composite	253
	6.1.2 Bone degradation	255
6.2	Numerical modeling of the degradation process of PLA/Mg composites	257
	6.2.1 Sample preparation	258
	6.2.2 Statistical reconstruction of samples	259
	6.2.3 Modeling corrosion	260
	6.2.4 Results	267
6.3	A numerical model of bone density loss in osteoporosis	274
	6.3.1 Model objectives	275
	6.3.2 3D reconstruction of the bone microstructure	275
	6.3.3 Determination of the bone quality in microgravity	276
	6.3.4 Theoretical model	277
	6.3.5 Optimization algorithm	279
	6.3.6 Results and discussion	281
6.4	Conclusion remarks	290
	References	291

7. Microstructure hull and design   **299**

7.1	Introduction	299
7.2	Practical approach to estimate microstructure hull and closures	306
	7.2.1 Sample 1 (hull space for SOFC microstructure)	306

	7.2.2	Sample 2 (hull space for infiltrated SOFC microstructures)	318
7.3	Practical approach for materials design of random heterogeneous materials		332
	7.3.1	Interpolation of random heterogeneous microstructures	332
	7.3.2	A framework for optimal microstructural design of random heterogeneous materials	344
7.4	Practical approach for materials design of periodic heterogeneous materials		363
	7.4.1	3D-PUC design using density-based topology optimization	363
	7.4.2	3D-PUC design using topology optimization and innovative building block	379
	7.4.3	Directional elastic modulus of the TPMS structures and a novel hybridization method to control anisotropy	391
7.5	Conclusion and remarks		410
References			410

Index **421**

About the authors

Dr. Majid Baniassadi is an associate professor at the School of Mechanical Engineering, University of Tehran, Iran. He holds a PhD in mechanics of materials from the University of Strasbourg (2011). He received his Master's degree from the University of Tehran (2007) and his undergraduate degree from Isfahan University of Technology (2004) in mechanical engineering. His research interests include multiscale analysis and micromechanics of heterogeneous materials, numerical methods in engineering, and electron microscopy image processing for microstructure identification. Dr. Baniassadi is also collaborating with ICube Laboratory in Strasbourg with activities in engineering science, computer science, and imaging. Thus far, he has published more than 150 scientific journal papers and is the reviewer of more than 10 international scientific journals, and he is often contacted for peer-reviewing submitted papers.

Dr. Mostafa Baghani is an associate professor at the School of Mechanical Engineering, College of Engineering, University of Tehran, Iran, since 2012. He has become the director of Smart Materials and Structures Laboratory at University of Tehran in 2016. In 2018 he has won the award for the best young researcher in the University of Tehran among more than 300 professors under 40. His research activities are focused on synthesis, design, and constitutive modeling of smart materials in the framework of continuum mechanics. He has also worked on computational modeling of the proposed constitutive models often developed via the nonlinear finite element method. He has published more than 170 journal papers and reviewed more than 20 international journals.

Prof. Yves Remond is currently a distinguished professor (exceptional class) at the University of Strasbourg in France. He is working at ICube—The Engineering Science, Computer Science, and Imaging Laboratory—which belongs both to the University of Strasbourg and to the CNRS. His teaching activity is conducted at the European Engineering School of Chemistry, Polymers and Materials Science (ECPM) in the field of continuum mechanics, mechanics of polymers, composite materials, and mechanobiology. He graduated in mechanics from Ecole Normale Supérieure of Cachan (now Paris-Saclay) and received his PhD degree from the University Paris VI in 1984 (Pierre et Marie Curie). Since 2012 he held a position of Scientific Deputy Director at CNRS Headquarters in Paris—INSIS, Institute for Engineering and Systems Sciences. He was also distinguished as an officer in the order of Academic Palms. He is a member of the International Research Center for

Mathematics and Mechanics of Complex Systems at the Universita dell'Aquila (Italy) and was the President of the French Association for Composite Materials (AMAC). He advised about 30 PhD and Habilitation students and published about 150 scientific papers in the field of mechanics of composite materials, polymers, and bioengineering.

Preface

The current book's primary targets are micromechanics and computational aspects of modeling complex microstructures and numerical characterization of heterogeneous materials. In this area's previously published books and numerical and theoretical approaches are devoted to only classical samples such as a periodic microstructure and composite composed of inclusion with a simple geometry. In most of the cases, the analysis has been performed on the materials with a linear constitutive law without any struggle for modeling complex geometry with nonlinear materials.

In this book, the main concepts of micromechanics and continuum modeling of different complex microstructures are explained. Various complex microstructures are chosen in our books, such as nanocomposites, multiphase composites, biomaterials, and biological materials. Detailed calculation of effective mechanical and thermal properties allows the audience to understand step-by-step modeling and homogenization of complex microstructures. Finally, a complementary chapter is added on microstructure hull and materials design to improve the knowledge of the reader. In this book, modeling of complex samples such as neural tissues, bone microstructures, liver tissues, and Voronoi structures, with nonlinear properties of the component, is explained and discussed.

The major features of the book that make it unique are listed as follows:

- reconstruction of the complex microstructure using experimental data,
- numerical modeling of microstructure with a complex geometry,
- nonlinear homogenization of microstructure using the finite element method,
- microstructure and property hull and closures with applied samples,
- materials design procedure for periodic and nonperiodic microstructures, and
- numerical modeling of the degradation process for biomaterials and biological materials.

Finally, we should mention that this book has been reproduced based on the research of the authors, which were published in prestigious journals in the past five years.

Majid Baniassadi,
Mostafa Baghani,
Yves Remond

Acknowledgments

We would like to thank our collaborators, Said Ahzi, Rene Muller, Hamid Garmestani, Daniel George, Masoud Safdari, Azadeh Sheidaei, Kui Wang, Mehran Tehrani, Karen Abrinia, Ali Hasanabadi, Akbar Ghazavizadeh, Alireza Moshki, David Ruch, MA Khaleel, Bohayra Mortazavi, Hoda Amani Hamedani, Ghader Faraji, Dongsheng Li, Xin Sun, Mostafa Mahdavi, Ensieh Yousefi, Morad Karimpour, Morteza Kazempour, Jalil Jamali, Ali Asghar Ataei, Amirhossein Bagherian, Poorya Chavoshnejad, Ali Sharifian, Mohammad Saber Hashemi, Fayyaz Nosouhi Dehnavi, Mohmmad Riazat, Mahdi Tafazoli, Julien Bardon, Mehrzad Taherzadehboroujeni, Mohammad Hosein Zarei, Vahid Fadaei Naeini, Mihaela Banu, P. Askeland, Maryam Pahlavan Pour, Nick Kuuttila, Farhang Pourboghrat, Lawrence T. Drzal, Alireza Babaei, Mohsen Shakeri, Mohsen Mazrouei Sebdani, Mohammad Ahadiparast, Hamid Shahsavari, Mojtba Haghighi-Yazdi, Hossein Jokar, Ali Tabei, Frederic Addiego, Abolfazl Alizadeh Sahraei, Iman Zeydabadi-Nejad, Naeem Zolfaghari, Christine Chappard, Stanislav Patlazhan, Marwan Al-Haik, Hossein Izadi, Saeed Khaleghi, Mahmoud Mousavi Mashhadi, and Hossein Memarian. We are deeply indebted to all of them. They helped us over the years to publish enriched articles in prestigious journals, and finally, we had a chance to arrange them as this book.

Special thanks are reserved for Fayyaz Nosouhi Dehnavi and Poorya Chavoshnejad for carefully reading various portions of the manuscript and providing valuable criticisms and suggestions.

The authors would be grateful for reports of typographical and other errors to be sent electronically via the following email: m.baniassadi@ut.ac.ir

Introduction to computational modeling of complex microstructures

1

Abstract

In this chapter, we introduce the computational reconstruction and modeling of complex microstructures for the calculation of effective mechanical and thermal properties of heterogeneous materials.

Statistical correlation functions are defined as descriptors of heterogeneous microstructures. The reconstruction of heterogeneous materials based on statistical descriptors is briefly explained and used to realize the three-dimensional (3D) microstructure from its 2D scanning electron microscopy (SEM) image for a multiphase medium. The 3D reconstruction of heterogeneous materials can be used to calculate effective properties and predict percolation threshold of heterogeneous materials. We briefly discuss different homogenization techniques (such as finite element, mean field, statistical continuum mechanics, and asymptotic approach) for calculation of effective mechanical and thermal properties.

Finally, the basic concept of microstructure hull and materials design are discussed. This chapter gives the audience a basic concept to understand step-by-step modeling and homogenization of complex microstructures.

1.1 Statistical descriptor

The description and characterization of heterogeneous systems have become extremely important to scientists during the past decades [1−8]. Statistical continuum mechanics provides a robust alternative approach for the reconstruction and characterization techniques of heterogeneous systems [9,10].

1.1.1 Correlation functions

Statistical methods using correlation functions are one of the most practical and powerful approaches to estimate properties of heterogeneous materials. Properties of these materials can be characterized by exploiting different orders of statistical correlation functions. In multiphase materials, the first-order correlation functions represent volume fractions of different phases and do not describe any information about the distribution and morphology of phases [9].

Applied Micromechanics of Complex Microstructures. DOI: https://doi.org/10.1016/B978-0-443-18991-3.00003-9
© 2023 Elsevier Inc. All rights reserved.

1.1.2 Two-point correlation functions

The data for the two-point distribution functions are acquired by assigning a number of random vectors within the microstructure, determining the likelihood of the head and tail of each vector ($\vec{r}$) landing in a particular phase, and examining the number fraction of the sets (vectors) which satisfy the different states [9].

Now assign a vector $\vec{r}$ starting at each of the random points in a heterogeneous microstructure (see Fig. 1.1). Depending on whether the beginning and the end of these vectors fall within phase 1 or phase 2, there will be four different probabilities (P^{11}, P^{22}, P^{12}, and P^{21}) defined as [9]

$$P^{ij}(\vec{r}) = \frac{M_{ij}}{M}\bigg|_{M \to \infty} \left\{ \vec{r} = \vec{r}_2 - \vec{r}_1 | (\vec{r}_1 \in \varphi_i) \cap (\vec{r}_2 \in \varphi_j) \right\} \tag{1.1}$$

where M_{ij} is the number of vectors with the beginning in phase $i(\phi_i)$ and the end in phase $j(\phi_j)$. Eq. (1.1) defines a joint probability distribution function for the occurrence of events constructed by two points ($\vec{r}_i$ and $\vec{r}_j$) as the beginning and end of a vector $\vec{r}$ when it is randomly inserted in a microstructure M number of times. The two-point function can be defined based on two other probability functions such that [9]

$$P^{ij}(\vec{r}) = P\{(\vec{r}_1 \in \varphi_i)|(\vec{r}_2 \in \varphi_j)\} P(\vec{r}_2 \in \varphi_j) \tag{1.2}$$

The first term on the right hand side is a conditional probability function. At very large distances, $r \to \infty$, the probability of occurrence of the beginning point

Figure 1.1 TPCF vectors in a heterogeneous microstructure.

does not affect the endpoint and the two points become uncorrelated or statistically independent and the conditional probability function reduces to a one-point correlation function [9].

$$P\left\{(|\vec{r}| \to \infty)(\vec{r}_1 \in \varphi_i)|(\vec{r}_2 \in \varphi_j)\right\} = P(\vec{r}_1 \in \varphi_i) \tag{1.3}$$

The two-point function will then reduce to

$$P^{ij}(\vec{r}||\vec{r}| \to \infty) = P(\vec{r}_1 \in \varphi_i)P(\vec{r}_2 \in \varphi_j) \tag{1.4}$$

or

$$P^{ij} = \nu_i \nu_j \tag{1.5}$$

For the case of a two-point function in a two-phase composite, we have symmetry for a nonfunctionally graded material (non-FGM) microstructure:

$$p^{12}(\vec{r}) \equiv p^{21}(\vec{r}) \tag{1.6}$$

and the normality condition gives

$$\sum_{j=1}^{2}\sum_{i=1}^{2} p^{ij}(\vec{r}) \equiv 1 \tag{1.7}$$

$$\sum_{j=1}^{2} p^{ij}(\vec{r}) \equiv \nu_i \tag{1.8}$$

Statistical two-point correlation functions (TPCFs) can be calculated using Monte Carlo simulation or the fast Fourier transform (FFT) approach.

According to Debye et al. [11] the slope of the phase-TPCFs of any isotropically homogeneous medium at the origin must be equal to $-s/\alpha$, where s is the phase-specific surface defined as the phase interface area per unit volume of the phase and α is a constant coefficient which depends on the space dimension. Thus for the proposed estimation the following inequality must be satisfied [5]:

$$\left.\frac{dp^{ij}(\vec{r})}{dr}\right|_{\vec{r}=0} \leq 0 \tag{1.9}$$

As another necessary condition for the phase-TPCFs of a homogeneous medium the so-called triangular inequality must be satisfied. For the proposed estimation, we need to show that [5]

$$p^{ii}(\vec{u}) + p^{ii}(\vec{t}) - p^i \leq p^{ii}(\vec{r}) \tag{1.10}$$

where $\vec{u}$ and $\vec{t}$ are two arbitrary vectors such that $\vec{r} = \vec{t} - \vec{s}$.

According to Weiner−Khinchtine theorem [12] a necessary and sufficient condition for the existence of a statistically homogeneous covariance function of phase η,

$$\gamma_\eta(\vec{r}) = p^{\eta\eta}(\vec{r}) - (p^\eta)^2 \tag{1.11}$$

where $p^{\eta\eta}$ and p^η are, respectively, two- and one-point correlation functions of phase η, is that it has the spectral representation [5]

$$\gamma_\eta(\vec{r}) = \frac{1}{(2\pi)^d} \int \tilde{\gamma}_\eta(\vec{\omega}) \exp(i\vec{\omega} \cdot \vec{r}) \, d\vec{\omega} \tag{1.12}$$

where d represents the space dimension. One of the consequences of this theorem is that the spectral density of $\gamma_\eta(\vec{r})$ is positive [5]

$$\tilde{\gamma}_\eta(\vec{\omega}) = \frac{1}{(2\pi)^d} \int \gamma_\eta(\vec{r}) \exp(-i\vec{\omega} \cdot \vec{r}) \, d\vec{r} \geq 0 \tag{1.13}$$

This inequality reduces to the following integral form for an isotropic d-dimensional medium

$$\tilde{\gamma}_\eta(\omega) = \int_0^\infty r^{d-1} \gamma_\eta(r) \frac{J_{(d/2)-1}(\omega r)}{(\omega r)^{d/2-1}} \, dr \geq 0 \tag{1.14}$$

where $\omega = |\vec{\omega}|$ and J_v is the Bessel function of order v. Therefore for the proposed estimation, we need to show that

$$\int_0^\infty r^{d-1} \frac{\left(p^{ii}(r) - (p^i)^2\right) J_{d/2-1}(\omega r)}{(\omega r)^{d/2-1}} \, dr \geq 0 \tag{1.15}$$

1.1.3 N-point correlation functions

Statistical N-point correlation functions are used for calculating properties of heterogeneous systems. The strength and the main advantage of the statistical continuum approach is the direct link to statistical information of microstructures. Higher-order correlation functions must be calculated or measured to increase the precision of the statistical continuum approach [9].

Now assign a set of n-1 vectors $\{\vec{v}_1 \text{ to } \vec{v}_{n-1}\}$ starting at each of the random points in a heterogeneous microstructure. Depending on whether the beginning and the end of these vectors fall within k-phase heterogeneous materials, there will be k^n different probabilities. Calculation of N-point correlation functions is time-consuming and has a high computational cost [9].

To achieve this aim a new approximation methodology is utilized to obtain N-point correlation functions for non-FGM heterogeneous microstructures.

Introduction to computational modeling of complex microstructures

Conditional probability functions are used to formulate the proposed theoretical approximation. In this approximation, weight functions are used to connect subsets of (N-1)-point correlation functions to estimate the full set of N-point correlation functions. The general formulation of the approximation of n-point correlation functions $(n > 3)$ was derived as [9]

$$C_n(x_1, x_2, x_3, \ldots, x_n)$$

$$= \sum_{i=1}^{n} \left(W_i^n \frac{\prod_{l=1}^{\binom{n-1}{n-2}} C_{(n-1)}(x_i, \ldots, x_{(n-1)})}{\prod_{l=1}^{\binom{n-1}{n-3}} C_{(n-2)}(x_i, \ldots, x_{(n-2)})} \cdot \frac{\prod_{l=1}^{\binom{n-1}{n-4}} C_{(n-3)}(x_i, \ldots, x_{(n-3)})}{\prod_{l=1}^{\binom{n-1}{n-5}} C_{(n-4)}(x_i, \ldots, x_{(n-4)})} \cdots \right),$$

$$\tag{1.16}$$

where $C_n(x_1, x_2, x_3, \ldots, x_n)$ and W_i^n are N-point correlation and dependency weight functions, respectively. In the formulation above, $(x_m \ldots x_i \ldots x_p)$ is defined as the subset of $(n-1)$ points that include x_i as a member of the subset. The weight functions W_m^n can be calculated using the boundary limits [9].

$$W_m^n = \frac{\left(\frac{(-1)^{n-1}}{2^{n-2}((n-2)!)^2} \Gamma_{n-1} \left(\left\{ |x_1 x_2|, |x_1 x_3|, \ldots, |x_{(n-2)} x_{(n-1)}| \right\}_m \middle| n \neq m \right) \right)^{\frac{\alpha_m}{2}}}{\sum_{k=1}^{n} \left(\frac{(-1)^{n-1}}{2^{n-2}((n-2)!)^2} \Gamma_{n-1} \left(\left\{ |x_1 x_2|, |x_1 x_3|, \ldots, |x_{(n-2)} x_{(n-1)}| \right\}_k \middle| n \neq k \right) \right)^{\frac{\alpha_k}{2}}} \middle| n < 6,$$

$$\tag{1.17}$$

1.1.4 Other important statistical descriptors

Other statistical descriptors such as the lineal-path function and two-point cluster functions can be used to characterize microstructures [8]. The lineal path function is the probability of placing an entire segment line of length r within the target phase, as shown in Fig. 1.2 [13,14].

Two-point cluster functions are acquired by assigning a number of random vectors within the microstructure, determining the likelihood of the head and tail of each vector $(\vec{r})$ landing in a particular phase and cluster as shown in Fig. 1.3.

Defining a new statistical descriptor is not difficult and can be performed with additional criteria [13,14].

1.2 Representative volume element

The representative volume element (RVE) plays an important role to calculate the effective properties of random heterogeneous materials. Effective stiffness tensor

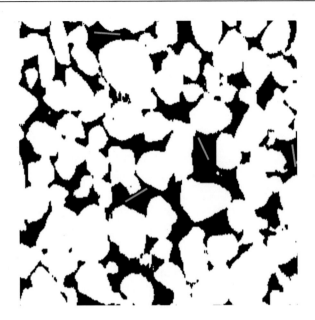

Figure 1.2 Random segment in a heterogeneous microstructure.

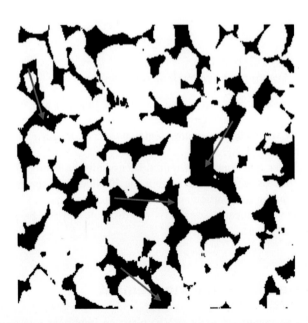

Figure 1.3 TPCCF or probability of happening of the head and tail of each vector in one phase and cluster of a heterogeneous microstructure.

Introduction to computational modeling of complex microstructures

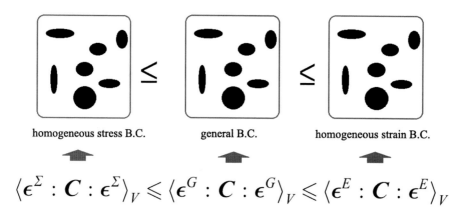

Figure 1.4 Universal inequalities.

can be calculated by the strain and stress fields within a sample. For different boundary conditions the effective stiffness tensor can vary depending on the applied displacement or traction boundary data on the boundary (see Fig. 1.4) [15].

The RVE is a minimum sample size of a heterogeneous material that can be used to determine the corresponding effective properties of a heterogeneous macroscopic sample. The size of the RVE should be large enough with respect to the heterogeneity in order to contain a sufficient number of inclusions for the apparent effective properties to be independent of the surface values of traction and displacement as long as these values are macroscopically uniform. Generally, two types of RVE can be defined; the first type is a microstructure containing a periodic unit cell (RUC), and the second involves random heterogeneous materials. The first type of microstructure can be homogenized by classical asymptotic homogenization for the material with periodic boundary conditions.

The second type involves random heterogeneous materials; the RVE size can be calculated based on the convergence range of effective properties for different sample sizes of microstructures [16].

1.3 Numerical reconstruction of heterogeneous materials

Microstructure reconstruction from limited information is a demanding inverse problem that has gained much attention lately. The main reason that 3D reconstructing methodologies are popular is due to their capability to evaluate macroscopic or effective properties of random materials through finite element analysis (FEA) while the sample experiences a range of loading conditions. The effective transport, mechanical, electrical, and electromagnetic properties of random heterogeneous materials are correlated with statistical functions that can describe a microstructure properly called statistical correlation functions. In addition, some crucial metrics of microstructures, including orientation, grain size, particle size, shape, and anisotropic and geometrical traits, can be investigated using statistical correlation functions. Acquiring a 3D representation of a microstructure from initial 2D micrographs can be done using both experimental and computational methods. Even though experimental methods such as X-ray computed tomography (CT) and focused ion beam/scanning electron microscopy (FIB-SEM) provide invaluable information about a microstructure's properties, generating novel computer-aided strategies have been more desirable recently due to the costly tools associated with experimental procedures, including sample preparation time and the need of expert operators to name but a few.

The reconstruction procedure can be expressed as an optimization problem, where through this problem a cost function would be optimized. This cost function can be defined by the correlation functions in a way that by evolving the microstructure through the reconstruction process the statistical correlation functions match the target functions [17]. It should be noted that the main aim of reconstruction is not to reproduce the exact replica of the initial sample but to generate a microstructure (or a series of cut sections while dealing with 3D reconstruction) that demonstrates the desired effective properties of the sample micrograph. These properties can vary from an optimization problem to the other, but we can mention porosity, permeability, specific surface, and volume fraction of each phase as some of the most important properties that the reconstructed microstructure needs to capture with a low amount of error. Yeong and Torquato proposed a methodology via low-order correlation functions and a stochastic optimization technique called simulated annealing to reconstruct random microstructures from its initial cut sections [14]. Li et al. presented a transfer learning-based methodology that is able to stochastically reconstruct a variety of microstructures and evaluate their respective properties. They used a model pruning approach to build a rational connection between microstructure characteristics and a deep convolutional network [18]. Tran et al. proposed a framework that can successfully reconstruct RGB and grayscale images with the aid of any physics-based descriptor. Their approach is not limited to binary microstructures, so the quality of reconstruction is reasonably high; also the sample size can be varied through the reconstruction process [19]. Xu and his co-workers developed an algorithm that can offer a 3D microstructure reconstruction using 2D reference images of an anisotropic battery separator. They evaluated the porosity and shape of the voids of the RVE and implemented FEA in order to run a uniaxial tensile simulation [20]. Agyei et al. offered a method for reconstruction of short fiber-reinforced composites (SERCs) using 2D segmentation of X-ray tomograms. Their nondestructive technique, which brings new insights into composite damage analysis, is roughly based on positioning elliptical regions on the different stacks to produce a crude microstructure [21]. Habte et al. presented a simulated annealing algorithm to generate a 3D microstructural representation using spherical particles to evaluate the effect of morphology of some cathode materials on Li-ion battery performance [22]. Li et al. developed a workflow employing a simulated annealing (SA) algorithm and Voronoi-based algorithm to perform a pseudo-3D reconstruction for sheet molding compounds (SMC) chopped

Introduction to computational modeling of complex microstructures 9

composites. They conducted FEA on the final reconstructed RVE to evaluate their desired mechanical properties and validate the results by experimental tensile tests [23]. Gao et al. devised an ultrafast algorithm that uses a mixture random field (MRF) model, non-Gaussian random field, and Karhunen−Loève (K-L) expansion method to reconstruct various mediums such as 2D, 3D, multiphase, and anisotropic materials [24]. Yang et al. introduced a set of algorithms for the reconstruction of random media through nonuniform rational B-splines (NURBS) data extraction from micro-CT of reference images. They used two packing algorithms to check the overlap of bounding boxes and represent fiber centerlines during packaging via NURBS. Furthermore an optimizer such as a genetic algorithm (GA) is applied to eliminate some of the inclusions with high error [25]. Feng et al. proposed an accelerating reconstruction scheme based on deep learning or, more accurately, conditional generative adversarial networks (CGANs) to make a relative connection between the conditioning dataset and the final void-solid microstructure. Their statistically efficient approach works with only one reference cut section and is faster (lower CPU usage) than previous multipoint statistics (MPS)-based methods [26]. Li et al. presented a procedure for accurate reconstruction of porous media, while the initial bimodal microstructures are extracted from 2D optical micrographs and limited-angle X-ray tomographic radiographs. They implemented an SA algorithm based on Yeong−Torquato methodology, and low-order spatial correlations belong to the pore phase to successfully reconstruct several sandstone samples [27].

The SA algorithm is a metaheuristic search technique that aids the reconstruction process as an optimizer and is used in numerous works due to its compatibility with correlation functions. SA is model-independent and can be implemented to multiphase materials. If a material and its molecular bonding form at a lower level of energy, it will be formidable and hard to break. The SA algorithm's responses are in the high-temperature range (brittle behavior), and consequently the crystalline lattice structure of a medium has some defects; hence one has to discover a way to reduce the temperature slowly and carefully in order to attain an excellent state for a medium. The SA algorithm should not be trapped in local minima, and that is when hill-climbing moves come into play. The hill-climbing move is one of the most substantial issues in the SA algorithm, and it all boils down to accepting answers that increase the temperature (does not improve the objective function) with a specific probability. Taking this into account, we come to the conclusion that the SA algorithm does not depend on the initial starting temperature, and this feature is a significant factor that distinguishes it over other algorithms [7]. The function that accepts nonimproving solutions can be defined as follows:

$$p = e^{-\frac{\Delta E}{k_B T}} \tag{1.18}$$

This formula was first developed by Metropolis et al. [28], where $k_B = 1$ and T is a temperature that diminishes at each iteration gradually. Also, ΔE represents the variation of objective function, while the initial sample experiences some arbitrary or planned changes.

In microstructure reconstruction application of SA, the following steps should be taken care of:

1. Initiating an arbitrary binary model with the same volume fraction of the reference micrograph.
2. Considering a mathematical descriptor such as TPCF and evaluating it for reference and initial images.
3. Defining a cost function (CF) or error function that captures the correlation difference between microstructures generated in step 2. One can determine the cost function in many ways, but here we stick to the widely used square difference method like below:

$$CF = (TPCF_{reference} - TPCF_{reconstructed})^2 \qquad (1.19)$$

Here "reconstructed" represents the initial randomly generated micrograph that needs to be changed in each iteration.
4. Exchanging two random pixels in the initial image and evaluating CF. It should be mentioned that in this step, one must exchange a "zero" pixel with a "one" pixel to achieve a lower amount of CF based on the Metropolis rule.
5. Decreasing temperature slow enough and iterating through steps 3 and 4.
6. Reaching the predefined stop condition and terminating the algorithm.

While using SA as an optimizer for reconstruction-based problems, one can implement different statistical correlations or even a combination of them to reach better-reconstructed micrographs. As stated earlier, one of the most important attributes of SA is its capability to accept nonimproving solutions based on Metropolis rules. It turns out that when this feature combines with a proper correlation function, getting a suitable microstructure is almost not out of reach. However, for complicated anisotropic micrographs, SA might converge slowly and also takes a considerable amount of time to reach the algorithm stop condition. The stop condition can be defined in several ways, including but not limited to

1. Reaching a predefined amount for T (temperature)
2. Satisfying a predetermined value for CF
3. When CF does not get improved, for instance, in 2000 successive epochs.

One way to tackle the problem of computational cost is to evaluate the descriptor until the long-range order of it is satisfied.

By implementing the long-range order of TPCF, the time-consuming optimization process would be decreased significantly.

Another way for dwindling operation time is to evaluate the desired descriptor in orthogonal directions and not in all directions. This method is most useful while implementing more complicated descriptors like the two-point correlation cluster function (TPCCF).

In Fig. 1.5 we have two perpendicular cut sections that are used for reconstruction via the SA algorithm. We have applied TPCF as the mathematical descriptor, which means that CF is defined based on TPCF and has to be optimized.

Introduction to computational modeling of complex microstructures

Figure 1.5 Two initial perpendicular cut sections (left side: xz plane and right side: yz plane) of two-phase heterogeneous materials.

Figure 1.6 3D-reconstructed microstructure of the final RVE produced using the SA algorithm.

In order to carry on the six steps discussed earlier the stop condition is defined in a way that if CF reaches an amount that is less than 0.1, the algorithm would be stopped automatically. Another stipulation that is implemented is allowing the temperature parameter (T) to be decreased provided it is not equal to zero because according to the Metropolis formula, a zero value of T would not let the annealing procedure reach better solutions as iteration increases. Fig. 1.6 demonstrates the final RVE produced using the SA algorithm. In order to evaluate the reconstructed micrographs which construct the final RVE, one can compare the TPCF of initial microstructures and reconstructed ones. This comparison is depicted in Fig. 1.7 for the first and second stacks of reconstructed TPCF and reference TPCF. The reason behind considering only eight stacks is the simple fact that TPCF reached its long-range order in the 8'th pixel, where the initial microstructure has a volume fraction of 0.33 and converged to a number close to 0.1089.

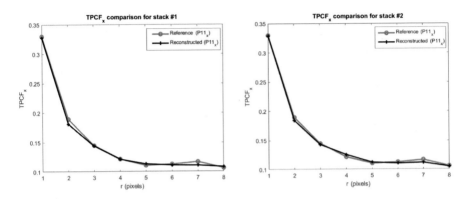

Figure 1.7 Comparison of TPCF between reconstructed and reference microstructures for the first and second stacks of reconstructed TPCF and reference TPCF.

1.4 Effective properties

Effective properties of heterogonous materials are the key issues in future applications and materials design. Due to difficulties in experimental characterizations, analytical and numerical simulations are getting more attractive as the experimental alternatives [29,30]. The analytical mechanics-based composite stiffness models, such as Mori−Tanaka (MT) [30,31] and Halpin−Tsai [32,33], have been widely used for the modeling of composite structures [34−38]. These methods cannot accurately consider the interactions between adjacent inclusions. Moreover, they cannot evaluate the microstresses involved with individual inclusions [39]. During the past decade, because of considerable improvement of computational facilities, numerical methods, such as the finite element method, have been widely used for the modeling of thermal conductivity and elastic properties of composite materials [38−46]. Using the finite element approach, it is possible to evaluate the microstresses, and we could also consider the effects of the adjacent inclusions on the effective properties. The main disadvantages of the finite element method in comparison with Mori−Tanaka and Halpin−Tsai methods are its higher computational costs and the complexities for the modeling of high-volume concentrations and aspect ratios for inclusions. Statistical continuum mechanics techniques also provide other tools for characterization and reconstruction of heterogeneous materials based on statistical correlation functions [14]. Statistical characterization can be implemented for different types of random heterogeneous materials [40,47−51]. A wide variety of effective properties of heterogeneous materials can be estimated by exploiting different types of statistical functions [14,52−55].

In this section the basic concept of three different groups of modeling techniques for the evaluation of the effective thermal conductivity and elastic modulus of random two-phase composite materials is discussed. These three groups are categorized as mean field methods (by Mori−Tanaka), numerical approaches (by finite element), and statistical continuum methods (by strong contrast). We present the details of 3D finite element, Mori−Tanaka, and strong contrast methods for the evaluation of thermal conductivity and elastic modulus of random two-phase composite materials. These approaches could be useful for the modeling and prediction of the elastic modulus and thermal conductivity of wide ranges of exciting composites, nanocomposites, biomaterials, and biological materials [55,56].

1.4.1 Calculation of effective properties

In this part the details of finite element, Mori–Tanaka, and strong contrast methods for the evaluation of effective thermal conductivity and elastic properties of two-phase composites containing randomly distributed and oriented inclusions are discussed. The inclusion shapes play an important role in the overall mechanical and thermal reinforcement. Therefore the modelings were carried out in ways to take into account different geometries for inclusions, such as cylinder, sphere, and plate. In the current study the inclusion geometry is introduced by the definition of the aspect ratio. For inclusions with a platelet shape the aspect ratio is the diameter to thickness ratio of the plate. In the case of cylindrical inclusions the aspect ratio corresponds to the length to diameter ratio of inclusion. Accordingly the aspect ratio of 1 is representative of spherical inclusions. In finite element and strong contrast modeling the hard-core models are used to generate particles without intersection or contact. In all the cases studied in this work the Poisson's ratio of the fillers and matrix was assumed to be 0.2 [55].

1.4.1.1 Finite element modeling

Now-a-days the finite element method is considered as a versatile tool for the modeling and simulation of a wide range of engineering problems. Computational limitations and modeling complexities of the finite element method impose limits on the maximum number of elements, which are used for introducing the geometries in the model. In this way the simulations of composite materials are limited to modeling of an RVE of the system. The finite element simulations in this study have been performed using the ABAQUS (Version 6.10) package. To reproduce the RVE in a status closer to those in experimentally fabricated random composites the 3D inclusions were randomly distributed and oriented in the RVE [57–60]. It is worth mentioning that while 2D finite element models could be geometrically acceptable for platelet [37,46] inclusions, they could not accurately describe the geometry for cylindrical and spherical inclusions. Furthermore, Hbaieb et al. [46] have shown that in the case of randomly distributed clay particles, 2D finite element models considerably underestimate the predictions by 3D models. In Fig. 1.7, samples of 3D cubic RVE with different geometries for inclusions are illustrated. The size of the RVE is adjusted based on the geometry, volume fractions, and numbers of perfect particles inside the RVE. To create the desired volume concentration of fillers inside the RVE accurately the RVE is constructed in a way to satisfy the periodicity criterion. This means that if a filling particle passes one boundary side of the RVE the remaining part of that particle continues from the opposite side. This accordingly means that if one puts these RVE cubes together side by side, no discontinuity will be observed in the constructed sample and all of the particles will have equal geometries. This way the developed RVE could also be considered as the representation of a bulk composite material. We should note that we have not applied the periodic boundary conditions in our modeling [55].

The random RVEs were constructed in ABAQUS by developing Python scripts as input files. We developed numerical tools (in C++ language) for creating randomly distributed and oriented inclusions inside the RVE observing the periodicity criterion. The modeling parameters are adjusted in C++ programs. The provided information by the C++ programs are written in a text file that is later used as an input file for Python scripts for the construction of final finite element RVE. We found that the combination of Python and C++ is a faster approach than exclusively relying on Python scripting [55].

As shown in Fig. 1.8D–F the specimens were meshed using four-node linear tetrahedron shape elements. For evaluation of mechanical properties, 3D stress (C3D4) elements are used [61–63]. The heat transfer (DC3D4) elements are used in the modeling of thermal conductivity. For the evaluation of elastic modulus a small uniform displacement was applied on one of the surfaces along the Z-direction and the remaining surfaces where fixed only in their normal directions. The reaction forces on each surface were calculated in order to obtain the corresponding stresses. Using the Hooke's law for an isotropic material the effective elastic modulus and Poisson's ratio of the RVE were estimated. For the evaluation of thermal conductivity a constant heating surface heat flux ($+q$) was exerted on one of the surfaces along the

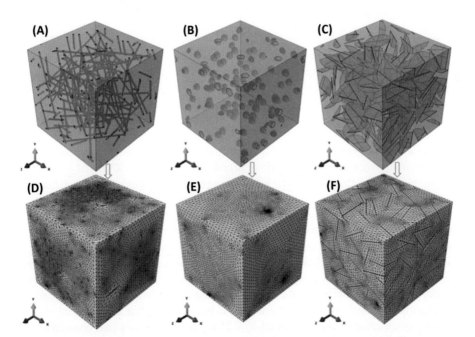

Figure 1.8 Developed 3D cubic periodic RVEs containing 100 perfect particles with different geometries, (A) tubes with an aspect ratio of 60, (B) spherical inclusions, and (C) platelets with an aspect ratio of 100. Under each RVE sample, the meshed specimen is also illustrated (D–F) [55].

Z-direction and on the opposite surface the same magnitude cooling surface heat flux $(-q)$ was applied. This introduces a heat flow along the specimen by transferring a constant quantity of heat flux. The total energy of the simulation RVE remains constant, and steady-state heat conduction is observed. In response to the energy redistribution, thermal energy moves from the hot reservoir to the cold reservoir and a temperature gradient is established in the system. The temperature gradient $\frac{dT}{dz}$ was evaluated by calculating the temperature differences between the hot and cold surfaces. The effective thermal conductivity of the nanocomposite was obtained using the 1D form of the Fourier law $\left(K = q\frac{dz}{dT}\right)$ [55].

1.4.1.2 Mori–Tanaka method

Mean field micromechanics methods provide simple tools for the evaluation of the effective thermal and mechanical properties of composites. A typical mean field formulation is based on the inclusion concentration tensors that connect inclusion averaged fields in reinforcements or matrixes with the corresponding macroscopic fields. One of these approaches, Mori–Tanaka, approximates the behavior of composites that contain reinforcements at nondilute volume fractions via dilute inhomogeneities that are subjected to effective matrix fields rather than the macroscopic fields [31]. These effective fields account for the perturbations caused by all other reinforcements in a mean-field sense [49]. Benveniste [64] expressed the Mori–Tanaka methods for elastic composites by the following dual relations [55]:

$$\langle\varepsilon\rangle^i = \overline{\mathbf{A}}^i_{dil}\langle\varepsilon\rangle^m = \overline{\mathbf{A}}^i_{dil}\overline{\mathbf{A}}^m_{MT}\langle\varepsilon\rangle \tag{1.20}$$

$$\langle\sigma\rangle^i = \overline{\mathbf{B}}^i_{dil}\langle\sigma\rangle^m = \overline{\mathbf{B}}^i_{dil}\overline{\mathbf{B}}^m_{MT}\langle\sigma\rangle \tag{1.21}$$

where $\langle\varepsilon\rangle^i, \langle\varepsilon\rangle^m, \langle\sigma\rangle^i$, and $\langle\sigma\rangle^m$ are the averaged strain and stress tensors of the inclusion and matrix phases, respectively. $\langle\varepsilon\rangle$ and $\langle\sigma\rangle$ are the macroscopic second-order strain and stress tensors, respectively; $\overline{\mathbf{A}}^m_{MT}$ and $\overline{\mathbf{B}}^m_{MT}$ stand for the Mori–Tanaka matrices; and $\overline{\mathbf{A}}^i_{dil}$ and $\overline{\mathbf{B}}^i_{dil}$ stand for the inclusion dilute, fourth-order strain and stress concentration tensors, respectively. The analogous form of the Mori–Tanaka formulation for the thermal conduction has the form [34,55]

$$\langle\nabla T\rangle^i = \overline{\mathscr{A}}^i_{dil}\langle\nabla T\rangle^m = \overline{\mathscr{A}}^i_{dil}\overline{\mathscr{A}}^m_{MT}\langle\nabla T\rangle \tag{1.22}$$

$$\langle q\rangle^i = \overline{\mathscr{B}}^i_{dil}\langle q\rangle^m = \overline{\mathscr{B}}^i_{dil}\overline{\mathscr{B}}^m_{MT}\langle q\rangle \tag{1.23}$$

where ∇T and q are the thermal gradient and heat flux vectors, respectively, and $\mathscr{A}$ and $\mathscr{B}$ stand for the thermal gradient and heat flux concentration second-order tensors, respectively. The Mori–Tanaka stress concentration tensors of the matrix and inclusions can be expressed by [55]

$$\overline{\mathbf{A}}^m_{MT} = \left[(1-v_f)\mathbf{I} + v_f\overline{\mathbf{A}}^i_{dil}\right]^{-1} \tag{1.24}$$

$$\overline{\mathbf{A}}_{MT}^{i} = \overline{\mathbf{A}}_{dil}^{i} \left[(1-v_f)\mathbf{I} + v_f \overline{\mathbf{A}}_{dil}^{i} \right]^{-1} \tag{1.25}$$

where $\mathbf{I}$ is the fourth-order identity tensor and v_f stands for the inclusion volume fractions. The dilute inclusion concentration tensor, $\overline{\mathbf{A}}_{dil}^{i}$, can be obtained in analogy to Hill's [65] expressions as [55]

$$\overline{\mathbf{A}}_{dil}^{i} = \left[\mathbf{I} + \mathbf{S}(\mathbf{C}^m)^{-1}(\mathbf{C}^i - \mathbf{C}^m) \right]^{-1} \tag{1.26}$$

where $\mathbf{C}^i$ and $\mathbf{C}^m$ stand for the stiffness tensor of the inclusion and matrix, respectively, and $\mathbf{S}$ is the Eshelby tensor [66]. The analogous relations link the inclusion gradient concentration tensors $\overline{\mathbf{A}}_{dil}^{i}$ by considering the $\mathbf{I}$ as the second-order identity tensor; inclusion and matrix conductivity tensors by K^i and K^m, respectively; and $\mathbf{S}$ as the Eshelby tensor of the diffusion problem [67].

Due to random orientations of particles inside the RVE, the Mori–Tanaka model results must be representative of the integral of all directions. For the second- and fourth-order tensors the volume average can be respectively expressed by [55]

$$[\![\blacksquare]\!] = \frac{1}{4\pi} \int_0^{2\pi} \int_0^{\pi} Q_{mi} Q_{nj} \blacksquare_{mnpq} Q_{pk} Q_{ql} \sin\theta d\theta d\psi \tag{1.27}$$

$$\{\blacksquare\} = \frac{1}{4\pi} \int_0^{2\pi} \int_0^{\pi} Q_{mi} \blacksquare_{mn} Q_{nj} \sin\theta d\theta d\psi \tag{1.28}$$

Here the braces denote the average over all possible orientations, θ and Ψ are Euler's angles between the local and global coordinate systems (as shown in Fig. 1.9), and Q denotes the corresponding transformation matrices. Finally the Mori–Tanaka macroscopic effective stiffness (C^e) and conductivity (K^C)

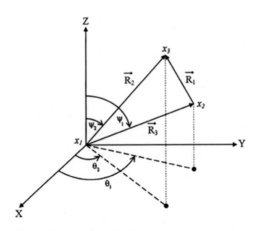

Figure 1.9 Vector representation in spherical coordinates [55].

Introduction to computational modeling of complex microstructures 17

tensors of the composite with randomly oriented inclusions can be expressed by [55,68]

$$C^c = \left[(1 - v_f)C^m + v_f \, [\![C^i \overline{\mathbf{A}}^i_{dil}]\!] \right] \left[(1 - v_f)I + v_f \, [\![\overline{\mathbf{A}}^i_{dil}]\!] \right]^{-1} \tag{1.29}$$

$$K^C = \left[(1 - v_f)K^m + v_f \left\{ K^i \overline{\mathscr{A}}^i_{dil} \right\} \right] \left[(1 - v_f)I + v_f \left\{ \overline{\mathscr{A}}^i_{dil} \right\} \right]^{-1} \tag{1.30}$$

1.4.1.3 Strong Contrast Method

In the current study, we used statistical continuum theory of strong contrast to predict effective properties of two-phase composite structures. As has been mentioned in the previous section, correlation functions are used as the statistical description of the microstructure. The expression of N-point correlation functions for a given phase i can be written as follows [14,55]:

$$P^i_N(x_1, x_2, \ldots, x_N) = Probability(x_1 \in Phase(i) \cap x_2 \in Phase(i) \cap \ldots \cap x_N \in Phase(i)) \tag{1.31}$$

where x_i is the position of random points (vector) in the microstructure.

In this work, Monte Carlo simulation is used to calculate TPCFs. In this method, 3D randomly oriented inclusions (spherical, cylindrical, and platelet) are generated and used to calculate the statistical TPCFs [47,48]. Higher-order correlation functions can provide more precise morphological details of heterogeneous systems. In this paper, three-point correlation function $P^i_3(x_1, x_2, x_3)$ is estimated using the TPCF by the following relation [39,55]:

$$P^i_3(x_1, x_2, x_3) \approx W^3_1 \left(\frac{P^i_2(x_1, x_2)P^i_2(x_1, x_3)}{P^i_1(x_1)} \right) + W^3_2 \left(\frac{P^i_2(x_1, x_2)P^i_2(x_2, x_3)}{P^i_1(x_2)} \right)$$
$$+ W^3_3 \left(\frac{P^i_2(x_3, x_2)P^i_2(x_1, x_3)}{P^i_1(x_3)} \right) \tag{1.32}$$

the weight coefficients could be defined as [55]

$$W^3_1 = \frac{|\vec{R}_1|}{|\vec{R}_1| + |\vec{R}_2| + |\vec{R}_3|}, \; W^3_2 = \frac{|\vec{R}_2|}{|\vec{R}_1| + |\vec{R}_2| + |\vec{R}_3|}, \; W^3_3 = \frac{|\vec{R}_3|}{|\vec{R}_1| + |\vec{R}_2| + |\vec{R}_3|} \tag{1.33}$$

where R_1, R_2, and R_3 are the position vectors (as illustrated in Fig. 1.8). In the strong contrast formulation presented in this study, the subscripts m, i, and c stand for the reference phase (matrix), inclusion phase, and composite effective properties, respectively [55].

1.4.1.3.1 Strong contrast solution for thermal conductivity

Assuming isotropic properties for the matrix and inclusion the effective thermal conductivity matrix, K^C, for 3D heterogeneous materials is calculated using the strong-contrast formulation for [31,55]:

$$\{K^C - K^m I\}^{-1} \cdot \{K^C + 2K^m I\}$$

$$= \frac{1}{\beta_{im} P_1^i(x_1)} I - 3K^m \int \left[\frac{P_2^i(x_1, x_2) - P_1^i(x_1) P_1^i(x_2)}{P_1^i(x_1) P_1^i(x_2)} \right] M^m(x_1, x_2) dx_2$$

$$- 9(K^m)^2 \beta_{im} \iint \left[\frac{P_3^i(x_1, x_2, x_3)}{P_1^i(x_1) P_1^i(x_2)} - \frac{P_2^i(x_1, x_2) P_2^i(x_2, x_3)}{P_1^i(x_1) P_1^i(x_2) P_1^i(x_3)} \right] M^m(x_1, x_2) \cdot M^m(x_2, x_3) dx_2 dx_3 - \ldots$$

$$(1.34)$$

In Eq. (1.34) I is the second-order identity tensor and β_{im} is the polarizability [14] scalar which is calculated using the inclusion and matrix thermal conductivity values, K_i and K_m, respectively, as follows {Mortazavi, 2013 #240}:

$$\beta_{im} = \frac{K_i - K_m}{K_i + 2K_m} \tag{1.35}$$

The M, second-order tensor, in Eq. (1.34) is defined as follows:

$$M^m(x_1, x_2) = \frac{1}{\Omega} \frac{1}{K_m} \frac{3t \otimes t - I}{|x_1 - x_2|^3} \tag{1.36}$$

where Ω is the total solid angle contained in a 3D sphere and $t = \frac{(x_1 - x_2)}{|x_1 - x_2|}$.

1.4.1.3.2 Strong contrast solution for elastic modulus

The strong-contrast method has been used to determine the effective elastic modulus of two-phase composites [9,54]. In this method, N-point correlation functions show up in the final equations that characterize the microstructure. The general equation for isotropic composites of the expansion for a reference phase m is written as follows [14,54] {Mortazavi, 2013 #240}:

$$v_f^2 \left[\frac{k_{im}}{k_{cm}} \Lambda_h + \frac{\mu_{im}}{\mu_{cm}} \Lambda_s \right] = v_f I - \sum_{n=2}^{\infty} B_n^{(i)} \tag{1.37}$$

As a common assumption the calculations have been performed for the first and second terms of $B_n^{(i)}$ and other terms have been neglected [14,54]:

$$v_f^2 \left[\frac{k_{im}}{k_{cm}} \Lambda_h + \frac{\mu_{im}}{\mu_{cm}} \Lambda_s \right] = v_f I - B_2^{(i)} - B_3^{(i)} \tag{1.38}$$

where v_f is the inclusion volume fraction and Λ_h and Λ_s are the fourth-order hydrostatic projection and shear projection tensors, respectively [14]. k_{mn} and μ_{mn} are introduced as bulk and shear moduli polarizabilities, respectively, where the subscripts m and n represent c, i, or m, which respectively stand for effective, inclusion, and matrix properties. The polarizabilities are expressed in terms of the effective or phase bulk and shear moduli, F and G, as follows [55]:

$$k_{mn} = \frac{F_m - F_n}{F_m + \frac{4}{3}G_n} \tag{1.39}$$

$$\mu_{mn} = \frac{G_m - G_n}{G_m + \dfrac{G_n[3F_n/2 + 4G_n/3]}{F_n + 2G_n}} \tag{1.40}$$

In Eq. (1.38) the tensor coefficients $B_2^{(i)}$ and $B_3^{(i)}$ are given by the following integrals:

$$B_2^{(i)} = \int dx_2 U^{(m)}(x_1, x_2)\left[P_2^{(i)}(x_1, x_2) - v_f^2\right] \tag{1.41}$$

$$B_3^{(i)} = \left(\frac{1}{v_f}\right) \int dx_2 \int dx_3 U^{(m)}(x_1, x_2):U^{(m)}(x_2, x_3)\Delta_3^{(i)}(x_1, x_2, x_3) \tag{1.42}$$

In Eq. (1.42) the determinant tensor of correlation functions $\Delta_3^{(i)}$ for the inclusion phase [14] is defined as follows [55]:

$$\Delta_3^{(i)}(x_1, x_2, x_3) = \begin{vmatrix} P_2^{(i)}(x_1, x_2) & P_1^{(i)}(x_2) \\ P_3^{(i)}(x_1, x_2, x_3) & P_2^{(i)}(x_2, x_3) \end{vmatrix} \tag{1.43}$$

The U tensors in Eqs. (1.41) and (1.42) for the reference phase [14] are based on the position-dependent fourth-order tensor H(r) and the related tensor for the reference phase:

$$U_{\alpha\beta\kappa\lambda}^{(m)}(r) = L_{\alpha\beta\gamma\eta}^{(m)} H_{\gamma\eta\kappa\lambda}^{(m)}(r)$$

$$= [3F_m + 4G_m]\left\{\left[k_{im} - \frac{5G_m}{3(F_m + 2G_m)}\mu_{im}\right]\frac{\delta_{ij}}{3}H_{\gamma\gamma\kappa\lambda}^{(m)}(r) + \frac{5G_m}{3(F_m + 2G_m)}\mu_{im}H_{\alpha\beta\kappa\lambda}^{(m)}(r)\right\} \tag{1.44}$$

H(r) is the symmetrized double gradient tensor [14]. The fourth-order tensor $L^{(m)}$ for the reference phase is expressed as

$$L^{(m)} = [3F_m + 4G_m]\left[k_{im}\Lambda_h + \frac{5G_m}{3(F_m + 2G_m)}\mu_{im}\Lambda_s\right] \tag{1.45}$$

1.5 Asymptotic homogenization

The homogenization method by asymptotic expansion is very helpful for periodic structures [69−72].

The main steps of the method are as follows: Let L represent the characteristic dimension of the overall structure; then if η is a small parameter, ηL will define the size of the RVE. We denote by X a space variable reflecting the variation in the magnitude of the properties at the structure scale. This variable is named "slow variable." We denote by Y the local space variable describing the variation of the properties inside the RVE. Y is named the "rapid variable."

Thus X and Y are related together through

$$\eta = \frac{X}{Y} \tag{1.46}$$

If $u(X.Y)$ is the displacement field solution of the problem, it may be expressed in terms of infinite expansion series:

$$u(X.Y) = u_0(X.Y) + \eta u_1(X.Y) + \eta^2 u(X.Y) + \ldots \tag{1.47}$$

The elastic deformation—written here in small deformations—can be decomposed as a function of slow and rapid variables (it can also be developed for large deformation):

$$\varepsilon_{ij} = \varepsilon_{Xij} + \frac{1}{\eta}\varepsilon_{Yij} \quad or \quad \varepsilon = \varepsilon_X + \frac{1}{\eta}\varepsilon_Y \tag{1.48}$$

where ε_X and ε_Y are the strain tensors corresponding to the slow and rapid space variables, respectively.

Knowing that $\sigma = C{:}\varepsilon$ (Hooke's law with the fourth-order stiffness tensor C representing the local elastic properties) the stress tensor σ depends also on the two space variables X and Y. Then the stress differential operator $div\sigma\,(X.Y)$ may be written as follows:

$$div(C{:}\varepsilon) = div_X(C{:}\varepsilon) + \frac{1}{\eta}div_Y(C{:}\varepsilon) \tag{1.49}$$

where div_X and div_Y are the divergence operators corresponding to slow and rapid variables, respectively. The equilibrium equation of the periodic medium is equal to

$$div\sigma(X.Y) + f = 0 \tag{1.50}$$

where f is the volume force. It induces the three following equations:

$$div_Y(C{:}\varepsilon_Y(u_0)) = 0 \tag{1.51}$$

$$div_Y(C{:}\varepsilon_Y(u_1)) + div_Y(C{:}\varepsilon_X(u_0)) + div_X(C{:}\varepsilon_Y(u_0)) = 0 \tag{1.52}$$

$$div_Y(C{:}\varepsilon_Y(u_2)) + div_Y(C{:}\varepsilon_X(u_1)) + div_X(C{:}\varepsilon_Y(u_1)) + div_X(C{:}\varepsilon_X(u_0)) + f = 0 \tag{1.53}$$

Solving (1.51), (1.52), and (1.53) gives u_0, u_1, and u_2, which are the solutions of the following variational formulation:

Find

$$\vartheta \in \vartheta_{per}. \ \forall \ \vartheta^* \in \vartheta_{per}. \int_M Tr[C{:}\varepsilon_Y(\vartheta){:}\varepsilon_Y(\vartheta^*)d\Omega = \int_m g\vartheta^* d\Omega \tag{1.54}$$

with $g = -div(C{:}\varepsilon_Y(u))$ and $\vartheta_{per} = \{u.\, u^+ = u^-\}$, where u^+ and u^- represent the values of the displacement u on two opposite faces of the RVE and are equal. M represents the unit cell (RVE).

From the solution of (1.51), (1.52), and (1.53), it can be concluded that

1. u_0 depends only on the space variable X:

$$u_0 = u_0(X)$$

2. u_0 and u_1 are linearly dependent,

$$\varepsilon_Y(u_1) = H{:}\varepsilon_X(u_0) \tag{1.55}$$

where H is a fourth-order symmetric linear operator similar to C. The equilibrium equation of the periodic medium becomes

$$div_X \int_M (H + C)\varepsilon_X(u_0)d\Omega + f \int_M d\Omega = 0 \tag{1.56}$$

Note that u_0 represents the homogenized displacement over the RVE, while u_1 represents the perturbation of the displacement. Finally, using the macroscopic (RVE) Hooke's law $\sigma^* = C^*{:}\varepsilon^*$, relating the macroscopic stress tensor σ^* to the macroscopic elastic strain tensor ε^*, along with (1.54) and (1.55), we get the homogenized behavior of the periodic medium. This leads to the following expression of the effective elastic stiffness tensor:

$$C^* = \frac{1}{Vol(M)} \int_M (H + C)dM \tag{1.57}$$

1.6 Hull space and materials design

In the traditional design process, material and geometry are usually varied iteratively to meet design objectives. An analogous approach for the design of materials

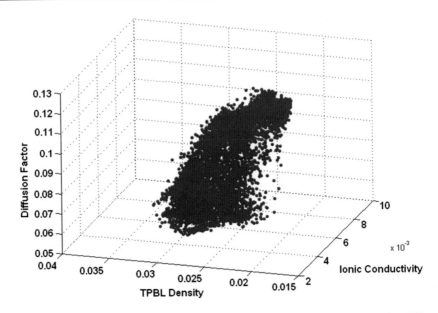

Figure 1.10 Microstructure hull; each axis represents one of the desirable properties [73].

requires a microstructure hull that consists of a set of possible realization of microstructures. The realization can be performed using different algorithms and can be implemented using a computer code to generated 3D microstructures. We can define input parameters to generate a variety of microstructures, and then we can characterize the microstructure to calculate effective properties. In other words a closure in this sense includes all possible effective property values predicted by sweeping input parameters defining these distribution functions, and generally a solution is a subset of such closure. For example for solid oxide fuel cell anode microstructures a sample property closure obtained for wide-range generated realization is shown in Fig. 1.10. Each axis represents one of the properties and its feasible range within the selected design space. Three-phase boundary density, ionic conductivity, and gas diffusivity are critical design objectives. Mathematically speaking a property closure for these parameters can be obtained using an arbitrary analytical or approximate method, and their boundaries represent constraints for the optimization process [73].

Design of new materials that have effective properties not usually seen in nature has become one of the challenging areas of materials science. In microstructural design of heterogeneous materials, periodic and random heterogeneous materials can be optimized to design a microstructure with desirable effective properties.

Any periodic structure can be represented by its unit cell (PUC); therefore we can optimize. Such a form of representation allows for scaling the problem of overall microstructural design down to the topology optimization of the corresponding PUC based on the prescribed effective properties.

The continuing advances in AM processes have provided greater degrees of freedom to fabricate a new generation of structures with intricate patterns in a variety of scientific and technological applications, such as architectured porous biomaterials [74], stretching-dominated cellular truss structures [75], and multiscale cellular structures [76]. Taking the capacity of AM technologies for granted the primary focus of the current study is presenting an efficient and easy-to-implement algorithm for optimal design of porous and periodic microstructures that are manufacturable by 3D printing and will have the mechanical properties specified beforehand.

In microstructural design of heterogeneous materials, topology optimization is therefore a useful tool for designing such periodic microstructures that can be constructed from the juxtaposition of identical unit cells. Properly speaking, in the context of periodic microstructures, topology optimization makes the generation of almost any manufacturable configuration within the design space possible. Working with the topology concept provides higher degrees of freedom (compared to sizing and shape optimization), thus enabling fabrication of lightweight materials of improved structural properties such as stiffness.

1.7 Conclusion remarks

In this chapter a crucial basic concept and fundamental definition of heterogeneous materials have been explained to improve our understanding of this knowledge. The basic concept of a statistical descriptor such as N-point correlation functions has been explained and used to reconstruct and homogenize microstructures. Design and manufacturing a new microstructure can open a window to design an efficient machine with extraordinary capability. Realization and reconstruction of heterogeneous materials help us to understand how we could generate the microstructure. Initial reconstructed and realized microstructures could be homogenized, and the effective properties can be predicted using a numerical method such as the mean field or finite element homogenization approach. Optimal design of microstructures is the next step to achieve our target materials. On the other hand, realization, homogenization, and optimization are the three steps that we need to follow repeatedly to design an optimal microstructure with target properties. Property hull and microstructure hull (in the case of existence) can be exploited to predict the materials design procedure's efficiency.

References

[1] Hasanabadi A, Baniassadi M, Abrinia K, Safdari M, Garmestani H. Efficient three-phase reconstruction of heterogeneous material from 2D cross-sections via phase-recovery algorithm. J Microsc 2016;264(3):384–93. Available from: https://doi.org/10.1111/jmi.12454.

[2] Baniassadi M, Safdari M, Garmestani H, Ahzi S, Geubelle PH, Remond Y. An optimum approximation of n-point correlation functions of random heterogeneous material systems. J Chem Phys 2014;140(7):074905. Available from: https://doi.org/10.1063/1.4865966.

[3] Chavoshnejad P, Razavi MJ. Effect of the interfiber bonding on the mechanical behavior of electrospun fibrous mats. Sci Rep 2020;10(1):7709. Available from: https://doi.org/10.1038/s41598-020-64735-5.

[4] Riahipour R, Sahraei AA, Werken Nvd, Tehrani M, Abrinia K, Baniassadi M. Improving flame-retardant, thermal, and mechanical properties of an epoxy using halogen-free fillers. Sci Eng Compos Mater 2017;25:939−46.

[5] Ghazavizadeh A, Soltani N, Baniassadi M, Addiego F, Ahzi S, Garmestani H. Composition of two-point correlation functions of subcomposites in heterogeneous materials. Mech Mater 2012;51:88−96.

[6] Rishi AM, Kandlikar SG, Rozati SA, Gupta A. Effect of ball milled and sintered graphene nanoplatelets−copper composite coatings on bubble dynamics and pool boiling heat transfer. Adv Eng Mater 2020;1901562.

[7] Haghverdi A, Baniassadi M, Baghani M, Sahraei AA, Garmestani H, Safdari M. A modified simulated annealing algorithm for hybrid statistical reconstruction of heterogeneous microstructures. Computat Mater Sci 2021;197:110636. Available from: https://doi.org/10.1016/j.commatsci.2021.110636.

[8] Bagherian A, Famouri S, Baghani M, George D, Sheidaei A, Baniassadi M. A new statistical descriptor for the physical characterization and 3D reconstruction of heterogeneous materials. Transp Porous Media 2022;142(1):23−40. Available from: https://doi.org/10.1007/s11242-021-01660-9.

[9] Baniassadi M, Ahzi S, Garmestani H, Ruch D, Remond Y. New approximate solution for N-point correlation functions for heterogeneous materials. J Mech Phys Solids 2012;60(1):104−19.

[10] Nosouhi Dehnavi F, Safdari M, Abrinia K, Hasanabadi A, Baniassadi M. A framework for optimal microstructural design of random heterogeneous materials. Computat Mech 2020;66(1):123−39. Available from: https://doi.org/10.1007/s00466-020-01844-y.

[11] Debye P, Anderson Jr HR, Brumberger H. Scattering by an inhomogeneous solid. II. The correlation function and its application. J Appl Phys 1957;28(6):679−83.

[12] Priestley M.B. Spectral analysis and time series: probability and mathematical statistics. vol. 04; QA280, P7. 1981.

[13] Rémond Y., Ahzi S., Baniassadi M., Garmestani H. Applied RVE reconstruction and homogenization of heterogeneous materials. Wiley Online Library; 2016.

[14] Torquato S. Random heterogeneous materials: microstructure and macroscopic properties. New York: Springer Science & Business Media; 2013.

[15] Hori M, Nemat-Nasser S. On two micromechanics theories for determining micro−macro relations in heterogeneous solids. Mech Mater 1999;31(10):667−82. Available from: https://doi.org/10.1016/s0167-6636(99)00020-4.

[16] Gitman IM, Askes H, Sluys LJ. Representative volume: Existence and size determination. Eng Fract Mech 2007;74(16):2518−34. Available from: https://doi.org/10.1016/j.engfracmech.2006.12.021.

[17] Torquato S. Necessary conditions on realizable two-point correlation functions of random media. Ind Eng Chem Res 2006;45(21):6923−8.

[18] Li X, Zhang Y, Zhao H, Burkhart C, Brinson LC, Chen W. A transfer learning approach for microstructure reconstruction and structure-property predictions. Sci Rep 2018;8(1):13461. Available from: https://doi.org/10.1038/s41598-018-31571-7.

[19] Tran A, Tran H. Data-driven high-fidelity 2D microstructure reconstruction via nonlocal patch-based image inpainting. Acta Mater 2019;178:207−18. Available from: https://doi.org/10.1016/j.actamat.2019.08.007.

[20] Xu H, Bae C. Stochastic 3D microstructure reconstruction and mechanical modeling of anisotropic battery separators. J Power Sources 2019;430:67−73. Available from: https://doi.org/10.1016/j.jpowsour.2019.05.021.

[21] Agyei RF, Sangid MD. A supervised iterative approach to 3D microstructure reconstruction from acquired tomographic data of heterogeneous fibrous systems. Compos Struct 2018;206:234−46. Available from: https://doi.org/10.1016/j.compstruct.2018.08.029.

[22] Habte BT, Jiang F. Microstructure reconstruction and impedance spectroscopy study of LiCoO2, LiMn2O4 and LiFePO4 Li-ion battery cathodes. Microporous Mesoporous Mater 2018;268:69−76. Available from: https://doi.org/10.1016/j.micromeso.2018.04.001.

[23] Li Y, Chen Z, Su L, Chen W, Jin X, Xu H. Stochastic reconstruction and microstructure modeling of SMC chopped fiber composites. Compos Struct 2018;200:153−64. Available from: https://doi.org/10.1016/j.compstruct.2018.05.079.

[24] Gao Y, Jiao Y, Liu Y. Ultra-efficient reconstruction of 3D microstructure and distribution of properties of random heterogeneous materials containing multiple phases. Acta Mater 2021;204:116526. Available from: https://doi.org/10.1016/j.actamat.2020.116526.

[25] Yang M, Nagarajan A, Liang B, Soghrati S. New algorithms for virtual reconstruction of heterogeneous microstructures. Computer Methods Appl Mech Eng 2018;338:275−98. Available from: https://doi.org/10.1016/j.cma.2018.04.030.

[26] Feng J, Teng Q, He X, Wu X. Accelerating multi-point statistics reconstruction method for porous media via deep learning. Acta Mater 2018;159:296−308. Available from: https://doi.org/10.1016/j.actamat.2018.08.026.

[27] Li H, Chen P-E, Jiao Y. Accurate reconstruction of porous materials via stochastic fusion of limited bimodal microstructural data. Transp Porous Media 2018;125 (1):5−22. Available from: https://doi.org/10.1007/s11242-017-0889-x.

[28] Metropolis N, Rosenbluth AW, Rosenbluth MN, Teller AH, Teller E. Equation of state calculations by fast computing machines. J Chem Phys 1953;21(6):1087−92.

[29] Khatibi AA, Mortazavi B. A study on the nanoindentation behaviour of single crystal silicon using hybrid MD-FE method. Adv Mater Res Trans Tech Publ 2008;259−62.

[30] Mortazavi B, Khatibi AA, Politis C. Molecular dynamics investigation of loading rate effects on mechanical-failure behaviour of FCC metals. J Computat Theor Nanosci 2009;6(3):644−52.

[31] Mori T, Tanaka K. Average stress in matrix and average elastic energy of materials with misfitting inclusions. Acta Metallurg 1973;21(5):571−4.

[32] Halpin JC. Primer on composite materials analysis, (Revised). CRC Press; 1992.

[33] Halpin J. Stiffness and expansion estimates for oriented short fiber composites. J Compos Mater 1969;3(4):732−4.

[34] Fornes T, Paul D. Modeling properties of nylon 6/clay nanocomposites using composite theories. Polymer 2003;44(17):4993−5013.

[35] Li X, Gao H, Scrivens WA, Fei D, Xu X, Sutton MA, et al. Reinforcing mechanisms of single-walled carbon nanotube-reinforced polymer composites. J Nanosci Nanotechnol 2007;7(7):2309−17. Available from: https://doi.org/10.1166/jnn.2007.410.

[36] Luo J-J, Daniel IM. Characterization and modeling of mechanical behavior of polymer/clay nanocomposites. Compos Sci Technol 2003;63(11):1607−16.

[37] Sheng N, Boyce MC, Parks DM, Rutledge G, Abes J, Cohen R. Multiscale micromechanical modeling of polymer/clay nanocomposites and the effective clay particle. Polymer 2004;45(2):487−506.

[38] Zeng Q, Yu A, Lu G. Multiscale modeling and simulation of polymer nanocomposites. Prog Polym Sci 2008;33(2):191−269.

[39] Luo D, Wang W-X, Takao Y. Effects of the distribution and geometry of carbon nanotubes on the macroscopic stiffness and microscopic stresses of nanocomposites. Compos Sci Technol 2007;67(14):2947−58.

[40] Baniassadi M, Mortazavi B, Hamedani HA, Garmestani H, Ahzi S, Fathi-Torbaghan M, et al. Three-dimensional reconstruction and homogenization of heterogeneous materials using statistical correlation functions and FEM. Computat Mater Sci 2012; 51(1):372−9.

[41] Fisher F, Bradshaw R, Brinson L. Fiber waviness in nanotube-reinforced polymer composites—I: modulus predictions using effective nanotube properties. Compos Sci Technol 2003;63(11):1689−703.

[42] Kanit T, Forest S, Galliet I, Mounoury V, Jeulin D. Determination of the size of the representative volume element for random composites: statistical and numerical approach. Int J Solids Struct 2003;40(13−14):3647−79.

[43] Tu S-T, Cai W-Z, Yin Y, Ling X. Numerical simulation of saturation behavior of physical properties in composites with randomly distributed second-phase. J Compos Mater 2005;39(7):617−31.

[44] Fertig III RS, Garnich MR. Influence of constituent properties and microstructural parameters on the tensile modulus of a polymer/clay nanocomposite. Compos Sci Technol 2004;64(16):2577−88.

[45] Hbaieb K, Wang Q, Chia Y, Cotterell B. Modelling stiffness of polymer/clay nanocomposites. Polymer. 2007;48(3):901−9.

[46] Jafari A, Khatibi AA, Mashhadi MM. Comprehensive investigation on hierarchical multiscale homogenization using representative volume element for piezoelectric nanocomposites. Compos Part B: Eng 2011;42(3):553−61.

[47] Baniassadi M, Laachachi A, Hassouna F, Addiego F, Muller R, Garmestani H, et al. Mechanical and thermal behavior of nanoclay based polymer nanocomposites using statistical homogenization approach. Compos Sci Technol 2011;71(16):1930−5.

[48] Baniassadi M, Laachachi A, Makradi A, Belouettar S, Ruch D, Muller R, et al. Statistical continuum theory for the effective conductivity of carbon nanotubes filled polymer composites. Thermochim acta 2011;520(1−2):33−7.

[49] Hamedani HA, Baniassadi M, Khaleel M, Sun X, Ahzi S, Ruch D, et al. Microstructure, property and processing relation in gradient porous cathode of solid oxide fuel cells using statistical continuum mechanics. J Power Sources 2011;196(15):6325−31.

[50] Li D, Saheli G, Khaleel M, Garmestani H. Quantitative prediction of effective conductivity in anisotropic heterogeneous media using two-point correlation functions. Computat Mater Sci 2006;38(1):45−50.

[51] Nikolov S, Petrov M, Lymperakis L, Friák M, Sachs C, Fabritius HO, et al. Revealing the design principles of high-performance biological composites using ab initio and multiscale simulations: the example of lobster cuticle. Adv Mater 2010;22(4):519−26.

[52] Mortazavi B, Bardon J, Bomfim JAS, Ahzi S. A statistical approach for the evaluation of mechanical properties of silica/epoxy nanocomposite: verification by experiments. Computat Mater Sci 2012;59:108−13.

[53] Fullwood DT, Adams BL, Kalidindi SR. A strong contrast homogenization formulation for multi-phase anisotropic materials. J Mech Phys Solids 2008;56(6):2287−97.

[54] Sen AK, Torquato S. Effective conductivity of anisotropic two-phase composite media. Phys Rev B Condens Matter 1989;39(7):4504−15. Available from: https://doi.org/10.1103/physrevb.39.4504.

[55] Mortazavi B, Baniassadi M, Bardon J, Ahzi S. Modeling of two-phase random composite materials by finite element. Mori−Tanaka Strong Contrast Methods Compos Part B: Eng 2013;45(1):1117−25.

[56] Nosouhi Dehnavi F, Safdari M, Abrinia K, Sheidaei A, Baniassadi M. Numerical study of the conductive liquid metal elastomeric composites. Mater Today Commun 2020;23:100878. Available from: https://doi.org/10.1016/j.mtcomm.2019.100878.

[57] Chavoshnejad P, Razavi MJ. Effect of the Interfiber Bonding on the Mechanical Behavior of Electrospun Fibrous Mats. Sci Rep 2020;10(1):7709. Available from: https://doi.org/10.1038/s41598-020-64735-5.

[58] Chavoshnejad P, Alsmairat O, Ke C, Razavi MJ. Effect of interfiber bonding on the rupture of electrospun fibrous mats. J Phys D Appl Phys 2020;54(2):025302.

[59] Mahdavi M, Yousefi E, Baniassadi M, Karimpour M, Baghani M. Effective thermal and mechanical properties of short carbon fiber/natural rubber composites as a function of mechanical loading. Appl Therm Eng 2017;117:8−16.

[60] Yousefi E, Sheidaei A, Mahdavi M, Baniassadi M, Baghani M, Faraji G. Effect of nanofiller geometry on the energy absorption capability of coiled carbon nanotube composite material. Compos Sci Technol 2017;153:222−31.

[61] Chavoshnejad P, Ayati M, Abbasspour A, Karimpur M, George D, Remond Y, et al. Optimization of Taylor spatial frame half-pins diameter for bone deformity correction: Application to femur. Proc Inst Mech Eng H 2018;232(7):673−81. Available from: https://doi.org/10.1177/0954411918783782.

[62] Yarali E, Baniassadi M, Baghani M. Numerical homogenization of coiled carbon nanotube reinforced shape memory polymer nanocomposites. Smart Mater Struct 2019; 28(3):035026.

[63] Safdari M, Baniassadi M, Garmestani H, Al-Haik MS. A modified strong-contrast expansion for estimating the effective thermal conductivity of multiphase heterogeneous materials. J Appl Phys 2012;112(11):114318.

[64] Benveniste Y. A new approach to the application of Mori-Tanaka's theory in composite materials. Mech Mater 1987;6(2):147−57.

[65] Nogales S, Böhm HJ. Modeling of the thermal conductivity and thermomechanical behavior of diamond reinforced composites. Int J Eng Sci 2008;46(6):606−19.

[66] Hill R. A self-consistent mechanics of composite materials. J Mech Phys Solids 1965;13(4):213−22.

[67] Eshelby J.D. The determination of the elastic field of an ellipsoidal inclusion, and related problems. 1226 ed: R Soc., 1957, p. 376−396.

[68] Hiroshi H, Minoru T. Equivalent inclusion method for steady state heat conduction in composites. Int J Eng Sci 1986;24(7):1159−72.

[69] Abdel Rahman R, George D, Baumgartner D, Nierenberger M, Rémond Y, Ahzi S. An asymptotic method for the prediction of the anisotropic effective elastic properties of the cortical vein: superior sagittal sinus junction embedded within a homogenized cell element. J Mech Mater Struct 2012;7(6):19.

[70] Devries F, Dumontet H, Duvaut G, Lene F. Homogenization and damage for composite structures. Int J Numer Methods Eng 1989;27(2):285−98. Available from: https://doi.org/10.1002/nme.1620270206.

[71] Dumont JP, Ladeveze P, Poss M, Remond Y. Damage mechanics for 3-D composites. Compos Struct 1987;8(2):119−41. Available from: https://doi.org/10.1016/0263-8223 (87)90008-0.

[72] Sanchez-Palencia E. Comportements local et macroscopique d'un type de milieux physiques heterogenes. Int J Eng Sci 1974;12(4):331−51. Available from: https://doi.org/10.1016/0020-7225(74)90062-7.

[73] Riazat M, Tafazoli M, Baniassadi M, Safdari M, Faraji G, Garmestani H. Investigation of the property hull for solid oxide fuel cell microstructures. Computat Mater Sci 2017;127:1−7.

[74] Montazerian H, Davoodi E, Asadi-Eydivand M, Kadkhodapour J, Solati-Hashjin M. Porous scaffold internal architecture design based on minimal surfaces: a compromise between permeability and elastic properties. Mater Des 2017;126:98−114.

[75] Kaur M, Yun TG, Han SM, Thomas EL, Kim WS. 3D printed stretching-dominated micro-trusses. Mater Des 2017;134:272−80.

[76] Wang Y, Zhang L, Daynes S, Zhang H, Feih S, Wang MY. Design of graded lattice structure with optimized mesostructures for additive manufacturing. Mater Des 2018;142:114−23.

Numerical characterization of micro- and nanocomposites

2

Abstract

Additive inclusion significantly can improve thermomechanical properties of matrixes. Many researchers have focused their studies on determining the effective thermal and mechanical properties of micro- and nanocomposites. Modeling mechanical and thermal properties of these types of composites is one of the most important challenges in materials science. In this chapter, at first, different approaches to realize and reconstruct representative volume elements (RVEs) are discussed, and in the second step, numerical characterization is applied on RVEs to calculate effective thermal properties (such as thermal conductivity) and mechanical properties. Different types of samples with linear and nonlinear constitutive material models are discussed, and interphase zone are taken into account in the modeling of nanocomposites. Different types of RVEs which include different categories of inclusions like cylindrical, penny, helical, and ellipsoidal are generated, and effective mechanical and thermal properties are calculated.

2.1 Introduction

In recent years, there has been significant progress in fabricating nanomaterials such as graphite nanplateles [1−3], carbon nanocones [4,5], carbon nanotubes [6−8], coiled carbon nanotubes (CCNTs) [9−11], nanoclay [12,13], and metal oxide nanoparticles with extraordinary mechanical, electrical, and thermal properties.

Cement composites or the associated materials in construction of buildings have brittle nature, low tensile strength, and durability. Therefore extensive research has been conducted to improve the mechanical properties of cement composites. Different types of reinforcement, such as fibers or steel bars, have traditionally been employed to improve the mechanical properties of cement and resist microcrack growth [14−20].

Various nanoparticles such as carbon nanotubes, TiO_2, silica, graphite nanoplatelets (GNPs), and so on have been exploited to reinforce cement composites. For example, graphene can improve mechanical properties by creating a bridge between microcracks. Nanomaterials such as graphene, graphene oxide (GO), or carbon nanotubes have a high specific surface and extraordinary properties. Therefore nanotechnology can be applied to develop a cement composite with optimum properties [14,21−30].

Some particular applications of nanocomposites could be found in the aerospace, automotive, electronics, oil, and similar industries. For a while, researchers have

---

Applied Micromechanics of Complex Microstructures. DOI: https://doi.org/10.1016/B978-0-443-18991-3.00008-8
© 2023 Elsevier Inc. All rights reserved.

focused their attention on carbon nanotubes (CNTs) due to their excellent mechanical properties, strength, modulus [31–38], and electrical and thermal conductivities along with a low density. Therefore CNTs are great fibers for making nanocomposites with exceptional multifunctional properties [6,39]. In early 1990s Ihara et al. claimed the existence of CCNTs [40]; nearly a year after, Zhang et al. synthesized regular CCNTs [41]. CCNTs were imaged by scanning electron microscopy (SEM) in 1998 [42]. One of the most popular methods for making multiwall carbon nanotubes is chemical vapor deposition (CVD) [43], and in the synthesis of CNTs, some helical-shaped carbon nanotubes were observed by accident in the jungle of CNTs [44]. Later, CVD became the most efficient method to produce CCNTs [11]. Recently, researchers have shown particular interest to nanocomposites reinforced by CCNTs for exceptional properties and specific geometry of CCNTs [11,45]. Volodin et al. [46] produced CCNTs by thermal decomposition of hydrocarbon gas and then evaluated local elasticity by FMM (force modulation microscopy). Their results showed that coiled multiwall nanotubes have a high module equal to 0.7 TPa. In addition to experimental work, some researchers used numerical methods to predict CCNT properties [47–52]. Feng et al. [53] used molecular mechanics (MM) simulation to study the tensile behavior of carbon nanosprings (CNSs). In this research the spring stiffness of three-turn CNS was calculated to be 0.36 N/m. CCNTs have a helical structure; in other words, CCNT is similar to a CNT which is turned into a spring with a uniform pitch and rise angle. CCNT has the following geometric characteristics (Fig. 2.1): tube diameter (d_c), coil diameter (D_c), helical angle (θ), length of coil tube (l_c), coiled length (L_c), pitch length (p), and number of pitches (N_c) [54].

Li et al. [55] produced CCNTs using the CVD method and made a polymer CCNT nanocomposite. They showed that by adding 5 wt% CCNTs to the epoxy resin, the elastic modulus increased almost 40%, but toughness was reduced [54].

To understand the mechanics of nanocomposites in the nanoscale, experimental approaches are often difficult and impractical. Therefore computational and analytical techniques have been developed to study the mechanics of nanocomposites. Eshelbi, Mori–Tanaka, self-consistent, and Halpin–Tsi are analytical methods. Molecular dynamics (MD), MM, finite element, and multiscale modeling are numerical methods. The MD method for a complex model is computationally expensive. The MD method is suitable for studying the interface between fibers and matrixes [56–58]. Chen et al. [59] used MD to study interaction forces between single-wall carbon nanotubes (SWCNTs) and polyphenylacetylene (PPA). Some studies [60,61] assumed noncovalent bonds between CNTs and matrixes. Frankland et al. [62] used MD simulation to model the interface between CNTs and matrixes

Figure 2.1 Geometrical parameters of the solid CCNT model [54].

and showed that by adding less than 1 wt% CNTs to polyethylene, the shear strength increased by 10-fold. Almasi et al. [63] investigated the effect of thickness and stiffness of the interphase between clay and epoxy nanocomposite; their results showed that the elastic modulus of the nanocomposite reduces by increasing the interphase thickness [54].

Multiscale modeling facilitates bridging the gap between the nanoscale and macroscale by considering the physical and mechanical properties of nanocomposites, and hence, data from MD modeling and a suitable multiscale model can be integrated to describe the phenomena at each scale. There are some studies [64,65] that used multiscale modeling (nanoscale, microscale, mesoscale, and macroscale) to explore nanocomposite properties. Effects of CNT length, waviness, and volume fraction were studied on properties of nanocomposites. One of the issues encountered in modeling is understanding the damage and fracture mechanism of nanocomposites. In theoretical and numerical methods the interface of particles and the matrix is usually assumed to be strong or perfect. However, the interface behavior can significantly affect the properties of nanocomposites. Therefore the assumption of perfect or strong bonding is not realistic for nanocomposites. There are some studies about growing crack in nanocomposites [66−71]. Most debonding occurs around corners of clays because of discrepancy between the matrix and nanoparticle elastic modulus [72,73]. Hamedia et al. [74] studied the total failure energy of nanocomposites reinforced by clays. Perfect bonding was considered between matrixes and nanoparticles, and failure of matrixes has been investigated. Toughness and fracture energy are two of the important material properties that many researchers paid attention to; for instance, in [75] fracture toughness and tensile strength were obtained by using the J integral method [54].

Khani et al. [76] used the finite element method (FEM) to study the effect of geometrical parameters of CCNTs on mechanical properties of nanocomposites. Their result showed that nanocomposites reinforced by CNTs had a higher elastic modulus compared to nanocomposites reinforced by CCNTs [54].

Safaie et al. [70] developed an interfacial debonding-induced damage model to study the damage initiation in GNPs and polyethylene. They showed that for a weakly bonded interface, nanocomposites had lower stress−strain response compared to the pure matrix. Therefore interfacial strength of nanofillers and matrixes is an important parameter that has a high impact on overall mechanical performance of nanocomposites. Zhang et al. [77] used a three-dimensional analytical solution for modeling a nanocomposite reinforced with CNT and modeled the interphase with a gradient elastic modulus. In their study a van der Waals-based cohesive law was set for the connection between CNTs and the interphase. The results showed that the stiffness of the nanocomposite is highly dependent on interphase and cohesive parameters [54].

In this chapter, different types of RVEs with nanoparticles are realized and reconstructed using a statistical approach. Generated RVEs are exploited to calculate effective mechanical and thermal properties. In this chapter a variety of samples for linear and nonlinear behavior of polymer nanoconposites (PNCs) have been simulated, and the finite element method is used to homogenize thermomechanical properties of microstructures.

2.2 Realization of composites with different types of inclusions

2.2.1 Composites loaded with ellipsoidal inclusions

Numerical simulations have been carried out inside a cubic cell with a side length of 1000 nm. Nanoparticles are modeled as simple ellipsoids of oblate, prolate, and spherical shapes. Dispersion of nanoparticles should be uniform (both the spatial position and orientation) inside the unit cell. In oblate and prolate ellipsoids, one semiaxis is held at 100 nm, and the other varies according to the aspect ratio. The ellipsoids should not penetrate each other. This requires checking the minimum distance between each generated ellipsoid with neighboring elements. An auxiliary envelope is then generated surrounding each inclusion representing the interface distance [78].

Ellipsoid generation was done using the following method. Two of the semiaxes were assumed to be equal in length, and the normal vector was assumed to be aligned with a semiaxis. Points representing ellipsoid centers were distributed using the Monte Carlo technique, and the direction of the normal vector was determined using random homogeneous functions [Eq. (2.1)] [78].

$$\begin{cases} \theta = 2\pi v \\ \varphi = \arccos(2u - 1) \end{cases} \tag{2.1}$$

In the above equation, θ and φ indicate spherical coordinate system angles (shown in Fig. 2.2), and u and v are random variables between 0 and 1 [78].

The following criterion was checked after generation of each ellipsoid to ensure that none of ellipsoids penetrate each other. If the newly generated ellipsoid did not satisfy the conditions described in Eq. (2.2), it would be dismissed. Generated geometries would only be accepted if they satisfied the aforementioned criterion [78].

$$\begin{aligned} &\text{if } D > d_{max} \Rightarrow \text{ No Contact.} \\ &\text{if } D \le d_{min} \Rightarrow \text{ Contact Exists.} \\ &\text{if } d_{min} \le D \le d_{max} \Rightarrow \begin{cases} \text{if } D > d_c \Rightarrow \text{No Contact.} \\ \text{if } D \le d_c \Rightarrow \text{Contact Exists.} \end{cases} \end{aligned} \tag{2.2}$$

where

$$\begin{aligned} d_{max} &= \max\{a, b, c\} \\ d_{min} &= \min\{a, b, c\} \\ d_c &= \text{distance of closest approach of two ellipsoids} \end{aligned} \tag{2.3}$$

a, b, and c are the semiaxes of the ellipsoid, D is the distance between centers of the two ellipsoids, and d_c is their minimum distance [79–81].

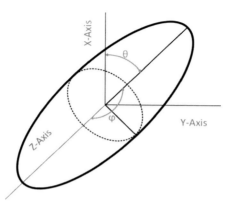

Figure 2.2 3D representation of the normal semiaxis angles with the Cartesian axes.

Figure 2.3 A cut section of a simulated oblate inclusion. The interphase is shown as a layer surrounding the core ellipsoid [78].

The interphase region is of particular interest when studying properties of PNCs [82]. In nanocomposites, inclusions have a high specific surface, resulting in a significant interphase region [78,82].

With increasing number of inclusions, the possibility of occurrence of penetrating interphase regions forming percolating networks increases. These percolating networks can affect the properties of the nanocomposite [82].

To study the effect of interphase thickness, three thickness values were prescribed based on values reported by other researchers [83,84]. The interphase layer was obtained by offsetting each ellipsoid by the thickness value (illustrated in Fig. 2.3). It should be noted that this interphase region was allowed to penetrate neighboring interphase regions, forming percolated networks throughout the RVE [78].

Percolation is defined as a group of connected clusters in a nonconducting medium [85]. Percolation algorithms are used for testing the continuity of touching inclusions. The criterion for determination of percolation occurrence is existence of at least one cluster that makes a path between the free surfaces of the RVE [78,86].

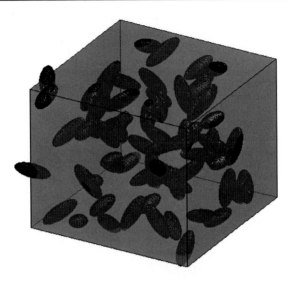

Figure 2.4 Isotropic random distribution of prolate inclusions [78].

In the present study, percolation occurrence check has been performed after each ellipsoid generation. If clustering is detected a cluster number is assigned to it, and if the newly created cluster is connected to existing clusters, it is merged with the existing cluster(s) with a unified cluster number. It should be noted that the percolation occurrence check algorithm becomes unstable for small interphase thickness values. A sample of the isotropic random distribution of inclusions and the generated clusters within the RVE is illustrated in Fig. 2.4 [78].

2.2.2 Composites loaded with helical inclusions

A 3D RVE consisting of a polymer and CCNTs was constructed. The RVE is the smallest material volume element of the composite for which the usual spatially constant "overall modulus" macroscopic constitutive representation is a sufficiently accurate model to represent mean constitutive response [87]. In Fig. 2.5 stress−strain curves of the nanocomposite with different sizes were plotted. For all of the samples, the RVE lengths were 3- to 5-fold compared to the length of CNTs [54].

Optimum distribution and volume fraction of CNTs in nanocomposites are very important [88]. Nanofillers were distributed randomly by the Monte Carlo method. A computational code has been developed in MATLAB® to avoid any intersection between nanofillers. To determine the direction of a nanotube a vector along the tube axis is defined by random homogeneous Monte Carlo functions that are described by Eq. (2.1) [89].

A convergence study was performed to determine the RVE size. RVE size is the smallest volume where by increasing that volume the overall properties of the

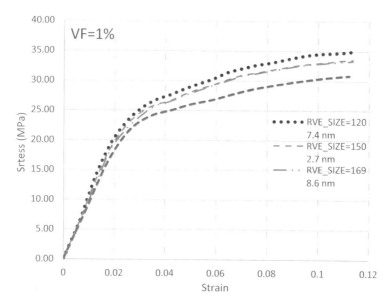

Figure 2.5 Effect of RVE size on the stress–strain curve of HDPE reinforced by 1% volume fraction carbon nanotubes [54].

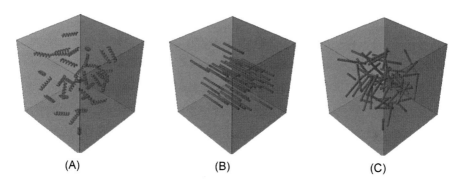

Figure 2.6 Three samples of RVE with VF = 1% for different shapes, (A) CCNT (isotropic), (B) aligned carbon nanotube, and (C) carbon nanotube (isotropic) [54].

material do not change significantly. The RVEs were generated in ABAQUS 6.13 for different volume fractions and different shapes of nanofillers. RVEs are shown in Fig. 2.6 [54].

CCNT's morphology plays an important role in the overall properties of the nanocomposites. The geometrical parameters for each CCNT studied in this section are summarized in Table 2.1. All nanofillers were labeled (filler index), which can be used to later identify the type of nanofillers in the nanocomposite. In FEM analysis, CCNT can be an equivalent solid element, shell element, 3D truss element

Table 2.1 Types of nanofillers [54].

Filler index	No 1	No 2	No 3	No 4	No 5	No 6
Type of filler						
Θ (°)	90	50	15	15	15	15
L_c(nm)	500	388.97	167.13	167.13	334.26	334.26
d_c(nm)	40	40	40	40	40	30
D_c(nm)	–	90	90	100	90	90

[90], or beam element [91]. The equivalent continuum solid element has been used for CCNT in the finite element model. We will discuss homogenization of mechanical properties in the next part of this section [54].

2.3 Experimental reconstruction of inclusionary composites

2.3.1 Small-angle neutron scattering

Two-point or higher-order correlation functions describe the distribution, orientation, and shape of different phases in heterogeneous materials. These functions can be determined from appropriate microstructure measurements [92−95], which contain the morphological information. In nanotube-polymer nanocomposites the heterogeneity can be indicated by nanotube distribution in the polymer matrix, nanotube curvature, and the local heterogeneity of the nanotubes, the so-called dispersion state [96]. In fact the dispersion state most significantly affects material properties. Dispersion of the nanoparticles within the polymer matrix is therefore the key parameter to take into consideration in the statistical theory [97].

To achieve information about nanotube dispersion, electron microscopy (i.e., TEM and SEM) or scattering can be employed [98]. In the case of microscopy, however, images are only relevant when the entire dispersion gradients of the nanoparticles are indicated [99]. The size of nanoparticle aggregates depends on the nanoparticle size, the processing method, and the chemical interaction between nanoparticles and the matrix and can reach several hundred nanometers. In microscopy, correlation functions strongly depend on the magnification at which the images are recorded. If a high magnification is used the correlation function will be dictated by the position in the heterogeneous material, where the microscopic images are taken (e.g., whether the images are chosen to include aggregates or not) [100]. In other words a high-resolution image may not represent the representative area (or volume), which is much larger than the selected image [101]. On the contrary a lower-magnification image contains more statistical information of the morphology at the price of resolution. The dispersion state of nanoparticles can be also characterized by neutron scattering measurements. Small- and ultra-small-angle neutron scattering (SANS and USANS, respectively) are easy and fast measurement techniques that encompass detailed structural information over several cubic millimeters (high statistics) of the sample without compromising the resolution. The scattering signal illustrates the distribution, shape (form factor), and relative position (structure factor) of the scatterers (nanotubes) in the polymer matrix. SANS and USANS signals can be exploited to calculate two-point correlation functions (TPCFs) that represent the material morphology [97,102−105].

2.3.1.1 Mathematical theory of small-angle neutron scattering

SANS and USANS rely on scattering intensity from heterogeneities whose size typically ranges between a few nanometers to tens of microns, depending on the equipment configuration. The scattered intensity depends on the difference between a local scattering length density (SLD), ρ, from the scattered heterogeneities and its surrounding, which can be represented by an average density $\overline{\rho}$. The local fluctuation η of the SLD can be defined as follows [97]:

$$\eta = \rho - \overline{\rho} \tag{2.4}$$

Assuming a statistically isotropic system with no long-range order a correlation function that considers the amplitude of the SLD fluctuations can be defined as [97]

$$\gamma(r)\langle\eta^2\rangle = \langle\eta_A\eta_B\rangle \tag{2.5}$$

where A and B are two distinct points in the sample that are defined by the vectors r_a and r_b, respectively, and $\gamma(r)$ is the characteristic or autocorrelation function depending on the vector r, where $r = r_b\text{-}r_a$. $\gamma(r)$ can be defined as follows [97]:

$$\gamma(r) = \langle\eta(r_a)\eta(r_b)\rangle \tag{2.6}$$

For random distribution of heterogeneities the autocorrelation function $\gamma(r)$ must satisfy the following conditions: $\gamma(r = 0) = \eta^2$ and $\gamma(r \rightarrow \infty) = 0$. It is convenient to define the autocorrelation function of phase 1 for a statistically homogeneous medium as [97,106]

$$\gamma(r) = \langle\eta(r_a)\eta(r_b)\rangle = P_2^1(r) - \phi_1^2 \tag{2.7}$$

where ϕ_1 is the volume fraction of phase 1 (fillers) and $P_2^1(r)$ is the two-point probability function, that is the probability that phase 1 exists at both r_a and r_b. Recalling that $\rho(r)$ is the SLD per unit volume a volume element dV will have an SLD that is equal to $\rho(r) \times dV$ [97].

The intensity of the neutron scattering, I, as a function of the scattering vector $\vec{q}$ over the entire volume V is given by the following Fourier integral [97,107]:

$$I(q) = \iiint \iiint dV_1 dV_2 \rho(r_1) \rho(r_2) e^{-iqr} = \iint \rho(r_1) \rho(r_2) e^{-iqr} dr_1 dr_2 \tag{2.8}$$

An autocorrelation function can then be defined as

$$\tilde{\rho}^2(r) = \iiint dV_1 \rho(r_1) \rho(r_2) \tag{2.9}$$

which allows $I(q)$ to be rewritten as

$$I(q) = \iiint dV \tilde{\rho}^2(r) e^{-iqr} dr \tag{2.10}$$

The intensity distribution in reciprocal space is uniquely determined by the structure of the density field. Considering statistical isotropy, it can be proved that [104]

$$\langle e^{-iqr} \rangle = \frac{\sin(qr)}{qr} \tag{2.11}$$

As a result the average scattering intensity of an isotropic sample reduces to

$$I(qh) = \int \tilde{\rho}^2(r) \frac{\sin(qr)}{qr} 4\pi r^2 dr \tag{2.12}$$

Recalling the autocorrelation function, γ, the above equation can be rewritten:

$$I(q) = V n_0^2 \int 4\pi r^2 \, \gamma(r) \frac{\sin(qr)}{qr} dr \tag{2.13}$$

where n_0 is the mean scattering density of the sample. Therefore

$$\gamma(r) = \frac{1}{2\pi^2 V n_0^2} \int_0^\infty I(q) \frac{\sin(hr)}{hr} q^2 dq \tag{2.14}$$

Here, n_0 is a constant. Using Eq. (2.7), Eq. (2.14) can be rewritten as follows [97]:

$$\gamma(r) = P_2^1(r) - \phi_1^2 = \frac{1}{2\pi^2 V n_0^2} \int_0^\infty I(q) \frac{\sin(hr)}{hr} q^2 dq \tag{2.15}$$

where $P_2^1(r)$ represents the two-probability correlation function, which characterizes the spatial distribution of the heterogeneities (phase 1) in the matrix (phase 2). The intensity, $I(q)$, becomes very small at higher q values, and it is therefore safe to only integrate Eq. (2.15) for the SANS and USANS q ranges. The two-point probability function $P_2^1(r)$ should verify the following conditions [97]:

$$P_2^1(r) = \phi_1 \quad \text{when } r = 0$$

$$P_2^1(r) = (\phi_1)^2 \quad \text{when } r \rightarrow \infty \tag{2.16}$$

The second condition in Eq. (2.16) is an indicator of the degree of homogeneity of the distribution of heterogeneities in the matrix (i.e., if the second condition is not verified, then the distribution of the heterogeneities is not homogeneous in the matrix) [97].

2.3.1.2 Structural characterization using SANS materials and preparation of the nanocomposites

Nano-cPT-100 high-purity ($>97\%$) SWCNTs from Nano-C, Inc. were used. They have a length of $1-1.1$ microns and a diameter of $0.9-1.3$ nm according to Nano-C. PEDOT:PSS was in the form of a 1.3% solution in water under brand name Clevios PH 1000. While CNT dispersion in a polymer matrix is often burdensome, PEDOT:PSS is known to adsorb on (i.e., wrap around) and disperse CNTs, rendering the formation of a relatively well-dispersed CNT-PEDOT:PSS nanocomposite. Bath sonication for 48 hours was carried out to achieve a relatively good dispersion of SWCNTs in the solution. The samples were drop-cast from the solution of Clevios and SWCNTs, dried in the fume hood overnight, and postdried in a vacuum oven at 80°C for 2 hours [97].

2.3.1.2.1 SANS and USANS

SANS data were collected at the high flux isotope reactor (HFIR) at the Oak Ridge National Laboratory (ORNL), and USANS data were collected at the National Institute of Standards and Technology (NIST) center for neutron research. The combined SANS and USANS data for the SWCNT-PEDOT:PSS samples are plotted in Fig. 2.7. For this plot the logarithm of the intensity (cm^{-1}) minus the incoherent scattering background is plotted as a function of the logarithm of scattering vector, q (1/Å). SANS and USANS provide structural information over a range of length scales (1 nm to 30 μm) that span segmental polymer structures, polymer−nanotube interfaces, individual nanotubes, and aggregates of nanotubes. Based on the examined structural sizes, RVEs were constructed over a volume of $30 \times 30 \times 30$ microns^3 [97].

2.3.1.3 Reconstruction of nanostructures using the SANS approach

Drop casting from solutions will result in a randomly distributed CNT network. The building blocks for such networks are CNT bundles that are separated by the polymer matrix. To graphically reconstruct this network, bundle diameters and curvatures are required. Bundle diameters can be extracted from the analysis of the

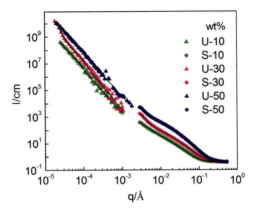

Figure 2.7 U (USANS) and S (SANS) samples. The symbols represent the data, and solid lines are the fits to the profiles. The difference of intensities of SANS (S) and USANS (U) data are attributed to the variation in the normalization of the data from the two instruments [97].

Figure 2.8 Representative SEM image of a 10 wt. % SWCNT -PEDOT:PSS sample [97].

mid-q range SANS data, and bundle curvatures were extracted from SEM micrographs of the cross-sections of the samples. The SEM micrograph for a 10wt% loaded CNT-PEDOT:PSS sample is shown in Fig. 2.8 [97].

The hierarchical Beaucage model was used to calculate the CNT bundle size [108,109]. This model has been useful in the analysis of light, X-ray, and neutron scattering of CNTs [110–112]. The Beaucage model assumes that the material is composed of N domains where each domain contains two components with the first representing the size of a scattering domain (R_g) and the second representing an interfacial scaling power (p) of the domain. The Beaucage function is correlates scattering intensity, I, to wave vector, q, as follows [97]:

$$I(q) = \sum_{i=1}^{N} \left[G_i exp\left(-q^2 R_{g,i}^2/3\right) + B_i \left[\frac{erf\left(qR_{g,i}/\sqrt{6}\right)^3}{q}\right]^{P_i} \right] + B \qquad (2.17)$$

where N is the hierarchical level, G_i and B_i are the prefactors for the two components in a given level, R_g is the radius of gyration of the level, P_i is the power of the interfacial term in the level, and B is the incoherent background. Following the analysis of the experimental mid-q range SANS curves, CNT bundle diameters of 8.7, 11.0, and 11.5 nm for the 10, 30, and 50wt% samples were calculated, respectively [97].

The first step in developing the RVE is to specify the RVE boundaries and then generate an initial structure of the random CNT nanocomposite. To simulate a long wavy nanotube, each nanotube was divided into many adjoined segments as shown in Fig. 2.9. Curvature in the CNT was determined by the number of segments and the orientation angle between the adjoining unit cylinders. As such a random point within the RVE is picked, and the first segment of the bundle is randomly generated. The second segment starts from the end point of the first segment and is bounded to be within a cone volume of cone angle ϕ. Cone angle determines CNT bundle curvatures and was roughly estimated from the SEM images. Cone angles were ~90 degrees for both 10% and 30% CNT samples and ~60 degrees for the 50% CNT sample. High CNT loadings will restrict the bundles to twist as much compared to the lower lodgings. Input parameters to generate bundles were the bundle diameter from the Beaucage analysis, number of segments in each bundle, and range of spatial angles between segments in each bundle. Using this approach, any curvature can be produced by adjusting the curvature angle and segment length. Bundles were assumed to be as large as the

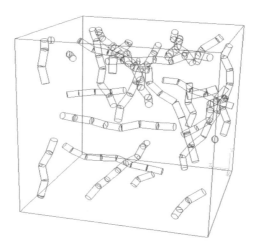

Figure 2.9 Adjoined segment structure used to model the curvy CNT network [97].

RVE size, that is, 30 microns. The employed method guarantees a random, isotropic distribution of carbon nanotubes. After the constructed model reached the objective volume fraction, its TPCF was calculated and compared against the experimental TPCF. Cone angles and segment numbers were then iteratively varied in the model structure to best approximate the experimental TPCF. A schematic of the construction algorithm is shown in Fig. 2.10 [97].

As explained in Section 1.1 the angular dependence of scattering that evolves from the scattering experiment is the Fourier transform of the density distribution function of the scattering heterogeneities in the sample [113]. The distribution, orientation, and shape of the nanotubes are described by two-point or higher-order correlation functions and can be determined from scattering experiments using Eq. (2.15). Experimental scattering curves of the PEDOT:PSS-CNT nanocomposites were used to calculate their TPCFs using Eq. (2.15) [97].

As shown in Fig. 2.11, there is good agreement between the TPCFs calculated for the RVE and from the scattering curves, specifically at small lengths ($r < 10$ m). In general, more than one structure can be responsible for a specific scattering curve. For the system of high-aspect ratio nanotubes, however, the scatterers are in the form of large cylinders, and therefore the scattering curves and their corresponding TPCFs should be very close to those of the actual nanostructure [97].

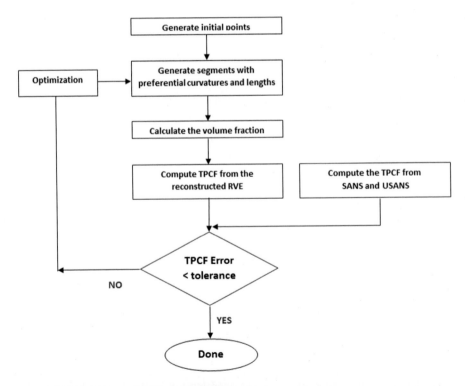

Figure 2.10 Reconstruction flow chart [97].

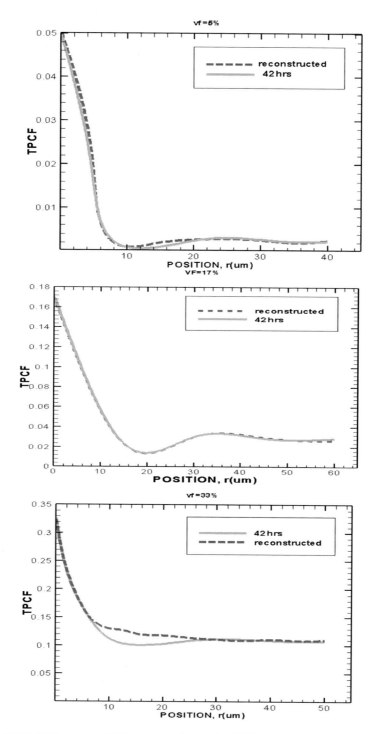

Figure 2.11 TPCF curves of RVE and experiments for CNT volume fractions of 5, 17, and 33% (10, 30, and 50 wt%, respectively) [97].

Fig. 2.12A and B show a reconstructed CNT structure based on the scattering curve of Fig. 2.7. The RVE and experimental TPCFs for this structure match well. The generated RVE can be meshed and implemented in an FEM simulation [97].

2.3.2 Reconstruction using SEM-FIB images

2.3.2.1 Serial sectioning of the nanocomposite using FIB-SEM

SEM was carried out on the Halloysite nanotube/ polypropylene composite sample with a pixel size of 12.17 nm. The dimensions of the slides were 1024×768 pixels, which represent 12.46×9.34 μm area. The distance between the slices was 50 nm.

Simultaneous sectioning and imaging of the nanocomposite was performed using a dual-column focused ion beam (FIB)-SEM (Carl Zeiss Auriga CrossBeam). The ion source was gallium, and the ion and electron columns were at a fixed angle of 36 degrees. The sample was tilted to 54 degrees so that the surface of the composite was perpendicular to the ion beam. After tilting the electron beam image was at 36 degrees, so a software-controlled tilt correction was enabled to ensure that the image is orthogonal.

Serial sectioning involved the removal of a known volume of material by the ion beam, followed by an incremental analysis with the electron beam. Because the sputtered material may redeposit onto the surface to be analyzed, significant in situ sample preparation was required. To begin a trapezoid was milled into the composite such that the shorter face was in a position to be imaged by the electron beam. The width of the shorter face was around 12 μm, and this width defined the width of the serial sectioning. The wider end of the trapezoid allowed for an unobstructed view of the analysis face. Two wings were cut as rectangles of 3×15 μm, approximately 6 μm on either side of the short face, such that after milling, a shape like that shown in Fig. 2.13A was observed. The wings were used as channels for the sputtered material to redeposit away from the surface of interest. A large beam (30 kV, 20 nA) was used to excavate the bulk of the material, and a smaller beam (30 kV, 4 nA) was used to square the edges. The trenches were milled to a depth of 20 μm. Water vapor was leaked into the chamber above the sample to assist the etching [114].

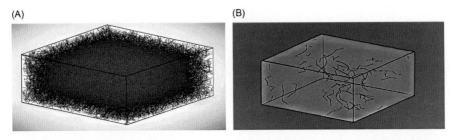

Figure 2.12 (A) Reconstructed RVE for a CNT volume fraction of 33%. (B) Diluted RVE showing the structure of curved CNTs [97].

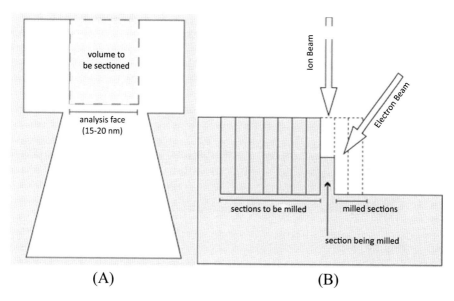

Figure 2.13 Schematic representation of serial sectioning: (A) front view and (B) top view [114].

A polished face was created by milling with a fine current (30 kV, 1 nA) to a depth of 20 μm. The 1 nA current, while larger than typically used for fine milling, was sufficient for producing an image that was adequate for the reconstruction. An area was selected on this polished face that was deemed representative of the entire face. The imaging electron beam was 2.0 kV and 1.2 nA. The area was chosen near the top and middle to avoid any contamination by redeposition. A volume was then established in the software (SmartSEM, Carl Zeiss) with a width and height larger than the viewing area. A milling current of 1 nA was used again. A smaller current could have been used to produce a cleaner surface but at the cost of a longer analysis time.

A schematic of the serial sectioning is shown in Fig. 2.13, and the real images recorded during the FIB procedure are presented in Fig. 2.14 [114].

The width of each slice was 50 nm; in other words, 50 nm of the nanocomposite would be milled away with the ion beam, followed by an image capture with the electron beam. A slow scan speed was chosen with the electron beam to ensure low noise in the image. Additionally the image contrast was turned slightly higher than what would normally be used to acquire a good image to accentuate the Halloysite nanotube from the matrix and aid in the reconstruction. Around 60 to 100 slices were taken per sample, a process that took 2−3 hours [114].

2.3.2.2 *Computational models*

2.3.2.2.1 3D reconstruction using VCAT software

By importing the bitmap files obtained at each step in serial sectioning into VCAT software (released by V-CAD Program, RIKEN, Japan), the 3D object is represented with

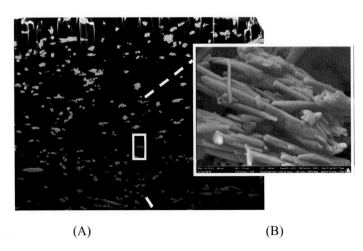

Figure 2.14 Multiscale imaging of the Halloysite nanotube composite; (A) slide generated through serial sectioning of the Halloysite nanotube composite and (B) Halloysite nanotubes imaged with Auriga SEM [114].

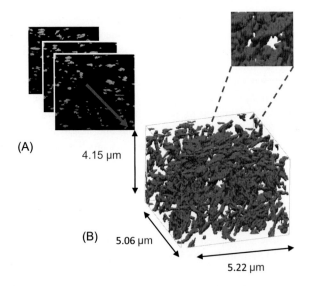

Figure 2.15 (A) 2D SEM images of the Halloysite nanotube polypropylene composite and (B) 3D reconstruction of the RVE based on serial sectioning [114].

gray levels in the range of colors 0–255 according to the image binarization mode 8-bit HSV (hue, saturation, value) color map (Fig. 2.15). By choosing a threshold of (0,0,100) the image part representation gives the best approximation of the dimensions of the cluster of Halloysite nanotubes inside the matrix. A mask property is associated with the matrix that will be a color value between 0 and 255. A noise reduction filter is applied in order to smooth the surface on which the mask property is applied. Each phase (matrix and filler) has a unique ID in order to distinguish between the phases inside the composite [114].

VCAT software offers a relatively simple function for mesh generation, refinement, simplification, volume data storage, and data transfer. Mask Property software is used to assign material properties to each phase of the nanocomposite prior to finite element analysis to predict the mechanical properties. The inclusion volume fraction of the reconstructed RVE has been calculated using Mask Property software to be 7.2%. This value is very close to the experimental data [114].

2.4 Numerical homogenization of thermomechanical properties

2.4.1 Finite element homogenization of polymer nanocomposites with the interphase zone

In this section the effect of percolation of the interphase on effective Young's modulus has been investigated using the Monte Carlo method and percolation check algorithms for spherical, oblate, and prolate inclusion shapes. The results suggest that an increased inclusion volume fraction increases the effective Young's modulus and that beyond percolation threshold a jump is observed. The significance of this jump varies with the shape and is most significant in the case of prolate-shaped inclusions. Furthermore the results show that a thicker interphase results in improved elastic properties.

In conclusion, nanocomposites with prolate inclusions and a high interphase thickness were found to exhibit superior elastic characteristics, which are postulated to be due to percolation occurring at lower inclusion volume fractions [78].

2.4.1.1 Methodology

RVE generation for composite with ellipsoidal inclusion with the interphase zone has been discussed in the section. Python scripts were used to set up a model in the commercial finite element package Abaqus based on the ellipsoid generation algorithm described in the previous section. The ellipsoid generation algorithm has been implemented in a MATLAB program. The output is then read by the Python script that generates the 3D structure in Abaqus. The resulting elements are then placed and integrated with the cubic domain using Boolean operations. Material properties of different regions of the model have been specified in accordance to those reported by Peng et al. [115] and are listed in Table 2.2 [78].

Specified boundary conditions are such that one side of the RVE is fixed, and a 10% strain has been applied on the opposite side. The model was meshed using tetrahedral 3D stress elements.

Table 2.2 Young's modulus and Poisson's ratio for the matrix, particle, and interphase [78,115].

	Young's modulus (GPa)	Poisson's ratio
Matrix	4.2	0.4
Particle	88.7	0.26
Interphase	8.4	0.4

Each model undergoes postprocessing in order to extract the stress, strain, and volume of individual elements, which is then used in homogenization. The homogenization method used in this study is described by [78]

$$\sum \sigma_i \times v_i = C \sum \varepsilon_i \times v_i \tag{2.18}$$

where $\sigma_i, \varepsilon_i, v_i$ denote the stress, strain, and volume of individual elements, respectively, and C is the elasticity tensor.

2.4.1.2 Effect of percolation of the interphase on mechanical properties

Elastic properties of the RVE have been studied for spherical, oblate, and prolate inclusions for 50, 75, and 100 nm of interphase thickness, respectively. Smaller values of the interphase thickness would cause the computational cost to rise exponentially for little gain, which is why they have been avoided. The process involved a gradual increase of the inclusion population until the percolation threshold was reached. Elastic properties of the resulting RVEs are presented in Figs. 2.16, 2.17, and 2.18 [78].

Fig. 2.16 shows that the Young's modulus increases with the volume fraction of the nanoparticles. It should be noted that the percolation occurs at 0.09, 0.1, and 0.11 inclusion volume fractions for the different interphase thicknesses; furthermore, this coincides with the onset of a sudden increase in the Young's modulus of the RVE. The results also suggest that higher interphase thicknesses yield superior elastic properties of the structure [78].

In the case of oblate inclusions, similar behavior is observed, and elastic properties improve with increasing inclusion volume fraction (shown in Fig. 2.17). An increased Young's modulus is observed at 0.0018, 0.0015, and 0.0012 corresponding to the percolation threshold as was expected. The difference in the onset of Young's modulus jump was found to be more profound in the case of oblate inclusions compared to that of the spherical inclusions [78].

The increased Young's modulus due to an increased inclusion volume fraction was observed in prolate inclusions as well, and the onset of the jump was found to happen at 0.007, 0.009, and 0.01 for the three interphase thickness levels (see Fig. 2.18).

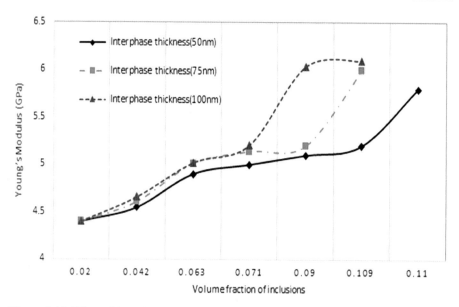

Figure 2.16 Effect of inclusion volume fraction on Young's Modulus for spherical-shaped inclusions [78].

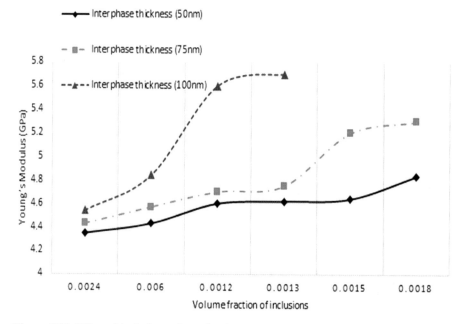

Figure 2.17 Effect of inclusion volume fraction on Young's Modulus for oblate-shaped inclusions [78].

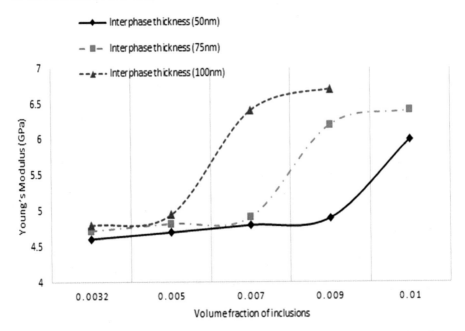

Figure 2.18 Effect of inclusion volume fraction on Young's modulus for prolate-shaped inclusions [78].

Comparing the three sets of results suggests that the sudden increase in the value of Young's modulus when percolation takes place can be due to the chain of inclusions connecting two sides of the RVE to one another. Furthermore, this jump is most evident in the case of prolate ellipsoids, and the elastic properties are least affected by spherical inclusions. It is postulated that the geometry of prolate inclusions facilitates chain formation at lower volume fractions [78].

2.4.2 Homogenization of nanocomposites with interfacial debonding using FEMs

RVEs with different types of nanofillers were generated using MATLAB and Python codes in ABAQUS. Nanofillers were distributed randomly in the RVEs. Mechanical response of the nanocomposites with both perfectly and cohesively bonded interfaces has been studied. The effect of volume fraction, shape, and orientation of nanofillers and types of interactions (perfect and cohesive bonding) on the strength of nanocomposites were studied. Elastic moduli of all samples are shown in Table 2.3. Table 2.4 shows the overall strain of the RVE when the damage parameter CSDMG reaches 1 nanofiller deboned from matrix. Table 2.4 shows that by increasing the volume fraction, damage starts in lower strain. Table 2.5 shows that nanocomposites reinforced by CCNT undergo

Table 2.3 Elastic modulus of nanocomposites (we can see the types of nanofillers in Table 2.1) [54].

Type of nanocomposite	Matrix	No 1/ HDPE	No 2/ HDPE	No 3/ HDPE	No 4/ HDPE	No 5/ HDPE	No 6/ HDPE
Volume fraction (%)	–	0.5 1 1.5	1	0.5 1 0.5	1	0.5	0.5
Elastic modulus (GPa)	0.8	0.848 0.990 1.092	0.916	0.826 0.909 0.945	0.849	0.898	0.925

Table 2.4 Strain of damage initiation in different volume fractions (we can see the types of nanofillers in Table 2.1) [54].

Volume fraction	e_{11} of No 1/HDPE (%)	e_{11} of No 3/HDPE (%)
0.5%	6.33%	10.85%
1%	5.49%	8.89%
1.5%	5.14%	5.60%

Table 2.5 Specific energy absorption in different volume fractions (we can see the types of nanofillers in Table 2.1) [54].

Volume fraction	Energy absorption of No 1/ HDPE (MJ/m^3)	Energy absorption of No 3/ HDPE (MJ/m^3)
0.5%	1.13	2.30
1%	1.012	1.84
1.5%	0.882	1.093

higher strain before complete damage compared to nanocomposites reinforced by CNT. Toughness of nanocomposites by calculating absorption energy (area upon stress−strain curve) was checked. The results show that nanocomposites reinforced by CCNT have higher toughness than nanocomposites reinforced by CNT. Therefore we can have nanocomposites with a high elastic modulus and good toughness. This is in agreement with the results of the recent experimental work presented by [54,116].

We can relate this property improvement to the shape of the coiled CNT which induces mechanical interlocking when the composites are subjected to loading. In fact the coiled configuration of the nanotubes enhances the fracture toughness as well as mechanical strength of the composites, even though there is no direct chemical bonding between the nanotubes and matrix [54].

In the future the crack propagation and its effect on nanocomposite toughness will be studied.

2.4.2.1 Modeling the interface between nanofillers and the matrix

The cohesive zone model has been used to model the damage in the interface between nanofillers and the matrix. In ABAQUS, there are two methods for modeling the interface, cohesive elements and cohesive surfaces. In this section, cohesive surfaces were used. In the interest of brevity, details and criteria of the model are skipped here. Those details can be found in our previous work [70].

The cohesive zone model's parameters are chosen from MD simulation conducted by Li and Seidel [57]. Figs. 2.19 and 2.20 show MD cohesive zone traction-

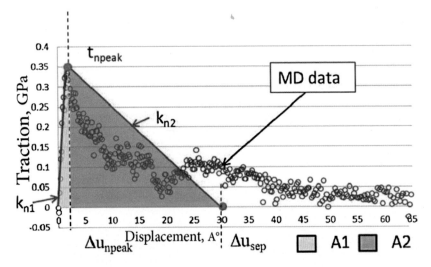

Figure 2.19 Traction-separation response in opening separation modes [54,57].

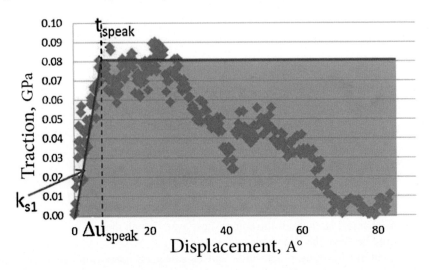

Figure 2.20 Traction-separation response in sliding separation modes [54,57].

displacement laws and corresponding bilinear cohesive zone laws in normal and sliding separation, respectively [117]. Peak traction is the top of the curve, and fracture energy is the area under the curve. Cohesive parameters obtained from Figs. 2.19 and 2.20 are listed in Table 2.6 [54].

2.4.2.1.1 Material

In this study, CNT [33] and CCNT [46,49] were assumed to be elastic isotropic materials. As mentioned before, nanofillers are assumed to be equivalent solid elements. Therefore the elastic modulus of nanofillers needs to be updated. For the equivalent solid fiber of CCNT, CCNT was assumed as a straight SWCNT and then the elastic modulus was calculated [76]. Effective moduli can be expressed by [54,118,119].

$$E_{eff} = \frac{A_{NT}}{A_{eff}} E_{NT} \qquad (2.19)$$

where E_{eff} and E_{NT} are the effective and average elastic moduli of CNT, respectively. A_{eff} and A_{NT} are the effective and real cross-sectional areas, respectively. E_{NT} is considered 1 TPa [120]. Following our previous work an elastoplastic material model was used for modeling high-density polyethylene (HDPE) [54,121].

2.4.2.1.2 Finite element model

RVEs generated in Section 2.1 were imported into ABAQUS software using a Python script. The Python code can automatically generate and assemble parts and assign an appropriate section. It can also identify all the surfaces and sets for the cohesive zone model and set the boundary conditions (Fig. 2.21 and Table 2.7).

Table 2.6 Opening and sliding parameters of the cohesive zone model [54].

Fracture mode	Peak traction (MPa)	Separated energy (MJ/m^2)
Normal mode	344	540.655
Shear mode	81	688.500

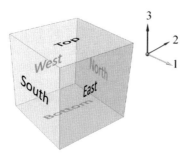

Figure 2.21 Notation of the faces for identifying lead cases [54].

Table 2.7 Boundary conditions [54].

East	Bottom	North	West	Boundary face
$U_1 \neq 0$	$U_3 = 0$	$U_2 = 0$	$U_1 = 0$	Boundary condition

Due to the high aspect ratio of the nanofiller, the RVE had at least 1,300,000 tetrahedron elements.

3D eight-node linear hexahedral elements C3D8R and 3D four-node linear tetrahedron elements C3D4 were used for CCNT and the matrix, respectively. All RVEs were subjected to the uniaxial displacement loading on one face, while the opposite face was fixed. The computational time for the explicit finite element simulation was reduced by mass scaling. In all simulations the kinetic energy was less than 5% of the total strain energy, indicating a quasistatic loading condition [54].

Homogenized variables at the macroscopic scale were obtained by volume averaging of variables in the RVE. The macroscopic stress and strain can be calculated from [54]

$$\langle \sigma \rangle = \frac{1}{\Omega_m} \int_{\Omega_m} \sigma_m d\Omega \tag{2.20}$$

$$\langle \varepsilon \rangle = \frac{1}{\Omega_m} \int_{\Omega_m} \varepsilon_m d\Omega \tag{2.21}$$

where $\langle \sigma \rangle$ and $\langle \varepsilon \rangle$ are the volume averaging (macroscopic) stress and strain, respectively, σ_m and ε_m are the local (microscopic) stress and strain in each element, respectively, and Ω_m is the volume of the RVE.

Fig. 2.22 shows the step-by-step procedure of microstructure reconstruction, material assignment, and finite element simulation [54].

2.4.2.2 Results and discussion

2.4.2.2.1 The effect of volume fraction and type of filler on cohesively and perfectly bonded nanocomposites

RVEs of nanocomposites with three different volume fractions (0.5%, 1%, and 1.5%) and two types of nanofillers (No 1 and No 3) were created. The results for both perfectly and cohesively bonded nanocomposites are shown in Figs. 2.23 and 2.24 [54].

Figs. 2.23A−C and 2.24A−C show that as the volume fraction increases the difference between cohesively bonded and perfectly bonded responses will increase. This means that in nanocomposites with a high volume fraction of nanofillers, more inclusions will be debonded compared to those with a low volume

Numerical characterization of micro- and nanocomposites

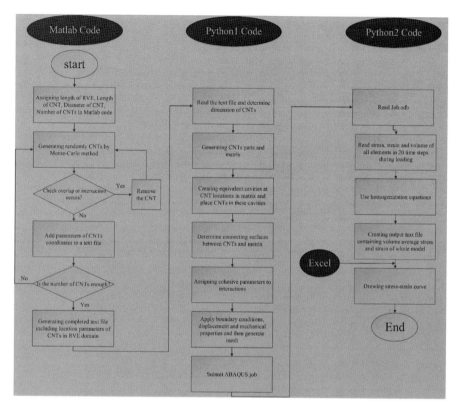

Figure 2.22 Flowchart of modeling and analysis of nanocomposites [54].

fraction. This effect was less pronounced in CCNT composites compared to CNT composites [54].

2.4.2.2.2 The effect of orientation of CNT in perfectly and cohesively bonded nanocomposites

Two RVEs, one with randomly dispersed CNTs (isotropic) and one with aligned CNTs (anisotropic), have been created. In the cohesively bonded composite the results show a huge reduction in the stress–strain response of the anisotropic composite compared to the isotropic composite (see Fig. 2.25) [54].

2.4.2.2.3 The effect of CCNT's helical angle in perfectly bonded and cohesively bonded nanocomposites

CCNT nanofillers, No 1, No 2, and No 3, with different helical angles have been considered. The effect of the coil shape of CCNT was studied and compared with that of straight CNT. As shown in Fig. 2.26, straight CNT is a better reinforcement compared to CCNT, and as the coil angle increases the overall stress–strain response improves as well [54].

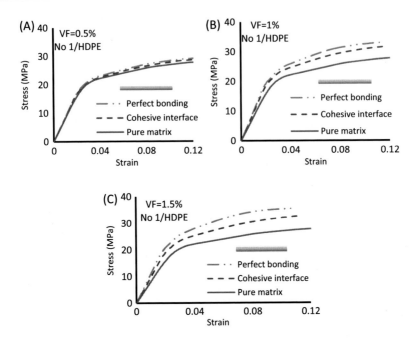

Figure 2.23 Comparing No 1/HDPE with perfectly bonded and cohesively bonded nanocomposites in different volume fractions [54].

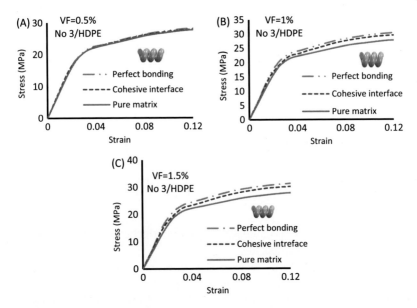

Figure 2.24 Comparing No 3/HDPE with perfectly and cohesively bonded nanocomposites in different volume fractions [54].

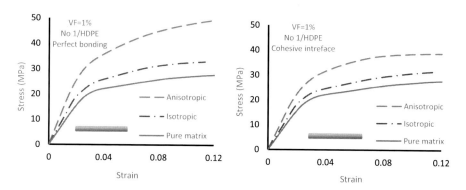

Figure 2.25 Effect of orientation of CNTs: (Left) perfectly bonded; (Right) cohesively bonded [54].

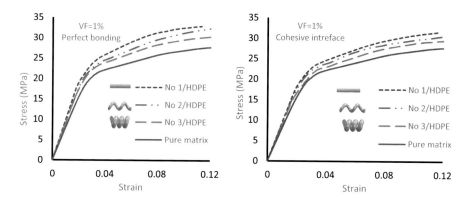

Figure 2.26 Effect of helical angle on properties of nanocomposites with (Left) perfectly bonded and (Right) cohesively bonded [54].

2.4.2.2.4 The effect of CCNT's pitch number in perfectly and cohesively bonded nanocomposites

To study the effect of the number of pitches in CCNT, nanofillers No 3 and No 5 were used in nanocomposites. Fig. 2.27 shows that the overall stress–strain also improves by increasing the number of pitches in CCNT [54].

2.4.2.2.5 The effect of CCNT's coil diameter in perfectly and cohesively bonded nanocomposites

In this section, nanocomposites reinforced by nanofillers No 3 and No 4 have been considered to study the effect of CCNT coil diameter on overall stress–strain response. As shown in Fig. 2.28, by increasing the coil diameter the overall stress–strain response decreases [54].

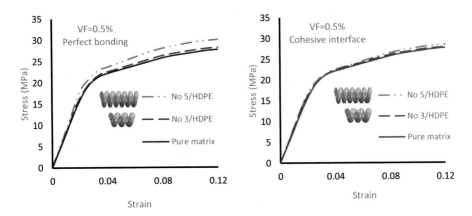

Figure 2.27 The effect of the number of pitches on properties of nanocomposites with (Left) perfectly bonded and (Right) cohesively bonded [54].

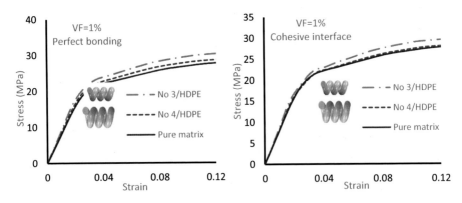

Figure 2.28 Effect of coil diameter on properties of nanocomposites with (Left) perfectly bonded and (Right) cohesively bonded [54].

2.4.2.2.6 The effect of CCNT's tube diameter in perfectly and cohesively bonded nanocomposites

The effect of CCNT's tube diameter on the stress−strain curve was studied. Fig. 2.29 compares two nanocomposites reinforced by No 5 and No 6. As shown, stress−strain curves improve in perfectly and cohesively bonded composites by reducing the tube diameter.

2.4.2.3 Damage in nanocomposites with cohesive interfaces

The CSDMG parameter in ABAQUS was used to show the amount of damage during the deformation. When this parameter reaches unity the interface fails completely. The zoom areas in Fig. 2.30 indicate the interface failure.

Numerical characterization of micro- and nanocomposites

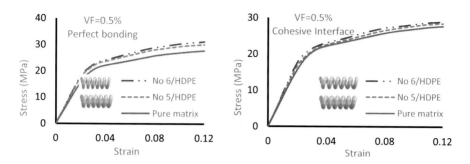

Figure 2.29 Effect of CCNT's tube diameter on properties of nanocomposites with (Left) perfectly bonded and (Right) cohesively bonded [54].

Figure 2.30 Damage in composites with CNT and CCNT fillers at e_{11} = 15% [54].

It was also observed that the nanocomposite reinforced by coiled CNTs reaches complete damage (all fibers debonded from the polymer) in a higher strain compared to the nanocomposite reinforced by straight CNTs. For example, for VF = 1%, nanocomposites No 1/HDPE and No 3/HDPE reach complete damage at 5.49% and 8.89% strain, respectively (see Fig. 2.31). This indicates that the CCNT composite has a better energy absorption capability compared to the CNT composite [54].

The areas under the curves in Fig. 2.31 were calculated to quantify the energy absorption of the nanocomposite. Once it is divided by total volume, it gives the specific energy absorption [54].

$$\frac{Energy}{Volume} = \int_0^{\varepsilon_f} \sigma d\varepsilon \tag{2.22}$$

where σ and ε are the stress and strain, respectively. ε_f is the strain upon failure. The specific energy absorption for nanocomposites No 1/HDPE and No 4/HDPE with VF = 1% is 1.012 MJ/m^3 and 1.842 MJ/m^3, respectively. This shows that nanocomposites reinforced with CCNTs have higher toughness compared to nanocomposites reinforced with CNTs [54].

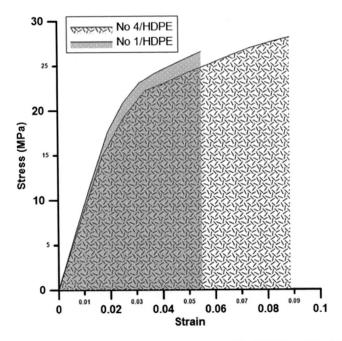

Figure 2.31 Comparing energy absorption of nanocomposites No 1/HDPE and No 4/HDPE [54].

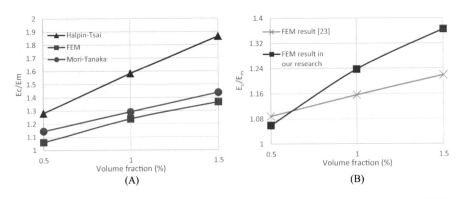

Figure 2.32 Comparison between FEM results: (A) analytical method and (B) another FEM simulation [54].

In Fig. 2.32A the results of finite element simulation were compared to those of Mori–Tanaka and Halpin–Tsai to verify the simulations. FEM results have a similar trend to the theoretical method. In Fig. 2.32B the results in different volume fractions were compared to another FEM result available in the literature [54].

In [76] E_c/E_m (Relative Young's modulus of CNT/nanocomposite and matrix) for the nanocomposite reinforced by CCNT with a helical angle of 50 degrees and a volume fraction of 1% was reported to be 1.09, while in this research, it was calculated 1.14. Additionally the elastic modulus of the nanocomposite declined with increasing helical angle [54].

2.4.3 Homogenization of nanocomposites with nonlinear matrix properties using finite element methods

In this part a finite element modeling for the characterization of the elastic properties of SMP nanocomposites is presented to take into account the effects of volume fraction and aspect ratio of nanofillers on the composite material. A user-defined material subroutine (UMAT) is used to implement 3D constitutive behavior of SMP into the nonlinear finite element software ABAQUS/Standard. Linear elastic orthotropic and isotropic stiffnesses are considered for GNPs inclusions and SMP matrix, respectively. To account for the limited size effects and possible boundary condition effects a series of finite element simulations are carries out to determine the suitable RVE size [122].

The effect of the imperfect GNP/SMP interface is also investigated, and it is concluded that for strain levels utilized in the current study the assumption of perfect bonding between GNP nanoparticles and the SMP polymer matrix is acceptable. In the next step a series of RVEs with varying inclusion volume fractions are created to apply the free-stress strain recovery cycle and observe the shape recovery behavior. Finally, three classes of GNPs with different aspect ratios for inclusions (10, 20, and 40) are considered. For each aspect ratio, several RVEs with different volume fractions (from 0.5% up to 3%) are modeled. The results indicated that elastic properties and the recovery stress increase with an increase in either volume fraction or aspect ratio of inclusions. The recovery stress increases by about 100% when using 1.7% or 3% GNP with an aspect ratio of 40 or 20, respectively [122].

2.4.3.1 Constitutive equations for SMP

In this section the shape memory effect in a stress−strain−temperature diagram is described. As is shown in Fig. 2.33, point $\boxed{a}$ is the start state in this diagram where

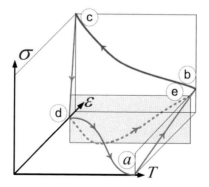

Figure 2.33 Stress−strain−temperature diagram showing the thermomechanical behavior of SMP under strain or stress recovery processes [122].

SMP holds its permanent shape. At this state a specified strain is applied on the SMP, and it demonstrates a rubbery behavior up to point $\boxed{b}$. Then the strain is fixed, and the temperature is decreased until the rubbery phase is converted into Glassy phase T_l (temporary shape at point $\boxed{c}$). Then the material is unloaded. Because of the high stiffness of the glassy phase polymer the strain slightly recovers (point $\boxed{d}$). Finally the temperature rises and SMP recovers its permanent shape at point $\boxed{a}$. This cycle is called stress-free strain recovery in SMP applications. On the other hand, SMP could have gone from point $\boxed{d}$ to point $\boxed{e}$ if the strain at point $\boxed{d}$ was kept fixed and the temperature was increased. In this way the cycle is called fixed-strain stress recovery (shown in Fig. 2.33 with a dotted line) [122].

Now a brief explanation of the small strain constitutive model proposed by Baghani et al. [123] is provided. In this model an equivalent RVE of SMP consisting of a frozen and an active phase has been used. According to the mixture rule the total strain is given by the following relation [122]:

$$\varepsilon = \varphi_a \varepsilon^a + \varphi_f \varepsilon^F + \varepsilon^T, \tag{2.23}$$

where ε^a and ε^F denote the elastic strain in the active and frozen phases, respectively, and ε^T denotes the thermal strain which is evaluated by $\alpha^T dT$, and α^T is the effective thermal expansion coefficient. In this relation, φ_a and φ_f denote the volume fraction of the active and frozen phases, respectively. Both φ_a and φ_f are functions of temperature. The strain in the frozen phase ε^F is decomposed into two parts such that [122]

$$\varphi_f \varepsilon^F = \varphi_f \varepsilon^f + \varepsilon^{is}, \tag{2.24}$$

where ε^f is the elastic strain in the frozen phase and ε^{is} is the inelastic stored strain. As a result the total strain is the weighted summation of the strains in each phase [122].

Assuming the temperature to decrease the strain in the newly generated glassy phase, already been in the rubbery phase, had experienced the ε^a previously. Then $\varphi_f \varepsilon^F$ is defined as [122]

$$\varphi_f \varepsilon^F = \varphi_f (\varepsilon^f + \bar{\varepsilon}^f) = \varphi_f (\varepsilon^f + \frac{1}{V_f} \int_{V_f} \varepsilon^a dv) = \varphi_f \varepsilon^f + \frac{1}{V} \int_{V_f} \varepsilon^a dv, \tag{2.25}$$

where V_f and V are the volume of the frozen phase and the total volume of the RVE, respectively. In Eq. (2.25), strain in the frozen phase is divided into two parts: strain in the old frozen phase, ε^f, and strain in the newly generated frozen phase, $\bar{\varepsilon}^f$. The term "$\varphi_f \varepsilon^f$" is called as the stored strain, and it is denoted by ε^{is}. We recast Eq. (2.25) to [122]

$$\varphi_f \varepsilon^F = \varphi_f \varepsilon^f + \varepsilon^{is}. \tag{2.26}$$

In the cooling process, ε^{is} is defined by

$$\varepsilon^{is} = \int \varepsilon^a d\varphi_f.$$

(2.27)

In the heating process the strain stored in the frozen phase should be relaxed. This is mathematically expressed by [122]

$$\varphi_f \varepsilon^F = \varphi_f(\varepsilon^f + \bar{\varepsilon}^f) = \varphi_f(\varepsilon^f + \frac{1}{V_f} \int_{V_f} \frac{\varepsilon^{is}}{\varphi_f} dv) = \varphi_f \varepsilon^f + \frac{1}{V} \int_{V_f} \frac{\varepsilon^{is}}{\varphi_f} dv,$$

(2.28)

which in a more compact form is

$$\varphi_f \varepsilon^F = \varphi_f \varepsilon^f + \varepsilon^{is}.$$

(2.29)

Thus the total strain could be recast to

$$\varepsilon = \varphi_a \varepsilon^a + \varphi_f \varepsilon^F + \varepsilon^T + \varepsilon^{is}.$$

(2.30)

In addition the stored strain obeys the following evolution law:

$$\dot{\varepsilon}^{is} = \varphi_f \left[k_1 \varepsilon^a + k_2 \frac{\varepsilon^{is}}{\varphi_f} \right]; \begin{cases} k_1 = 1, k_2 = 0, \dot{T} < 0 \\ k_1 = 0, k_2 = 1, \dot{T} > 0, \\ k_1 = 0, k_2 = 0, \dot{T} = 0 \end{cases}$$

(2.31)

Now, based on a first-order rule of mixture the convex free-energy density function Ψ is defined by [122]

$$\Psi\left(\varepsilon, T, \varphi_f, \varphi_a, \varepsilon^a, \varepsilon^f, \varepsilon^{is}\right) = \varphi_a \Psi^a(\varepsilon^a) + \varphi_f \Psi^f\left(\varepsilon^f\right) + \Psi^\lambda\left(\varepsilon, T, \varphi_f, \varphi_a, \varepsilon^a, \varepsilon^f, \varepsilon^{is}\right) + \Psi^T(T),$$

(2.32)

where Ψ^a and Ψ^f stand for the Helmholtz free-energy density function of the active and frozen phases, respectively. Also Ψ^T denotes the thermal energy and the term Ψ^λ enforces kinematic constraint in the following form [122]

$$\Psi^\lambda(\varepsilon, T, \varphi_a, \varphi_f, \varepsilon^a, \varepsilon^f, \varepsilon^{is}) = \lambda:[\varepsilon - (\varphi_a \varepsilon^a + \varphi_f \varepsilon^f + \varepsilon^{is}) - \varepsilon^T],$$

(2.33)

where λ is the Lagrange multiplier. Applying the second law of thermodynamics in the sense of the Clausius–Duhem inequality the following equation is obtained [122]

$$\sigma = \lambda = \frac{\partial \Psi^a}{\partial \varepsilon^a} = \frac{\partial \Psi^f}{\partial \varepsilon^f}.$$

(2.34)

Eq. (2.34) is a consequence of the basic assumption of simultaneous existence of the rubbery and glassy phases [122].

2.4.3.2 Three-dimensional modeling and numerical considerations

FEM is one of the most powerful numerical schemes that are used for the modeling and simulation of a wide range of engineering problems. Because of high computational costs and modeling complexities a perfect realistic modeling of a nanocomposite is extremely challenging or might be impossible. One way to bypass some of these limitations in the simulation of composite materials is to exchange the macroscopic model with a large enough representative volume element [122].

In this section, FEM has been utilized for the evaluation of effective elastic properties of two-phase SMP composites filled with randomly distributed and oriented GNPs. The geometry of inclusions plays an important role on the overall effective mechanical properties [124,125]. Thus in this work, we also study the effect of inclusion geometry in our finite element model where we change the aspect ratio of inclusions with platelet geometry. For a platelet, by aspect ratio, we denote the ratio between the diameter and thickness. In the modeling procedure, particles are subject to hard-core constraint where we avoid intersection or contact between them. The commercial finite element code ABAQUS (V6.10) is used to carry out the simulations [122].

In order to reconstruct the RVE similar to experimentally fabricated random composites, 3D inclusions are randomly distributed and oriented in the RVE. In Fig. 2.34, samples of 3D cubic RVE with different aspect ratios of inclusions are shown. To automate the model generation process, RVEs are constructed in ABAQUS using an in-house Python script. Also the position of the center of

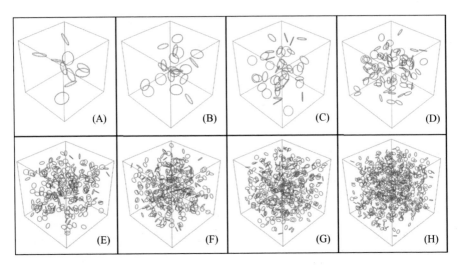

Figure 2.34 Different sizes of RVEs that were used in the size dependency checking simulations. The volume fraction for all the above models was 0.005. There were 10, 20, 40, 80, 240, 320, 400, and 600 inclusions in models (A), (B), (C), (D), (E), (F), (G), and (H), respectively [122].

particles and their orientation are first computed using an in-house C++ code [126]. The modeling parameters are adjusted in the C++ code [122].

To simulate the SMP matrix in the finite element model the 3D constitutive equations described in Section 2.2 [123] are implemented in the finite element software ABAQUS/Standard through a UMAT. The material parameters used in all simulations are listed in Table 2.8. The details of the material model used here are given elsewhere [123,127], where the model has been verified with several experimental data sets. It is noted that the elastic properties of GNPs are listed in Table 2.9 [122,128].

To account for the effect of size dependency and determine the proper size for RVEs, several simulations are carried out initially. In these preliminary simulations, different RVE sizes with the same inclusion volume fraction were created (see Fig. 2.34) to measure the size dependency of calculated results and determine an acceptable RVE size [122].

The intension of this research is to study the effect of the aspect ratio and the volume fraction on the mechanical behavior of SMP nanocomposites. Making a realistic model for a nanocomposite is beyond the scope of the current work. In all numerical studies on composites reinforced with platelet-shaped inclusions such as graphene, GNP, and clay nanoplatelets [129–132], inclusions have been modeled as thin circular disks and in a few cases as rectangular thin disks. This simplistic model could provide useful information about the above-mentioned parameters on the overall stress–strain behavior of the composite materials [122].

In this study the interface between nanoparticles and the matrix are assumed to be perfect. In practices the interface between nanoparticles and GNPs may undergo failure and debonding. To investigate the potential effect of interface debonding on the mechanical properties of the nanocomposite, we introduce imperfect interfaces with associated surface energy into our finite element model for one of the cases. For this purpose, we use cohesive zones to the interfaces between GNPs and the

Table 2.8 Material parameters of SMP modeling [122,123].

Material parameters	Values	unit
E^a, E^f	15.2, 2600	MPa
v^a, v^f	0.49, 0.4	–
T_l, T_g, T_h	25, 46, 62	$°C$
ρ	1100	kg/m^3
φ_f	$\frac{\tanh(\frac{T_h - T_g}{b}) - \tanh(\frac{T - T_g}{b})}{\tanh(\frac{T_h - T_g}{b}) - \tanh(\frac{T_l - T_g}{b})}, b = 4.817°C$	–

Table 2.9 The elastic constants of GNP reported by Blakslee [122,128].

	C_{11}	C_{12}	C_{13}	C_{33}	C_{44}	C_{66}
GNP elastic constants (GPa)	1060	180	15	36.5	4	440

matrix and assume a bilinear traction-separation law [133]. The cohesive model used here is parameterized by an initial stiffness, a peak traction value, and a critical surface energy for failure. For the GNP/SMP system, these cohesive zone parameters can be extracted from molecular dynamics studies or by experimental measurements, which is beyond the scope of the current study. An alternative is to find a material which has a similar structure to SMPs with known cohesive zone parameters. Using molecular dynamics, Awasthi et al. [134] have studied the interfacial interactions between graphene and polyethylene. The molecular structure of polyethylene is quite close to that of SMP; therefore here, we assume that the interaction between GNPs and SMP closely matches that of GNP and polyethylene reported in this study as listed in Table 2.10 [122].

The RVEs are meshed using four-node linear tetrahedral elements (C3D4 elements). To evaluate the elastic modulus a small uniform strain (7%) is applied on one side of the RVE cube where the opposite side is kept fixed in its normal direction and the other side is allowed to move freely. The corresponding stresses are calculated using reaction forces on each side. The effective elastic modulus and Poisson's ratio are estimated employing Hooke's law for isotropic materials. To ensure that the selected RVE obeys Hooke's law for isotropic materials the homogenized stiffness tensor of the RVE is extracted. An RVE loaded with 0.5% of GNPs with an aspect ratio of 10 is subjected to six different load cases, and each time only one normal or pure shear strain is applied to the faces. The effective stiffness tensor for the described RVE extracted by volumetric homogenization of the resultant stresses is examined as a measure of isotropy [122].

2.4.3.3 Numerical homogenization results

To check the convergence of the finite element results the dependency of the reported results to the RVE size are considered (Fig. 2.35). The size of RVE is closely related to the dimension and geometry of inclusions [126]. The size of RVE must be large enough eliminate the effect of boundary distribution of inclusions [135,136]. Fig. 2.35 shows the finite element results for elastic modulus of the SMP composite at T_h for different RVE sizes including GNP inclusions with the aspect ratio of 20. The volume fraction of particles is 0.5% for all cases. As an overall trend an increase in the number of inclusions enhances the accuracy of the finite element results. Nevertheless, utilizing larger RVE sizes is limited by the difficulties in the mesh generation step and it is also associated to higher computational

Table 2.10 Cohesive zone model parameters for interfacial interaction between GNP and polyethylene [122,134].

Fracture mode	Fracture energy (mJ/m^2)	Peak traction (MPa)
Shear mode	331.650	108.276
Normal mode	246.525	170.616

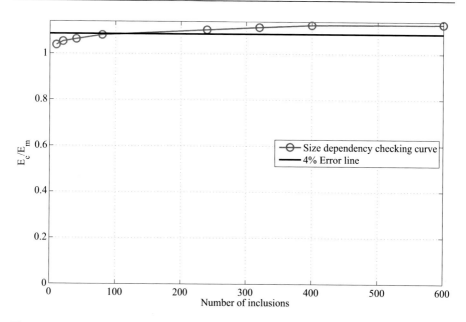

Figure 2.35 Effect of the RVE size on the SMP nanocomposite effective elastic modulus (E_c/E_m) for different GNP inclusions. The volume fraction of inclusions is 0.5% [122].

costs. Thus a maximum acceptable relative error was adopted in the simulation to determine the acceptable size of the RVE. As shown in Fig. 2.35, increasing the size of RVE or equally increasing the number of inclusions the estimated composite to matrix material elastic modulus ratio (E_c/E_m) converges to a constant value. Assuming this constant value as the accurate result the relative error is calculated. In addition, we assume the relative error to be less than 4%, and the results from a series of simulations show that the minimum size for RVEs with the volume fraction of 0.5% should be above of the demonstrated line depicted in Fig. 2.35. According to these results an RVE with a volume fraction of 0.5% and including 50 nm diameter inclusions is an acceptable RVE [122].

Using the cohesive zone model with parameters listed in Table 2.10 the behavior of an SMP/GNP nanocomposite system is computationally investigated under a simple tension test. The average stress/strain diagram obtained is shown in Fig. 2.36. Based on this diagram a linear response for stress–strain behavior is observed for strains less than 4%. At low strain levels, SMP/GNP interfaces show negligible failure, mostly limited to the stress concentration sites; therefore the assumption of perfect bonding for the interfaces is valid. For strain levels larger than 4%, the stress–strain response starts to deviate from linearity; however, the extent of nonlinearities remains small for up to 6% strains. It is worth noting that for all modeling efforts reported here the tensile strains never exceed 7%; therefore the assumption of a perfect interface should provide tight estimates for the properties reported. However, for strains larger than 7%, initial damage sites start to propagate along the interfaces and the total strain energy stored in the system will be

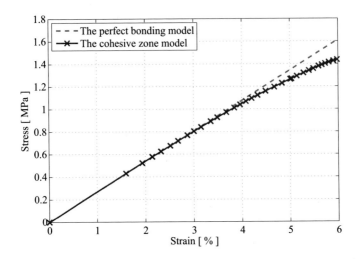

Figure 2.36 Homogenized stress/strain response for an RVE of the GNP/SMP nanocomposite system loaded with 0.5% GNPs under simple tension test (the temperature is set to T_h and remains constant) [122].

reduced by the formation of free (debonded) GNP/SMP surfaces in the system as shown in Fig. 2.37. As a result the total elastic stiffness of the GNP/SMP nanocomposite system diminishes [122].

Based on the initial observation presented, for the rest of the study a perfect bonding for interfaces is assumed and a proper RVE size is utilized. As is mentioned in Section 2.2 a specified strain was applied to the rubbery SMP sample. Then the temperature was decreased to transform SMP to the glassy phase. In the next step the constraints were removed, and the sample was unloaded. During this process, SMP released a negligible elastic strain. Finally the temperature was increased and SMP recovered its permanent shape. To apply this cycle on the SMP nanocomposite and observe effects of inclusions on the recovery behavior of SMP, three different RVEs with an inclusion aspect ratio of 20 and volume fractions of 1%, 2%, and 3% were considered. The results of these simulations are shown in Fig. 2.38 in the form of three stress-free strain recovery cycle diagrams. The results are presented as a function of the normalized temperature (T^*), which is defined by [122]

$$T^* = \frac{T - T_l}{T_h - T_l}, \qquad (2.35)$$

where T_l and T_h are identified in Table 2.8. The diagrams illustrate that the permanent shape was completely recovered after a stress-free strain cycle. The reported results are reasonable because the applied strain is in the range of small strains, and the elastic behavior of GNPs and SMP is therefore considered. Furthermore the results prove that by increasing the volume fraction of inclusions the effective

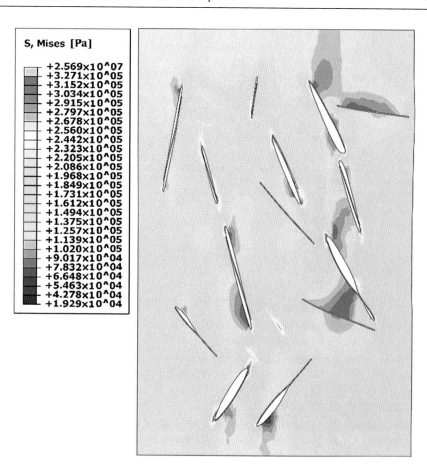

Figure 2.37 Partial debonding of interfaces in the GNP/SMP nanocomposite system. The plot shows distribution of effective stress (in von Mises) over a cross-section of the RVE undergoing strain levels above 7% [122].

elastic modulus of the composite increases. Also, this result is obtained in the fixed-strain stress recovery cycle [122].

To simulate the SMP nanocomposite in the fixed-strain stress recovery cycle, three configurations were considered, for which RVEs with different volume fractions of inclusions were presented. For the first configuration, five composites with an inclusion aspect ratio of 10 and volume fractions of 0.05%, 1%, 1.5%, 2%, and 2.5% were considered. For the second configuration the aspect ratio of inclusions was 20 and there were six different composites models (volume fractions of 0.5%, 1%, 1.5%, 2%, 2.5%, and 3%). For the third configuration, four different composites with an inclusion aspect ratio of 40, and volume fractions of 0.5%, 1%, 1.5%, and 2% were considered. For each of the above composites a fixed-strain stress recovery cycle was applied to RVE. In this cycle the temperature in the deformed RVE was decreased and the generated glassy phase stored the strain. Then the RVE

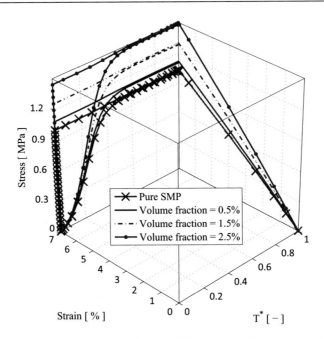

Figure 2.38 Temperature–strain–stress diagram of free-stress strain recovery cycles for three different SMP nanocomposites (volume fractions of 1%, 2%, and 3%) and pure SMP [122].

was unloaded to release the stored elastic strain. In the final step the strain remained constant and then the temperature was increased. In the last step a recovery stress was generated in the sample which imposed a recovery force on the supports [122].

The results of finite element simulation for the first, second, and third configurations are presented in Figs. 2.39, 2.40, and 2.41, respectively. As shown in these figures the recovery stress increases due to an increase in the volume fraction of GNP inclusions. Moreover, by comparing the reported results for different values of inclusion aspect ratio, it can be concluded that both the recovery stress and the elastic modulus increase as the aspect ratio of the particles increases for the same volume fraction [122].

As mentioned in the previous section a series of simulations were carried out to ensure that the reconstructed RVE for the nanocomposite obeys Hooke's law for isotropic materials. Eq. (2.36) shows the estimated stiffness tensor for a sample RVE. Comparing the estimated stiffness tensor with the stiffness tensor of an isotropic material, it is clear that for the selected RVE size, the assumption of isotropic material properties for the RVE/SMP nanocomposite system studied here is valid [122].

$$C = \begin{bmatrix} 2.7212 & 2.6163 & 2.6128 & 0.0040 & 0.0012 & 0.0004 \\ 2.6156 & 2.7225 & 2.6153 & 0.0021 & 0.0023 & 0.0020 \\ 2.6141 & 2.6166 & 2.7247 & 0.0002 & 0.0004 & 0.0024 \\ 0.0002 & 0.0001 & 0.0011 & 0.1068 & 0.0001 & 0.0007 \\ 0.0002 & 0.0002 & 0.0002 & 0.0001 & 0.1067 & 0.0001 \\ 0.0001 & 0.0002 & 0.0000 & 0.0007 & 0.0002 & 0.1068 \end{bmatrix} * 10^8 \text{Pa} \quad (2.36)$$

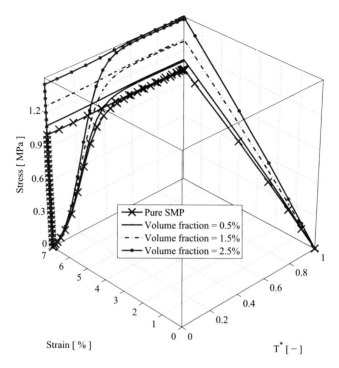

Figure 2.39 Temperature−strain−stress diagram of fixed-strain stress recovery cycles for three different SMP nanocomposites (an aspect ratio of 10 and volume fractions of 0.5%, 1.5%, and 2.5%) and pure SMP [122].

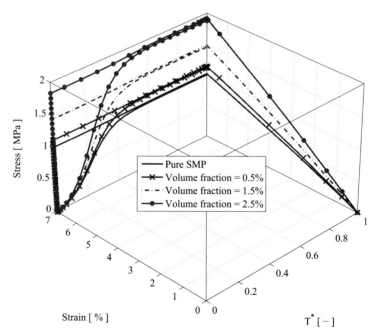

Figure 2.40 Temperature−strain−stress diagram of fixed-strain stress recovery cycles for three different SMP nanocomposites (an aspect ratio of 20 and volume fractions of 0.5%, 1.5%, and 2.5%) and pure SMP [122].

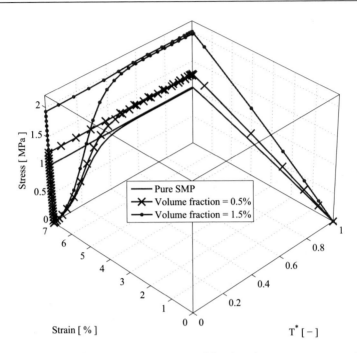

Figure 2.41 Temperature−strain−stress diagram of fixed-strain stress recovery cycles for three different SMP nanocomposites (an aspect ratio of 40 and volume fractions of 0.5% and 1.5%) and pure SMP [122].

Fig. 2.42 summarizes all the modeling results obtained. This figure illustrates the effect of the volume fraction and aspect ratio of inclusions on the elastic modulus of the SMP nanocomposite. From this figure, it is clear that an increase in either values of the aspect ratio or the volume fraction enhances the elastic properties of SMP nanocomposites. For instance, by increasing the volume fraction of a GNP, with an aspect ratio of 20, up to 3% the elastic modulus of the composite rises by 2-fold. The results of all modeling are listed in Table 2.11. In this table, ratios of the elastic modulus and the effective recovery stress of the composite to those for pure SMP are reported for a variety of SMP composites [122].

2.4.4 Finite element homogenization of coupling thermomechanical properties

In this part the effect of mechanical loading on the effective thermomechanical properties, elastic modulus, thermal conductivity, and Poisson's ratio of short carbon fibers (SCFs) in natural rubber is investigated. The Monte Carlo algorithm is exploited to generate a random distribution of SCFs in natural rubber. Three different volume fractions were used with five statistically equivalent samples generated for each volume fraction. The ABAQUS package is used to homogenize RVE models under eight mechanical loading scenarios. Two steps were prescribed for the

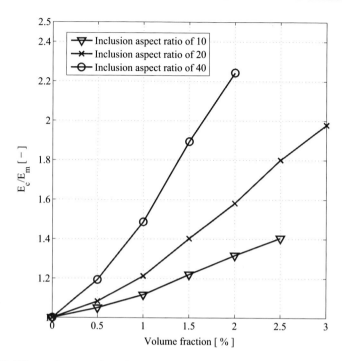

Figure 2.42 Effects of aspect ratio and volume fraction of inclusions on the elastic modulus [122].

Table 2.11 Elastic properties and recovery stress of SMP nanocomposites obtained by finite element simulations [122].

Aspect ratio of inclusions	Volume fraction (%)	$(E_c/E_{SMP})^a$	$(RS_c/RS_{SMP})^b$
10	0.5	1.0498 ± 0.0020	1.0838 ± 0.0034
	1	1.1169 ± 0.0047	1.1528 ± 0.0061
	1.5	1.2199 ± 0.0088	1.2586 ± 0.0103
	2	1.3171 ± 0.0127	1.3584 ± 0.0143
	2.5	1.4048 ± 0.0162	1.4485 ± 0.0179
20	0.5	1.0827 ± 0.0033	1.1176 ± 0.0047
	1	1.2119 ± 0.0085	1.2504 ± 0.0100
	1.5	1.4020 ± 0.0161	1.4455 ± 0.0178
	2	1.5820 ± 0.0233	1.6301 ± 0.0252
	2.5	1.8009 ± 0.0720	1.8541 ± 0.0342
	3	1.9787 ± 0.0391	2.0359 ± 0.0414
40	0.5	1.1933 ± 0.0077	1.2312 ± 0.0092
	1	1.4865 ± 0.0195	1.5321 ± 0.0213
	1.5	1.89390.0358	1.9488 ± 0.0380
	2	2.2433 ± 0.0497	2.3059 ± 0.0522

[a]Ratio of elastic modulus of the composite to that of pure SMP.
[b]Ratio of recovery stress of the composite to that of pure SMP.

simulations: mechanical loading followed by a thermal loading. A Python script is utilized for calculating the effective thermomechanical properties. The results show that the effective thermal conductivity of all samples increased during loadings, and the improvement of thermal conductivity in VF = 2% was 4.68%. The elastic modulus [137] increased with mechanical loading, and an improvement of 0.78% was observed for VF = 2% under maximum tensile strain. Poisson's ratio showed different behaviors for different samples which strongly depend on the alignment of SCFs. The results demonstrate that the alignment of SCFs changed under mechanical loading, which can explain the improvement of effective properties. The average effective thermal conductivity of RVEs with a volume fraction value of 2% was found to be greater than that of natural rubber. By applying mechanical loading the effective thermal conductivity of this sample increased by 4.68%, which illustrates the effect of mechanical loading on the effective thermal conductivity [138].

The properties of pitch-based short carbon fibers (M-2007S) were as follows: This filler has different aspect ratios (minimum 6.2) with a density value of 1.6 (g/cm^3), a modulus of elasticity of 35 GPa, and a thermal conductivity of 100 (W/m.K) [139,140]. These fillers were used to reinforce natural rubber as the matrix. Three different volume fractions of carbon fibers (1%, 1.5%, and 2%) were each subjected to eight loading scenarios. Natural rubber has a density of 0.93(g/cm^3), a modulus of elasticity of 1.5 MPa, and a thermal conductivity of 0.14 (W/m.K) [141,142]. Table 2.12 shows the process which has been done in this study [138].

2.4.4.1 Methodology

The RVEs have been simulated using the commercial FEA package ABAQUS. A MATLAB code was developed to generate models with volume fractions of 1%, 1.5%, and 2%. The Monte Carlo algorithm was used for generating random coordinates for the fillers resulting in isotropic RVEs [143]. Five statistically equivalent models were generated for each set of parameters of each volume fraction in order to ensure that the reported results are not affected by the randomness of the SCF distribution within the RVE [138].

In this study the effect of mechanical loading (at 3.33%, 5%, 6.66%, 8.34%, 10%, 11.67%, 13.33%, and 15% strain) on the effective thermal conductivity of models with three different volume fractions (five models were generated for each volume fraction) was investigated [138].

Based on the proposed algorithm a random point is selected as one end of SCF axis, and the second point is calculated using Eq. (2.1).

A Python script was utilized to set up the RVE in ABAQUS based on the output of the MATLAB code. SFCs were created based on their coordinates and radius, and the contact interaction was defined on their outer surface. The remaining model preparation steps such as loading definitions and mesh generation were done using ABAQUS CAE. Fig. 2.43 shows the generated RVE for 1% volume fraction. Perfect bonding was assumed between the SFCs and the rubber matrix [138].

Table 2.12 The process which has been done in this study [138].

Samples which have been used			The loadings which were applied to samples		Results
1% volume of SCFs in natural rubber 1.5% volume of SCFs in natural rubber 2% volume of SCFs in natural rubber	5 random distributions of SCFs in the matrix were generated for each volume fraction	15 samples for three volume fractions were generated	Mechanical loading Thermal loading	8 strains (3.33%, 5%, 6.66%, 8.34%, 10%, 11.67%, 13.33%, and 15% strain) were applied to each sample After applying mechanical loading, in each strain, thermal loading was applied	For each of the 15 samples the behavior of thermal conductivity, elastic modulus, and Poisson's ratio was illustrated in every strain

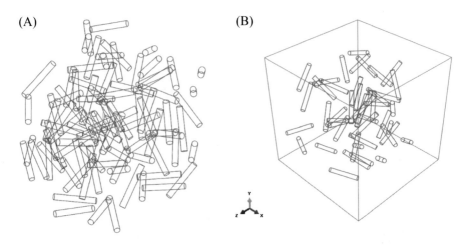

Figure 2.43 (A) Generated SCFs for one of the samples which are used in this study by the Monte Carlo algorithm. (B) Generated RVE in the finite element software (ABAQUS) [138].

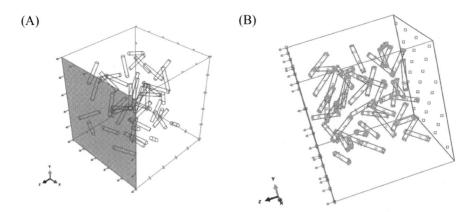

Figure 2.44 (A) Applying mechanical loading on the first step of simulation. (B) Applying thermal loading on the second step of simulation [138].

The simulation was broken down to two steps; the first step consisted of a coupled temperature−displacement analysis where mechanical loading was applied in one direction. This step is shown in Fig. 2.44A [138].

The second step was a coupled temperature−displacement analysis where a thermal loading was applied in the same direction as the mechanical loading in the previous step. For the thermal loading, one side of the RVE had a constant temperature, and its opposite side was specified to have free convection with air (Fig. 2.44B). In this study, eight tensile strain levels varying from 3.33% to 15% were applied on every sample. It should be noted that these tensile strains are in the elastic region of natural rubber [138].

RVEs were generated and studied with random distribution at three volume fractions of 1%, 1.5%, and 2%. Five statistically equivalent RVEs were generated for each volume fraction, and the average of their properties was reported as a function of mechanical loading. The effective thermal conductivity of RVEs prior to any tensile strain was determined to be used for comparison purposes.

Fig. 2.45 shows the RVE under 3.33% and 15% tensile strains where the contour plots show the displacement along the loading direction. Fig. 2.46A shows the distribution of heat flux within an RVE with the volume fraction of 1% in the direction of loading. Fig. 2.46B shows a cross-section of temperature distribution within the RVE [138].

The next samples contain 1.5% SCF (Fig. 2.47A shows the first sample of this volume fraction). Fig. 2.47B shows the first RVE of VF = 1.5% with the generated mesh. Fig. 2.48 shows the result of the first sample under minimum and maximum applied tensile strain. Fig. 2.48 shows the displacement of the SCFs for the minimum and maximum applied tensile strains (i.e., 3.33% and 15%, respectively). Fig. 2.48 shows that SCFs' alignment changes under mechanical loading, and their axes are oriented along the loading direction. This means that the effective thermal conductivity of the RVE increases along the loading direction [138].

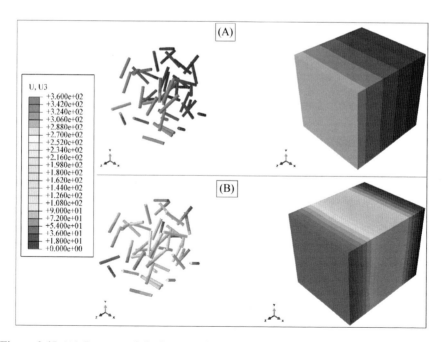

Figure 2.45 (A) Contours of displacement in deformed RVE with a tensile strain of 3.33% for the sample 1 of VF = 1% (B) Deformed RVE with a tensile strain of 15% for the sample 1 of VF = 1% (the RVE is dimensionless and RVE size is 2400 units) [138].

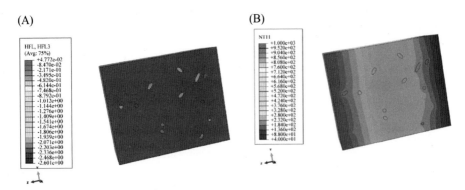

Figure 2.46 (A) Cross-section of the RVE for heat flux (W/m^2) in the direction of loading (sample 1 of VF = 1%). (B) Temperature (°C) contour throughout the RVE for sample 1 of VF = 1% [138].

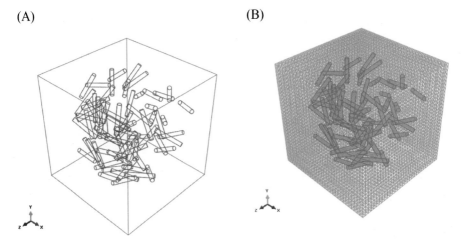

Figure 2.47 (A) Distribution of SCFs in the first RVE of VF = 1.5%. (B) Generated mesh for the first sample of VF = 1.5% [138].

For RVEs with VF = 2% the first sample is shown in Fig. 2.49A and its generated mesh is shown in Fig. 2.49B. Displacement of the first RVE and SCFs for VF = 2% is demonstrated in Fig. 2.50 [138].

Fig. 2.50 shows the RVE under minimum and maximum applied tensile strains (3.33% and 15%, respectively). Fig. 2.50 illustrates how the alignment of SCFs is changed under loading. At 15% tensile strain, SCFs are oriented along the loading direction which improves the effective properties of the sample [138].

For evaluating the effective thermal properties a Python script was used, which would calculate the average heat flux, stress, and strain for all the elements according to Eqs. (2.37)–(2.39). For this purpose, for calculating the effective properties

Numerical characterization of micro- and nanocomposites

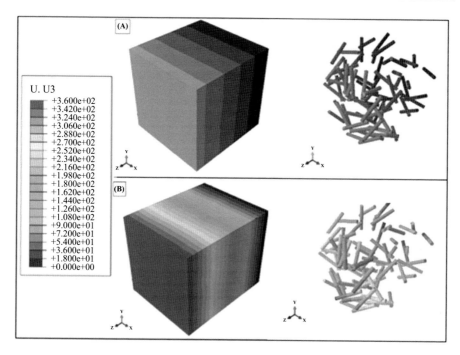

Figure 2.48 The first RVE with VF = 1.5% after (A) the first (3.33%) and (B) final (15%) tensile strains (the RVE is dimensionless and the RVE size is 2400 units) [138].

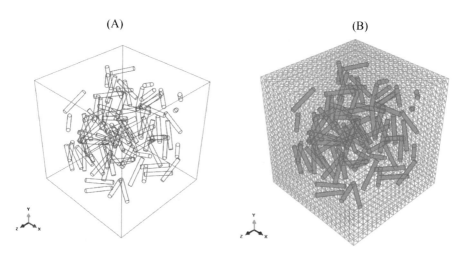

Figure 2.49 (A) Distribution of SCFs in the first RVE of VF = 2%. (B) Generated mesh for the first RVE of VF = 2% [138].

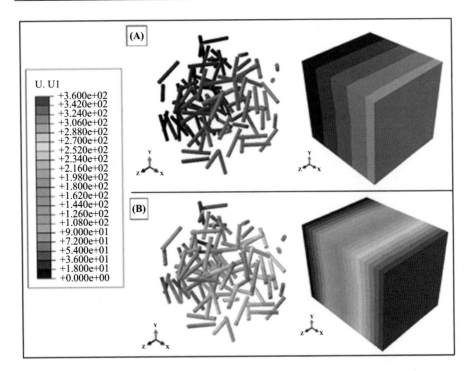

Figure 2.50 The first RVE with VF = 2% after the (A) 3.33% tensile strain and (B) 15% tensile strain (the RVE is dimensionless and the RVE size is 2400 units) [138].

the average of thermal and mechanical quantities of each element which can be extracted from output should be calculated by the following equations [138]:

$$HF_{ave} = \sum_1^N \frac{HF_i \times \nu_i}{V} \quad (2.37)$$

$$S_{ave} = \sum_1^N \frac{S_i \times \nu_i}{V} \quad (2.38)$$

$$E_{eff} = \sum_1^N \frac{e_i \times \nu_i}{V} \quad (2.39)$$

where HF denotes heat flux (W/m^2), ν_i is the volume of each element (m^3), S is the stress (pa), E is the strain, and V is the volume of the RVE (m^3). Then the mean temperature of nodes which had free convection with air is calculated as was discussed in Section 2.2 [138].

Then, using Eq. (2.37) the effective thermal conductivity was calculated:

$$k_{eff}\left(\frac{W}{m.K}\right) \times \frac{\Delta T}{L(m)} = HF\left(\frac{W}{m2}\right) \quad (2.40)$$

where k_{eff} is the effective thermal conductivity, ΔT is the temperature difference between the two surfaces of the RVEs exposed to constant temperature and free convection, L is the length of the RVE after applying load, and HF is the heat flux according to Eq. (2.37). The effective elastic modulus can be calculated using the following equation [138]:

$$\text{Elastic modulus} = \frac{S_{ave}}{E_{ave}} \quad (2.41)$$

The effect of mechanical loading on the effective thermal conductivity of all samples was investigated and is shown in Fig. 2.51 [138].

The average effective thermal conductivity for VF = 1% increased from 0.1621 W/m.K to 0.1644 W/m.k at 15% tensile strain along the direction of loading. This means that the effective thermal conductivity increased by 1.42% after 15% tensile strain. Fig. 2.51 explains that the average of the effective thermal conductivity for VF = 1.5% increases with load, and its maximum is at the 15% tensile strain. The effective thermal conductivity of samples increased by 3.38% under 15% tensile strain. As shown in Fig. 2.52 the average effective thermal conductivity of samples with VF = 2% increased from 0.1794 W/m.K at the initial state to 0.1878 W/m.K at the tensile strain level of 15%. This means that the average of the effective thermal conductivity of these samples increased by 4.68% during these loadings [138].

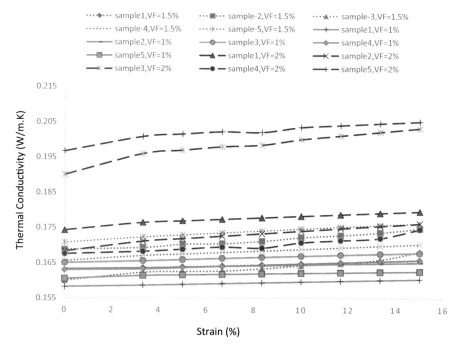

Figure 2.51 The effect of mechanical loading on the effective thermal conductivity of all samples was investigated [138].

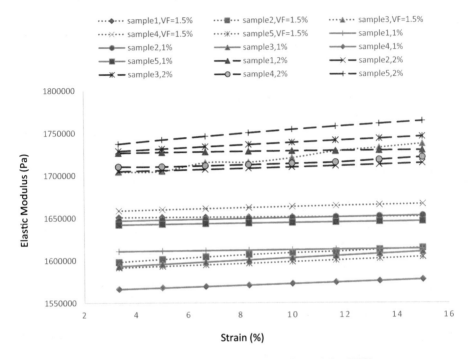

Figure 2.52 The effect of mechanical loading on the elastic modulus [138].

As shown in Figs. 2.45, 2.48, and 2.49, SCFs are reoriented during loading more toward the loading direction. The contrast between the thermal conductivity of SCFs is 100 W/m.K compared to the thermal conductivity of natural rubber (0.14 W/m.K) and plays a major role in the effective thermal conductivity of the composite. This leads to having more thermal channels (if we assume every SCF as a thermal channel due to the high thermal conductivity) in the direction of loading [138].

2.4.4.2 Comparison between analytical and numerical results

In this study the trend of effective properties as a function of mechanical loading was investigated; therefore the thermal resistance between SCFs and the matrix is assumed to be negligible. In order to verify this study, one simulation taking into account the thermal resistance between SCFs and the matrix was carried out with $R_k = 8.3 \times 10^{-8} \mathrm{m^2 K/W}$ reported by Nan et al. [143] for VF = 1.5%. The result of these simulations showed improvement in the effective thermal conductivity by 2.65%, and the results showed the same trend, while without considering thermal resistance, this improvement was 2.86%. We consider that the thermal resistance reduces the effective thermal conductivity in each strain, but it does not change the trend of improvement [138].

To compare this method to the analytical approach, which was mentioned in the introduction, the Mori−Tanaka method was used for samples with VF = 1.5%. Based on this method an increase of 4.05% was calculated in the effective thermal conductivity, whereas in the presented FEM simulation, this was found to be 3.38%, which suggests an agreement between the two techniques [138].

As illustrated in Fig. 2.52 the average effective elastic modulus for VF = 1% at 3.33% tensile strain was 1.6117 MPa, which increased to 1.62 MPa at 15% tensile strain. The average effective elastic modulus increased by 0.51%. The elasticity modulus was increased to its maximum at 15% tensile strain. This suggests that the SCFs are aligned along the direction of loading. According to Fig. 2.52 the average of effective elastic modulus for VF = 1.5% at 3.33% tensile strain was 1.6406 MPa, which increased to 1.6547 MPa at 15% tensile strain, which increased by 0.86%. This figure demonstrates that the average of elastic modulus of VF = 2% had increased by 0.78% during these loadings. As discussed for Fig. 2.52, by applying mechanical loading on these composites, alignment of SCFs was altered, and they were oriented along the loading direction which led to a higher thermal conductivity along this direction. The elastic modulus of SCFs is greater than that of natural rubber (35 GPa for SCFs compared to 1.5 MPa for natural rubber); therefore the alignment of SCFs in the composite is a significant factor for the effective thermal and mechanical properties. Consequently, having more SCFs in the direction of loading, improved effective properties are to be expected, which is in agreement with what was presented in the results [138].

The effect of mechanical loading on the Poisson's ratio, ν_{32}, is illustrated in Fig. 2.53, where direction 3 is the direction of loading [138].

The effect of mechanical loading on the Poisson's ratio ν_{32} was investigated. As shown in Fig. 2.53, Poisson's ratio varies for different samples because this parameter

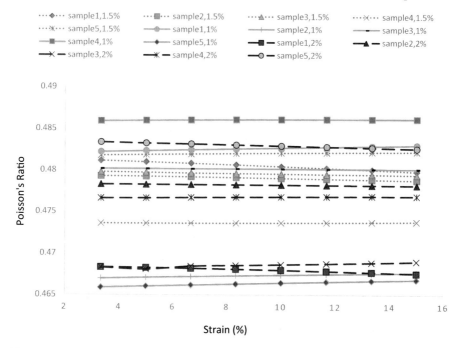

Figure 2.53 The effect of mechanical loading on the Poisson's ratio [138].

is highly dependent on the distributions of SCFs. As the generation of these RVEs was discussed in the methodology, SCFs were generated randomly in the RVEs using the Monte Carlo algorithm, and distribution of SCFs was completely random, which led to different Poisson's ratios for different samples [138].

Fig. 2.54 shows the average of thermal conductivity for the three volume fractions used in this study. Fig. 2.54 illustrates that the average of all volume fractions was increased during loading [138].

Fig. 2.55 shows the average of elastic modulus for the three volume fractions which were used in this study. Fig. 2.55 suggests that the average elastic modulus increases with loading, and it is greater for higher volume fractions [138].

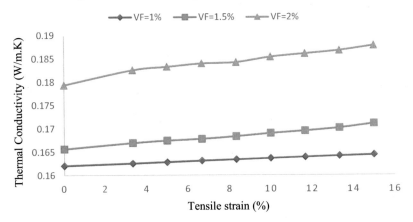

Figure 2.54 The average of thermal conductivity for three volume fractions used in this study [138].

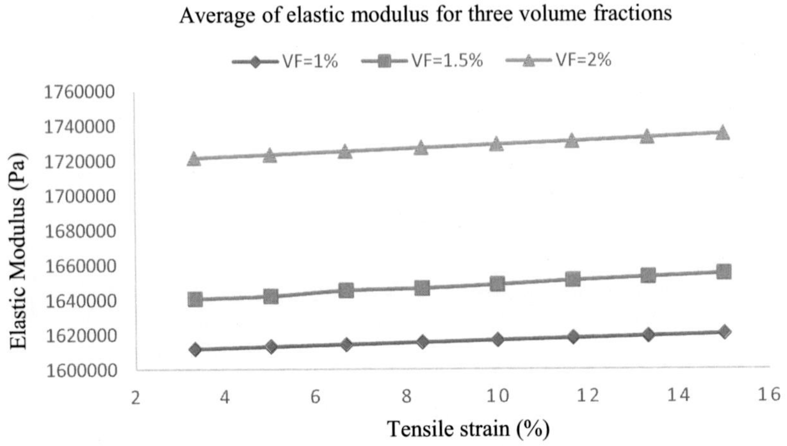

Figure 2.55 The average of elastic modulus for three volume fractions used in this study [138].

2.5 Conclusion and remarks

Modeling mechanical and thermal properties of the micro- and nanocomposites is one of the most important challenges in materials science. Realization and reconstruction of heterogeneous materials can be performed using experimental or numerical approaches. Experimental approaches such as Fib-SEM and X-ray computed tomography directly can output the micro- or nanostructure, but other experimental techniques such as the SAXS or SANS technique indirectly can provide statistical information to reconstruct heterogeneous materials. Numerical realization is the other simple alternative approach to generate RVE, and different categories of inclusions can be generated based on the size and distribution of particles.

Finally, numerical characterization is exploited to predict effective thermal properties (such as thermal conductivity), mechanical properties, and damage behavior of the micro- and nanocomposites.

References

[1] Drzal LT, Fukushima H. Exfoliated graphite nanoplatelets (xGnP): a carbon nanotube alternative. Proceedings of NSTI nanotechnology coneference and trade show. 2006.

[2] Fukushima H, Drzal LT, Rook BP, Rich MJ. Thermal conductivity of exfoliated graphite nanocomposites. J Therm Anal Calorim 2006;85(1):235−8. Available from: https://doi.org/10.1007/s10973-005-7344-x.

[3] Frank I, Tanenbaum DM, Van der Zande A, McEuen PL. Mechanical properties of suspended graphene sheets. J Vac Sci Technol B 2007;25(6):2558−61.

[4] Yang N, Zhang G, Li B. Carbon nanocone: a promising thermal rectifier. Appl Phys Lett 2008;93(24):243111.

[5] Wei J, Liew KM, He X. Mechanical properties of carbon nanocones. Appl Phys Lett 2007;91(26):261906.

[6] Thostenson ET, Ren Z, Chou T-W. Advances in the science and technology of carbon nanotubes and their composites: a review. Compos Sci Technol 2001;61(13):1899−912.

[7] Asiaei S, Khatibi AA, Baniasadi M, Safdari M. Effects of carbon nanotubes geometrical distribution on electrical percolation of nanocomposites: a comprehensive approach. J reinforced Plast Compos 2010;29(6):818−29.

[8] Nazem Salimi M, Alizadeh Sahraei A, Baniassadi M, Abrinia K, Ehsani M. Synergistic effect of carbon nanotubes and copper particles in an epoxy-based nanocomposite using electroless copper deposited carbon nanotubes: Part I−Mechanical properties. J Compos Mater 2016;50(14):1909−20.

[9] Liu L, Liu F, Zhao J. Curved carbon nanotubes: from unique geometries to novel properties and peculiar applications. Nano Res 2014;7(5):626−57. Available from: https://doi.org/10.1007/s12274-014-0431-1.

[10] Liu LZ, Gao HL, Zhao JJ, Lu JP. Superelasticity of carbon nanocoils from atomistic quantum simulations. Nanoscale Res Lett 2010;5(3):478−83. Available from: https://doi.org/10.1007/s11671-010-9545-x.

[11] Lau KT, Lu M, Hui D. Coiled carbon nanotubes: synthesis and their potential applications in advanced composite structures. Compos Part B Eng 2006;37(6):437−48.

[12] Alexandre M, Dubois P. Polymer-layered silicate nanocomposites: preparation, properties and uses of a new class of materials. Mater Sci Eng R: Rep 2000;28(1):1−63.

[13] Baniassadi M, Laachachi A, Hassouna F, Addiego F, Muller R, Garmestani H, et al. Mechanical and thermal behavior of nanoclay based polymer nanocomposites using statistical homogenization approach. Compos Sci Technol 2011;71(16):1930−5.

[14] Yang H, Cui H, Tang W, Li Z, Han N, Xing F. A critical review on research progress of graphene/cement based composites. Composites Part A Appl Sci Manuf 2017;102:273−96. Available from: https://doi.org/10.1016/j.compositesa.2017.07.019.

[15] Stynoski P, Mondal P, Marsh C. Effects of silica additives on fracture properties of carbon nanotube and carbon fiber reinforced Portland cement mortar. Cem Concr Compos 2015;55:232−40. Available from: https://doi.org/10.1016/j.cemconcomp.2014.08.005.

[16] Šmilauer V, Hlaváček P, Padevět P. Micromechanical analysis of cement paste with carbon nanotubes. Acta Polytech 2012;52(6).

[17] Said AM, Zeidan MS, Bassuoni M, Tian Y. Properties of concrete incorporating nanosilica. Constr Build Mater 2012;36:838−44.

[18] Reis JML. Fracture and flexural characterization of natural fiber-reinforced polymer concrete. Constr Build Mater 2006;20(9):673−8. Available from: https://doi.org/10.1016/j.conbuildmat.2005.02.008.

[19] Holschemacher K, Mueller T, Ribakov Y. Effect of steel fibres on mechanical properties of high-strength concrete. Mater Des (1980−2015) 2010;31(5):2604−15. Available from: https://doi.org/10.1016/j.matdes.2009.11.025.

[20] Ali M, Liu A, Sou H, Chouw N. Mechanical and dynamic properties of coconut fibre reinforced concrete. Constr Build Mater 2012;30:814−25. Available from: https://doi.org/10.1016/j.conbuildmat.2011.12.068.

[21] Sun J, Cao X, Xu Z, Yu Z, Zhang Y, Hou G, et al. Contribution of core/shell $TiO_2@SiO_2$ nanoparticles to the hydration of Portland cement. Constr Build Mater 2020;233:117127. Available from: https://doi.org/10.1016/j.conbuildmat.2019.117127.

[22] Jing G, Ye Z, Wu J, Wang S, Cheng X, Strokova V, et al. Introducing reduced graphene oxide to enhance the thermal properties of cement composites. Cement Concrete Compos 2020;109:103559. Available from: https://doi.org/10.1016/j.cemconcomp.2020.103559.

[23] Du M, Jing H, Gao Y, Su H, Fang H. Carbon nanomaterials enhanced cement-based composites: advances and challenges. Nanotechnol Rev 2020;9(1):115−35. Available from: https://doi.org/10.1515/ntrev-2020-0011.

[24] Yao Y, Zhang Z, Liu H, Zhuge Y, Zhang D. A new in-situ growth strategy to achieve high performance graphene-based cement material. Constr Build Mater 2022;335:127451. Available from: https://doi.org/10.1016/j.conbuildmat.2022.127451.

[25] Liu C, He X, Deng X, Wu Y, Zheng Z, Liu J, et al. Application of nanomaterials in ultra-high performance concrete: a review. Nanotechnol Rev 2020;9(1):1427−44. Available from: https://doi.org/10.1515/ntrev-2020-0107.

[26] Krystek M, Ciesielski A, Samorì P. Graphene-based cementitious composites: toward next-generation construction technologies. Adv Funct Mater 2021;31(27):2101887. Available from: https://doi.org/10.1002/adfm.202101887.

[27] Torabian Isfahani F, Li W, Redaelli E. Dispersion of multi-walled carbon nanotubes and its effects on the properties of cement composites. Cement Concrete Compos 2016;74:154−63. Available from: https://doi.org/10.1016/j.cemconcomp.2016.09.007.

[28] Rong Z, Sun W, Xiao H, Jiang G. Effects of nano-SiO2 particles on the mechanical and microstructural properties of ultra-high performance cementitious composites. Cement Concrete Compos 2015;56:25−31. Available from: https://doi.org/10.1016/j.cemconcomp.2014.11.001.

[29] Raki L, Beaudoin J, Alizadeh R, Makar J, Sato T. Cement and concrete nanoscience and nanotechnology. Materials 2010;3(2). Available from: https://doi.org/10.3390/ma3020918.

[30] Chuah S, Pan Z, Sanjayan JG, Wang CM, Duan WH. Nano reinforced cement and concrete composites and new perspective from graphene oxide. Constr Build Mater 2014;73:113−24. Available from: https://doi.org/10.1016/j.conbuildmat.2014.09.040.

[31] Li C, Chou T-W. Elastic moduli of multi-walled carbon nanotubes and the effect of van der Waals forces. Compos Sci Technol 2003;63(11):1517−24.

[32] Yu M-F, Files BS, Arepalli S, Ruoff RS. Tensile loading of ropes of single wall carbon nanotubes and their mechanical properties. Phys Rev Lett 2000;84(24):5552.

[33] Yu M-F, Lourie O, Dyer MJ, Moloni K, Kelly TF, Ruoff RS. Strength and breaking mechanism of multiwalled carbon nanotubes under tensile load. Available from: http://www.sciencemag.orgScience 2000;287(5453):637−40.

[34] Li D, Baniassadi M, Garmestani H, Ahzi S, Reda Taha M, Ruch D. 3D reconstruction of carbon nanotube composite microstructure using correlation functions. J Computat Theor Nanosci 2010;7(8):1462−8.

[35] Yarali E, Baniassadi M, Baghani M. Numerical homogenization of coiled carbon nanotube reinforced shape memory polymer nanocomposites. Smart Mater Struct 2019;28 (3):035026.

[36] Sharifian A, Baghani M, Wu J, Odegard GM, Baniassadi M. Insight into geometry-controlled mechanical properties of spiral carbon-based nanostructures. J Phys Chem C 2019;123(5):3226−38.

[37] Sahraei AA, Ayati M, Rodrigue D, Baniassadi M. A computational approach to evaluate the nonlinear and noisy DC electrical response in carbon nanotube/polymer nanocomposites near the percolation threshold. Computat Mater Sci 2020;173:109439.

[38] Sharifian A, Fadaei Naeini V, Baniassadi M, Wu J, Baghani M. Role of chemical doping in large deformation behavior of spiral carbon-based nanostructures: unraveling geometry-dependent chemical doping effects. J Phys Chem C 2019;123(31):19208−19.

[39] Ma PC, Kim J-K, Tang BZ. Effects of silane functionalization on the properties of carbon nanotube/epoxy nanocomposites. Compos Sci Technol 2007;67(14):2965−72.

[40] Itoh S, Ihara S, Kitakami J-i. Toroidal form of carbon C 360. Phys Rev B 1993;47 (3):1703.

[41] Zhang X, Zhang X, Bernaerts D, Van Tendeloo G, Amelinckx S, Van Landuyt J, et al. The texture of catalytically grown coil-shaped carbon nanotubules. EPL (Europhys Lett) 1994;27(2):141.

[42] Biró L, Gyulai J, Lambin P, Nagy J, Lazarescu S, Márk G, et al. Scanning tunnelling microscopy (STM) imaging of carbon nanotubes. Carbon 1998;36(5):689−96.

[43] Terrones M. Science and technology of the twenty-first century: synthesis, properties, and applications of carbon nanotubes. Annu Rev Mater Res 2003;33(1):419−501.

[44] Bai J, Hamon A-L, Marraud A, Jouffrey B, Zymla V. Synthesis of SWNTs and MWNTs by a molten salt (NaCl) method. Chem Phys Lett 2002;365(1):184−8.

[45] Wang J, Kemper T, Liang T, Sinnott SB. Predicted mechanical properties of a coiled carbon nanotube. Carbon 2012;50(3):968−76.

[46] Volodin A, Ahlskog M, Seynaeve E, Van Haesendonck C, Fonseca A, Nagy JB. Imaging the elastic properties of coiled carbon nanotubes with atomic force microscopy. Phys Rev Lett 2000;84(15):3342−5. Available from: https://doi.org/10.1103/PhysRevLett.84.3342.

[47] Tian L, Guo X. Fracture and defect evolution in carbon nanocoil − a molecular dynamics study. Computat Mater Sci 2015;103:126−33. Available from: https://doi.org/10.1016/j.commatsci.2015.03.026.

[48] Fakhrabadi MMS, Amini A, Reshadi F, Khani N, Rastgoo A. Investigation of buckling and vibration properties of hetero-junctioned and coiled carbon nanotubes. Computat Mater Sci 2013;73:93−112. Available from: https://doi.org/10.1016/j.commatsci.2013.02.020.

[49] Ju S-P, Lin J-S, Chen H-L, Hsieh J-Y, Chen H-T, Weng M-H, et al. A molecular dynamics study of the mechanical properties of a double-walled carbon nanocoil. Computat Mater Sci 2014;82:92−9.

[50] Ghaderi SH, Hajiesmaili E. Molecular structural mechanics applied to coiled carbon nanotubes. Computat Mater Sci 2012;55:344−9. Available from: https://doi.org/10.1016/j.commatsci.2011.11.016.

[51] Ghaderi SH, Hajiesmaili E. Nonlinear analysis of coiled carbon nanotubes using the molecular dynamics finite element method. Mater Sci Eng A 2013;582:225−34.

[52] Wu J, Nagao S, He J, Zhang Z. Nanohinge-induced plasticity of helical carbon nanotubes. Small 2013;9(21):3561−6. Available from: https://doi.org/10.1002/smll.201202830.

[53] Feng C, Liew KM, He P, Wu A. Predicting mechanical properties of carbon nanosprings based on molecular mechanics simulation. Compos Struct 2014;114:41−50. Available from: https://doi.org/10.1016/j.compstruct.2014.03.042.

[54] Yousefi E, Sheidaei A, Mahdavi M, Baniassadi M, Baghani M, Faraji G. Effect of nanofiller geometry on the energy absorption capability of coiled carbon nanotube composite material. Compos Sci Technol 2017;153:222−31.

[55] Li X-F, Lau K-T, Yin Y-S. Mechanical properties of epoxy-based composites using coiled carbon nanotubes. Compos Sci Technol 2008;68(14):2876−81.

[56] Gou J, Liang Z, Zhang C, Wang B. Computational analysis of effect of single-walled carbon nanotube rope on molecular interaction and load transfer of nanocomposites. Compos Part B Eng 2005;36(6):524−33.

[57] Li Y, Seidel GD. Multiscale modeling of the effects of nanoscale load transfer on the effective elastic properties of unfunctionalized carbon nanotube−polyethylene nanocomposites. Model Simul Mater Sci Eng 2014;22(2):025023.

[58] Odegard G, Clancy T, Gates T. Modeling of the mechanical properties of nanoparticle/polymer composites. Polymer 2005;46(2):553−62.

[59] Chen H, Xue Q, Zheng Q, Xie J, Yan K. Influence of nanotube chirality, temperature, and chemical modification on the interfacial bonding between carbon nanotubes and polyphenylacetylene. J Phys Chem C 2008;112(42):16514−20.

[60] Liao K, Li S. Interfacial characteristics of a carbon nanotube−polystyrene composite system. Appl Phys Lett 2001;79(25):4225−7.

[61] Rahmat M, Hubert P. Molecular dynamics simulation of single-walled carbon nanotube−PMMA interaction. J Nano Res Trans Tech Publ 2012;117−28.

[62] Frankland S, Caglar A, Brenner D, Griebel M. Molecular simulation of the influence of chemical cross-links on the shear strength of carbon nanotube-polymer interfaces. J Phys Chem B 2002;106(12):3046−8.

[63] Almasi A, Silani M, Talebi H, Rabczuk T. Stochastic analysis of the interphase effects on the mechanical properties of clay/epoxy nanocomposites. Compos Struct 2015;133:1302−12.

[64] Vu-Bac N, Silani M, Lahmer T, Zhuang X, Rabczuk T. A unified framework for stochastic predictions of mechanical properties of polymeric nanocomposites. Comput Mater Sci 2015;96:520−35.

[65] Vu-Bac N, Rafiee R, Zhuang X, Lahmer T, Rabczuk T. Uncertainty quantification for multiscale modeling of polymer nanocomposites with correlated parameters. Compos Part B Eng 2015;68:446−64.

[66] Benzaid R, Chevalier J, Saâdaoui M, Fantozzi G, Nawa M, Diaz LA, et al. Fracture toughness, strength and slow crack growth in a ceria stabilized zirconia–alumina nanocomposite for medical applications. Biomaterials 2008;29(27):3636–41.

[67] Hajshirmohammadi B, Khonsari M. On the entropy of fatigue crack propagation. Int J Fatigue 2020;133:105413.

[68] Hajshirmohammadi B, Khonsari M. A simple approach for predicting fatigue crack propagation rate based on thermography. Theor Appl Fract Mech 2020;107:102534.

[69] Bellemare SC, Bureau MN, Denault J, Dickson JI. Fatigue crack initiation and propagation in polyamide-6 and in polyamide-6 nanocomposites. Polym Compos 2004;25 (4):433–41.

[70] Safaei M, Sheidaei A, Baniassadi M, Ahzi S, Mashhadi MM, Pourboghrat F. An interfacial debonding-induced damage model for graphite nanoplatelet polymer composites. Comput Mater Sci 2015;96:191–9.

[71] Thompson AM, Chan HM, Harmer MP, Cook RE. Crack healing and stress relaxation in Al2O3 SiC "Nanocomposites". J Am Ceram Soc 1995;78(3):567–71.

[72] Silani M, Ziaei-Rad S, Talebi H, Rabczuk T. A semi-concurrent multiscale approach for modeling damage in nanocomposites. Theor Appl Fract Mech 2014;74:30–8.

[73] Silani M, Talebi H, Hamouda AM, Rabczuk T. Nonlocal damage modelling in clay/ epoxy nanocomposites using a multiscale approach. J Comput Sci 2016;15:18–23.

[74] Hamdia KM, Msekh MA, Silani M, Vu-Bac N, Zhuang X, Nguyen-Thoi T, et al. Uncertainty quantification of the fracture properties of polymeric nanocomposites based on phase field modeling. Compos Struct 2015;133:1177–90.

[75] Msekh MA, Silani M, Jamshidian M, Areias P, Zhuang X, Zi G, et al. Predictions of J integral and tensile strength of clay/epoxy nanocomposites material using phase field model. Compos Part B Eng 2016;93:97–114.

[76] Khani N, Yildiz M, Koc B. Elastic properties of coiled carbon nanotube reinforced nanocomposite: A finite element study. Mater Des 2016;109:123–32.

[77] Zhang Y, Zhao J, Jia Y, Mabrouki T, Gong Y, Wei N, et al. An analytical solution on interface debonding for large diameter carbon nanotube-reinforced composite with functionally graded variation interphase. Compos Struct 2013;104:261–9.

[78] Gharehnazifam Z, Baniassadi M, Abrinia K, Karimpour M, Baghani M. Elastic Percolation in Nanocomposites with Impenetrable Ellipsoidal Inclusion (Comprehensive Study of Geometry and Interphase Thickness). Int J Appl Mech 2016;8(4):1650055. Available from: https://doi.org/10.1142/s1758825116500551.

[79] Zheng X, Palffy-Muhoray P. Distance of closest approach of two arbitrary hard ellipses in two dimensions. Phys Rev E Stat Nonlin Soft Matter Phys 2007;75(6 Pt 1):061709. Available from: https://doi.org/10.1103/PhysRevE.75.061709.

[80] Zheng X, Iglesias W, Palffy-Muhoray P. Distance of closest approach of two arbitrary hard ellipsoids. Phys Rev E Stat Nonlin Soft Matter Phys 2009;79(5 Pt 2):057702. Available from: https://doi.org/10.1103/PhysRevE.79.057702.

[81] Gharehnazifam Z, Baniassadi M, Abrinia K, Rahimi M, Izadi M. Electrical percolation in nanocomposites with impenetrable ellipsoidal inclusion (comprehensive study of tunneling, geometry, anisotropy and mixing). J Computat Theor Nanosci 2015;12 (6):1010–16.

[82] Qiao R, Catherine, Brinson L. Simulation of interphase percolation and gradients in polymer nanocomposites. Compos Sci Technol 2009;69(3):491–9.

[83] Sevostianov I, Kachanov M. Effect of interphase layers on the overall elastic and conductive properties of matrix composites. Applications to nanosize inclusion. Int J Solids Struct 2007;44(3):1304–15.

[84] Petsi A, Burganos V. Interphase layer effects on transport in mixed matrix membranes. J Membr Sci 2012;421:247−57.

[85] Vigolo B, Coulon C, Maugey M, Zakri C, Poulin P. An experimental approach to the percolation of sticky nanotubes. Science 2005;309(5736):920−3. Available from: https://doi.org/10.1126/science.1112835.

[86] Baniassadi M, Safdari M, Ghazavizadeh A, Garmestani H, Ahzi S, Gracio J, et al. Incorporation of electron tunnelling phenomenon into 3D Monte Carlo simulation of electrical percolation in graphite nanoplatelet composites. J Phys D: Appl Phys 2011;44(45):455306.

[87] Drugan W, Willis J. A micromechanics-based nonlocal constitutive equation and estimates of representative volume element size for elastic composites. J Mech Phys Solids 1996;44(4):497−524.

[88] Ghasemi H, Brighenti R, Zhuang X, Muthu J, Rabczuk T. Optimal fiber content and distribution in fiber-reinforced solids using a reliability and NURBS based sequential optimization approach. Struct Multidiscip Optim 2015;51(1):99−112.

[89] Weisstein E, Lorenz Attractor M−A. Sphere Point Piking. MathWorld-A Wolfram web resource. 2010.

[90] Huang J, Rodrigue D. Equivalent continuum models of carbon nanotube reinforced polypropylene composites. Mater Des 2013;50:936−45. Available from: https://doi.org/10.1016/j.matdes.2013.03.095.

[91] Ghasemi H, Rafiee R, Zhuang X, Muthu J, Rabczuk T. Uncertainties propagation in metamodel-based probabilistic optimization of CNT/polymer composite structure using stochastic multi-scale modeling. Computat Mater Sci 2014;85:295−305.

[92] Jiao Y, Stillinger FH, Torquato S. Modeling heterogeneous materials via two-point correlation functions: basic principles. Phys Rev E Stat Nonlin Soft Matter Phys 2007;76(3 Pt 1):031110. Available from: https://doi.org/10.1103/PhysRevE.76.031110.

[93] Izadi H, Baniassadi M, Hasanabadi A, Mehrgini B, Memarian H, Soltanian-Zadeh H, et al. Application of full set of two point correlation functions from a pair of 2D cut sections for 3D porous media reconstruction. J Pet Sci Eng 2017;149:789−800.

[94] Hasanabadi A, Baniassadi M, Abrinia K, Safdari M, Garmestani H. Optimization of solid oxide fuel cell cathodes using two-point correlation functions. Computat Mater Sci 2016;123:268−76.

[95] Ghazavizadeh A, Soltani N, Baniassadi M, Addiego F, Ahzi S, Garmestani H. Composition of two-point correlation functions of subcomposites in heterogeneous materials. Mech Mater 2012;51:88−96.

[96] Torquato S. Necessary conditions on realizable two-point correlation functions of random media. Ind Eng Chem Res 2006;45(21):6923−8.

[97] Mahdavi M, Baniassadi M, Baghani M, Dadmun M, Tehrani M. 3D reconstruction of carbon nanotube networks from neutron scattering experiments. Nanotechnology 2015;26(38):385704. Available from: https://doi.org/10.1088/0957-4484/26/38/385704.

[98] Fullwood DT, Niezgoda SR, Kalidindi SR. Microstructure reconstructions from 2-point statistics using phase-recovery algorithms. Acta Mater 2008;56(5):942−8.

[99] Kashiwagi T, Fagan J, Douglas JF, Yamamoto K, Heckert AN, Leigh SD, et al. Relationship between dispersion metric and properties of PMMA/SWNT nanocomposites. Polymer 2007;48(16):4855−66.

[100] Baniassadi M, Laachachi A, Makradi A, Belouettar S, Ruch D, Muller R, et al. Statistical continuum theory for the effective conductivity of carbon nanotubes filled polymer composites. Thermochim acta 2011;520(1−2):33−7.

[101] Lingaiah S, Sadler R, Ibeh C, Shivakumar K. A method of visualization of inorganic nanoparticles dispersion in nanocomposites. Compos Part B: Eng 2008;39 (1):196−201.

[102] Guinier A, Fournet G, Walker CB, Yudowitch KL. Small-angle scattering of X-rays. 1955.

[103] Debye PAJH. J Appl Phys 1957;28:4.

[104] Debye P, Anderson Jr H, Brumberger H. Scattering by an inhomogeneous solid. II. The correlation function and its application. J Appl Phys 1957;28(6):679−83.

[105] Frisch H, Stillinger F. Contribution to the statistical geometric basis of radiation scattering. J Chem Phys 1963;38(9):2200−7.

[106] Torquato S, Hyun S, Donev A. Multifunctional composites: optimizing microstructures for simultaneous transport of heat and electricity. Phys Rev Lett 2002;89 (26):266601. Available from: https://doi.org/10.1103/PhysRevLett.89.266601.

[107] Mikdam A, Makradi A, Ahzi S, Garmestani H, Li D, Remond Y. Effective conductivity in isotropic heterogeneous media using a strong-contrast statistical continuum theory. J Mech Phys Solids 2009;57(1):76−86.

[108] Beaucage G. Approximations leading to a unified exponential/power-law approach to small-angle scattering. J Appl Crystallogr 1995;28(6):717−28. Available from: https://doi.org/10.1107/s0021889895005292.

[109] Beaucage G. Small-angle scattering from polymeric mass fractals of arbitrary mass-fractal dimension. J Appl Crystallogr 1996;29(2):134−46. Available from: https://doi. org/10.1107/s0021889895011605.

[110] Bauer BJ, Hobbie EK, Becker ML. Small-angle neutron scattering from labeled single-wall carbon nanotubes. Macromolecules 2006;39:2637−42.

[111] Chatterjee T, Jackson A, Krishnamoorti R. Hierarchical structure of carbon nanotube networks. J Am Chem Soc 2008;130(22):6934−5. Available from: https://doi.org/ 10.1021/ja801480h.

[112] Schaefer DW, Justice RS. How nano are nanocomposites? Macromolecules 2007;40 (24):8501−17. Available from: https://doi.org/10.1021/ma070356w.

[113] Higgins JS, Benoît H. Polymers and neutron scattering. Oxford series on neutron scattering in condensed matter. Oxford New York: Clarendon Press; Oxford University Press; 1994.

[114] Sheidaei A, Baniassadi M, Banu M, Askeland P, Pahlavanpour M, Kuuttila N, et al. 3-D microstructure reconstruction of polymer nano-composite using FIB−SEM and statistical correlation function. Compos Sci Technol 2013;80:47−54. Available from: https://doi.org/10.1016/j.compscitech.2013.03.001.

[115] Peng R, Zhou H, Wang H, Mishnaevsky Jr L. Modeling of nano-reinforced polymer composites: Microstructure effect on Young's modulus. Computational Mater Sci 2012;60:19−31.

[116] Lau K-t, Lu M, Liao K. Improved mechanical properties of coiled carbon nanotubes reinforced epoxy nanocomposites. Compos Part A: Appl Sci Manuf 2006;37 (10):1837−40.

[117] Mark S, Alger M. Polymer science dictionary. Springer; 1997.

[118] Thostenson ET, Chou T-W. On the elastic properties of carbon nanotube-based composites: modelling and characterization. J Phys D: Appl Phys 2003;36(5):573.

[119] Bhuiyan MA, Pucha RV, Worthy J, Karevan M, Kalaitzidou K. Understanding the effect of CNT characteristics on the tensile modulus of CNT reinforced polypropylene using finite element analysis. Comput Mater Sci 2013;79:368−76. Available from: https://doi.org/10.1016/j.commatsci.2013.06.046.

[120] Lu JP. Elastic properties of carbon nanotubes and nanoropes. Phys Rev Lett 1997;79 (7):1297.

[121] Kwon H, Jar P-Y. On the application of FEM to deformation of high-density polyethylene. Int J Solids Struct 2008;45(11):3521−43.

[122] Taherzadeh M, Baghani M, Baniassadi M, Abrinia K, Safdari M. Modeling and homogenization of shape memory polymer nanocomposites. Compos Part B: Eng 2016;91:36−43. Available from: https://doi.org/10.1016/j.compositesb.2015.12.044.

[123] Baghani M, Naghdabadi R, Arghavani J, Sohrabpour S. A constitutive model for shape memory polymers with application to torsion of prismatic bars. J Intell Mater Syst Struct 2012;23(2):107−16. Available from: https://doi.org/10.1177/1045389x11431745.

[124] Hou X, Hu H, Silberschmidt V. Numerical analysis of composite structure with in-plane isotropic negative Poisson's ratio: effects of materials properties and geometry features of inclusions. Compos Part B: Eng 2014;58:152−9. Available from: https://doi.org/10.1016/j.compositesb.2013.10.030.

[125] Sevostianov I. On the shape of effective inclusion in the Maxwell homogenization scheme for anisotropic elastic composites. Mech Mater 2014;75:45−59. Available from: https://doi.org/10.1016/j.mechmat.2014.03.003.

[126] Mortazavi B, Baniassadi M, Bardon J, Ahzi S. Modeling of two-phase random composite materials by finite element, Mori−Tanaka and strong contrast methods. Compos Part B: Eng 2013;45(1):1117−25.

[127] Baghani M, Naghdabadi R, Arghavani J, Sohrabpour S. A thermodynamically-consistent 3D constitutive model for shape memory polymers. Int J Plasticity 2012;35:13−30.

[128] Blakslee OL, Proctor DG, Seldin EJ, Spence GB, Weng T. Elastic constants of compression-annealed pyrolytic graphite. J Appl Phys 1970;41(8):3373−82. Available from: https://doi.org/10.1063/1.1659428.

[129] Dai G, Mishnaevsky L. Damage evolution in nanoclay-reinforced polymers: a three-dimensional computational study. Compos Sci Technol 2013;74:67−77. Available from: https://doi.org/10.1016/j.compscitech.2012.10.003.

[130] Doll K, Ural A. Mechanical evaluation of hydroxyapatite nanocomposites using finite element modeling. J Eng Mater Technol 2013;135(1):011007. Available from: https://doi.org/10.1115/1.4023187.

[131] Safdari M, Al-Haik MS. Synergistic electrical and thermal transport properties of hybrid polymeric nanocomposites based on carbon nanotubes and graphite nanoplatelets. Carbon. 2013;64:111−21.

[132] Safdari M, Baniassadi M, Garmestani H, Al-Haik MS. A modified strong-contrast expansion for estimating the effective thermal conductivity of multiphase heterogeneous materials. J Appl Phys 2012;112.

[133] Alger MS. Polymer science dictionary. Springer Science & Business Media; 1997.

[134] Amnaya PA, Dimitris CL, Daniel CH. Modeling of graphene−polymer interfacial mechanical behavior using molecular dynamics. Model Simul Mater Sci Eng 2009;17 (1):015002.

[135] Scheider I, Chen Y, Hinz A, Huber N, Mosler J. Size effects in short fibre reinforced composites. Eng Fract Mech 2013;100:17−27. Available from: https://doi.org/10.1016/j.engfracmech.2012.05.005.

[136] Qian C, Harper LT, Turner TA, Li S, Warrior NA. Establishing size effects in discontinuous fibre composites using 2D finite element analysis. Computat Mater Sci 2012;64:106−11. Available from: https://doi.org/10.1016/j.commatsci.2012.05.067.

[137] Jiang Z, Gyurova LA, Schlarb AK, Friedrich K, Zhang Z. Study on friction and wear behavior of polyphenylene sulfide composites reinforced by short carbon fibers and sub-micro TiO 2 particles. Compos Sci Technol 2008;68(3):734−42.

[138] Mahdavi M, Yousefi E, Baniassadi M, Karimpour M, Baghani M. Effective thermal and mechanical properties of short carbon fiber/natural rubber composites as a function of mechanical loading. Appl Therm Eng 2017;117:8−16.

[139] Zhang H, Zhang Z, Breidt C. Comparison of short carbon fibre surface treatments on epoxy composites: I. Enhancement of the mechanical properties. Compos Sci Technol 2004;64(13):2021−9.

[140] Khun N, Zhang H, Sun D, Yang J. Tribological behaviors of binary and ternary epoxy composites functionalized with different microcapsules and reinforced by short carbon fibers. Wear 2016;.

[141] Jones DR, Ashby MF. Engineering materials 2: an introduction to microstructures, processing and design. Butterworth-Heinemann; 2005.

[142] Ashby MF, Cebon D. Materials selection in mechanical design. Le J de Phys IV 1993;3(C7):C7−1-C7-9.

[143] Ghazavizadeh A, Baniassadi M, Safdari M, Atai A, Ahzi S, Patlazhan S, et al. Evaluating the effect of mechanical loading on the electrical percolation threshold of carbon nanotube reinforced polymers: A 3D Monte-Carlo study. J Computational Theor Nanosci 2011;8(10):2087−99.

Numerical realization and characterization of random heterogeneous materials

3

Abstract

Multiphase heterogeneous materials currently serve a wide range of engineering applications. The effective properties of these engineered materials strictly depend on the volume fraction and morphological characteristics of the phases. The realization of heterogeneous materials is a basic step for design and characterization of new materials with specific properties. Accurate evaluation of the macroscopic properties for the heterogeneous materials, for example, elastic, thermal, electrical, and transport properties, usually requires a detailed realization of the full 3D microstructure. Historically in many practical applications such as materials engineering, petroleum engineering, biology, and medicine, only 2D information about the internal microstructures is provided obtained from 2D images. 3D reconstruction of 2D images is highly sought for many applications in homogenization and characterization and in general to establish structure-to-property relationships.

3.1 Introduction

An effective reconstruction procedure enables one to generate accurate image-based digital models for a subsequent analysis step to obtain macroscopic properties [1,2]. Direct reconstruction techniques such as stitching digitized serial section images acquired by focused ion beam scanning electron microscopy (SEM), X-ray computed tomography (micro-CT), and scanning laser confocal microscopy are not well-suited to the routine engineering applications, mostly due to expensive technology, lack of skilled operators, and many other technical issues involved [3–5].

The first stage in an accurate reconstruction of a microstructure is to determine an appropriate tool to describe it. Many statistical measures can be used for this purpose, including surface correlation functions [1], lineal measures [1,6], pore-size functions [1], two-point cluster functions (TPCFs) [7], and n-point correlation functions [8–11]. Among these microstructure descriptors, n-point correlation functions have attracted a great deal of attention due to many reasons, including their coordination with spectral and material knowledge frameworks [12–15] and their direct application in estimating effective properties of random heterogeneous materials [1,16]. It has been shown that some of the effective properties of the random heterogeneous materials are strongly correlated to n-point statistics. Effective mechanical, thermal, electrical, and permeability properties of a wide

Applied Micromechanics of Complex Microstructures. DOI: https://doi.org/10.1016/B978-0-443-18991-3.00005-2
© 2023 Elsevier Inc. All rights reserved.

range of heterogeneous materials can be calculated directly based on their intrinsic properties and their n-point correlation functions [1,3,12,17].

The simplest form of the n-point correlation functions ($n = 1$) provides an estimate for the volume fraction of the phases in the media, and higher-order correlation functions carry more information about the geometry of the microstructure, where in the asymptotic limits, every microstructure can be uniquely reconstructed from infinite-order correlation functions [18−20]. The application of the correlation functions for the reconstruction has been proposed in the literature. Yeong and Torquato [21] first reported the possibility of the reconstruction of a full 3D medium by the utilization of morphological information obtained from a 2D slice of the initial microstructure. In this work, they considered reconstructing a 3D specimen of Fontainebleau sandstone by using the two-point probability function and lineal-path function estimated from 2D cut sections [21]. Later, Cule and Torquato [22] introduced a stochastic optimization technique that enables one to generate realizations of a heterogeneous material from a prescribed set of correlation functions. The reconstruction method was based on the minimization of the summation of the squared differences between the calculated and reference correlation functions using a simulated annealing method [22,23]. Another method that utilizes two-point probability functions for the purpose of the reconstruction is the Gaussian filtering method [24−26]. The Gaussian random field method uses linear and nonlinear filters to match the correlation functions in the reconstruction process. It is worth to mention that these methods are computationally costly and are mostly applicable to two-phase isotropic media. Because of these inherent limitations, it is impossible to extend these methods to general multiphase anisotropic media [3,25,27,28].

A well-known example of heterogeneous materials is a solid oxide fuel cell (SOFC) anode. SOFC is an energy transformation device that converts the chemical energy of a fuel (such as hydrogen or methane) into electricity based on electrochemical reactions. Many research works have been carried out on SOFCs to improve their efficiency by engineering their 3D microstructure. Many parameters involving the morphology [28], the shape and the size of the constituents [29], microscopic topology, and spatial connectivity of SOFCs [30] are studied for this purpose. Lanzini et al. [28] used the truncated-Gaussian method for the reconstruction of the 3D microstructures from 2D phase images using two-point statistics to analyze the microstructure of a fuel cell quantitatively based on the phase fraction, grain size, granulometry law, constituent shape factors, and the phase spatial distribution. Shi and Xue [29] employed a genetic-algorithm method to design electrodes for anode-supported planar solid oxide fuel cells. To maximize the cell performance, they considered more important specifications in their method, such as the level of porosities and particle size distribution for both anode and cathode electrodes. Baniassadi et al. [11] developed a Monte Carlo methodology, based on two-point statistical functions, for 3D reconstruction of the microstructure of a three-phase anode used in SOFCs. They considered TPCFs of a 2D SEM micrograph as the target. The realization of the microstructure was carried out using a hybrid stochastic methodology based on the colony and kinetic algorithms and optimization techniques. 3D reconstructed heterogeneous microstructures can be exploited to evaluate or homogenize properties [3].

In order to homogenize mechanical properties, it is appropriate to use the homogenization scheme with the highest possible speed alongside acceptable accuracy. Analytical homogenization theories [31,32] are relying on simplifying assumptions about the shape, distribution, and dispersions of the phases. Most of these methods, however, are only applicable to a dilute mixture of the phases with simple inclusion shapes [33]. Numerical methodologies such as finite element (FE) [34], finite difference (FD) [35], and fast Fourier transform (FFT) methods [36,37] can overcome these limitations, among which FFT methods are famous for their efficiency [38]. Moreover, FFT methods relax the need for costly preprocessing steps, such as mesh/grid generation needed for FD and FE methods. The accuracy of the FFT method compared to analytical solutions [37,39] and the computational efficiency of the method compared to FE homogenization methods are studied in the literature [38,40]. Further, due to the efficiency the FFT method was also extended to nonlinear materials with a reasonable computational cost [41−43].

Following our previous work [44], in this chapter a powerful approach for the reconstruction of three-phase microstructures based on a few cut sections is presented. The method consists of the approximation of TPCFs for all 3D vectors and reconstructing the representative volume element (RVE) using an adapted phase recovery algorithm. In the reconstruction of a three-phase microstructure using an approximation to TPCFs, some contradictions may occur, mainly originating from the approximate property of the method. For instance after the reconstruction of the first phase and applying interdependencies, there may be some points belonging to more than one phase. To illustrate the capabilities of the proposed method an SOFC anode microstructure is reconstructed using a few 2D cut sections and the accuracy of the method is discussed based on the microstructure properties such as tortuosity, percolation, and three-phase boundary (TPB) length (TPBL). Homogenization of random heterogeneous microstructures can be performed using a different numerical approach. In this chapter the FE and FFT methods are discussed to homogenize effective thermomechanical properties [3].

For this purpose an FE method was applied to study thermomechanical properties of random heterogeneous materials. This study also taps on the efficiency of the FFT methods for random heterogeneous materials relying on the latest development of FFT methods [43].

3.2 Realization of multiphase random heterogeneous materials using a Monte Carlo approach

In the current study a Monte Carlo approach is developed to realize various virtual 3D microstructures. The realization approach is stochastic, and it is composed of three major steps: generation, distribution, and growth of cells. Hereafter, by cells (or alternately grains or particles), we refer to the initial geometries assigned to each phase before the growth step. In each realization, for different phases, first a number of initial cells are placed at some random nucleation points. Upon assignment of the initial cell geometries the growth of cells starts based on a cellular automaton algorithm, as depicted in Fig. 3.1. Realization steps are repeated continuously until a microstructure with desired volume fractions for each phase is achieved. In fact the procedure of the growth of the cells is continued until they meet each other and the grid is filled with three phases. It should be mentioned that

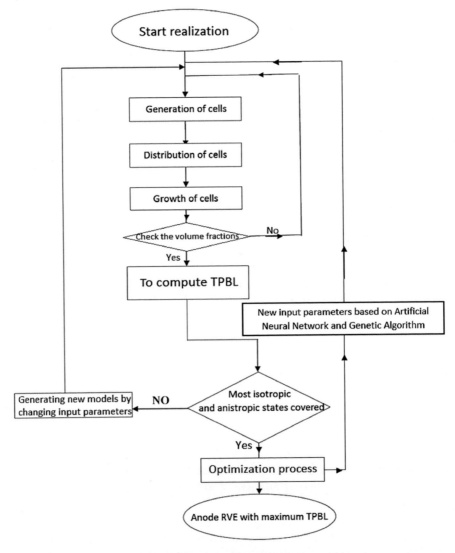

Figure 3.1 Adopted algorithm for the realization of virtual microstructures with optimal TPBL [45].

a hard-core (nonpenetrating) condition is always maintained throughout the initial distribution and growth of the cells. Additionally, depending on the desired characteristics of the final microstructure, during various steps of realizations, several control parameters are introduced and modulated [45].

Fig. 3.2 illustrates the step-by-step growth of three cells in a 2D cut section during eight evolutionary stages of growth. Starting with an empty 3D grid, during the cell generation procedure, each cell is rotated and transformed randomly and finally placed in a random position in the RVE. For generation of initial cells a 2D cell can be extruded to form an extrusion shape function. In this work, however, cubic basic cells are used. The cells can be transformed through a shrinkage function, S, defined as

$$S = \begin{bmatrix} f_1(x, y, z, \beta, p_1) \\ f_2(x, y, z, \beta, p_2) \\ f_3(x, y, z, \beta, p_3) \end{bmatrix} \tag{3.1}$$

where x, y, and z are Cartesian coordinates and f_i can be a simple polynomial function. In the transformation matrix above, $0 < \beta < 1$ is a random variable, where p_i is the optimization parameter. In this work, simple forms are chosen that only scale the initial cell randomly. The number of initial cells generated in each step can be controlled by a diffusion factor (or DF), which is defined to be $0 < DF < 1$. A minimum number of cell generations in each step occur when DF = 1 [45].

In the developed algorithm a Cellular Automata (CA) approach is used to implement an Eden fractal model [11] for simulating the kinetic growth of cells. It should be mentioned that grain boundaries in heterogeneous materials can be assumed as fractal geometries [46].

In the current study the CA algorithm is exploited on a grid of sites with a finite number of states. By assigning an initial state to each site of the grid the growth process is guided by the states of the neighboring sites and is governed by a few growth rules that are usually similar for all sites [45].

A cellular automaton is a collection of sites of the grid with specified states with a neighborhood relation, a set of states, and a local transition function that evolved through several discrete time steps according to the local transition function of the neighboring sites. In this study the Neumann neighborhood relation is adopted. For a lattice a 3D Neumann neighborhood has six neighboring sites on the right, left, front, top, bottom, and back of a central site, which are considered during the evolution and subsequent growth of the grid [11,45].

In this study, stochastic transition rules are used for every site of the Neumann neighborhood in the following way:

$$\psi_i(\beta, p_i) = p_i - \beta > 0, \tag{3.2}$$

where $i = 1\ldots6$, β $(0 < \beta < 1)$ is a random variable and p_i is an optimization parameter $(0 \leq p_i \leq 1)$. The model is updated synchronously, and if condition (2) is satisfied, the growth continues in that direction of empty sites. The SOFC anode microstructure

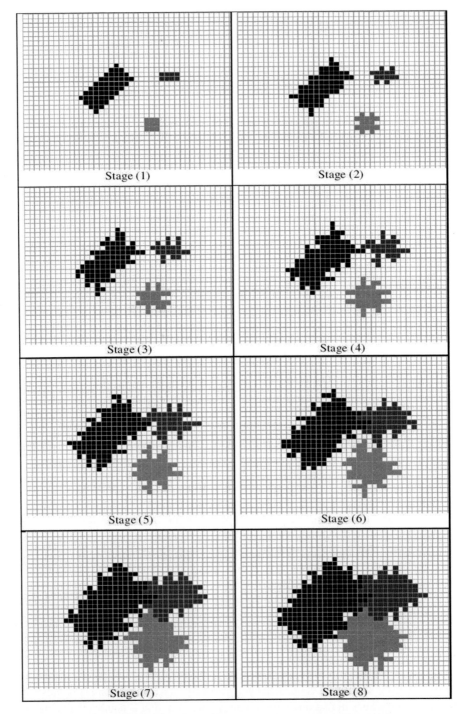

Figure 3.2 Step-by-step growth of three phases illustrated in a 2D plane [11,45].

is one the most complex microstructures that can significantly alter the performance of SOFC; therefore the realization method is implemented to the SOFC anode microstructure to design a 3D microstructure with maximum TPBL [45].

3.2.1 Realization of the SOFC anode

Microstructural attribution of the anode can significantly alter the performance of SOFC. Two main factors that affect TBPL of the anode microstructure are volume fractions of the phases (Ni, YSZ, and void) and the 3D microstructural attributions of the phases. In conjunction with experimental observations the optimal values for the volume fractions were utilized to design an optimal 3D microstructure of the anode with maximal TPBL. The best microstructure with the largest TPBL is achieved using volume fraction values of %26, %33, and %41 for Ni, YSZ, and pores, respectively [45].

A wide variety of 3D realizations are generated. Growth factors in the x-, y-, and z-directions are adopted as the optimization parameters and varied aiming to achieve maximal TPBL. The cycle of realization and optimization (artificial neural network and genetic algorithm) to obtain the highest value in TPBL is repeated until the OF is led to maximum TPBL. This strategy adopted here to optimize parameters is based on optimizing growth parameters and the diffusion factor as the most important parameter for designing anode microstructures. More details about the adopted 3D realization approach are provided elsewhere [11,45,47].

For this purpose, over 400 3D realizations were generated based on crystal growth and diffusion aiming to cover a wide range of possible isotropic and anisotropic configuration. A novel algorithm for calculating TPBL was discussed, which leads to accurate values for TPBL. Finally the largest TPBL was identified employing a coupled artificial neural network and genetic algorithm optimization scheme. It is observed that with the proposed scheme, 300% and 500% enhancement for the TPBL can be achieved for isotropic and anisotropic microstructures, respectively. Additionally, important factors leading to maximal values for TPBL were investigated. The current study results can be used as a guide for future experimental investigations [45].

There are seven factors to control the growth rates for the Ni and YSZ cells. Six of these parameters are growth rates of Ni and YSZ in x-, y-, z-directions, and the last parameter is the diffusion factor. The diffusion and crystal growth factors correspond to cell growth in our models. The ranges of growth rates and diffusion are selected to be $0.001-0.1$ and $0.005-0.99$, respectively, with relative units. All RVEs represent anode microstructures, and the surfaces of $z = 0$ and $z = 150$ are selected as the entrance surface of the fuel and the connected surface to electrolyte, respectively. A suitable RVE size is selected [11], and the real size for every voxel is related to grain sizes of Ni and YSZ and to the fabrication method. In the beginning, 25 samples for isotropic states are generated with a diffusion factor of 0.01, and then TPBL is calculated. Results for isotropic models in maximum, minimum, and average TPBL are displayed in Table 3.1. To study diffusion effects, three of the best isotropic models are generated again with a different diffusion factor, as

Table 3.1 Results for isotropic models in maximum, minimum, and average TPBL. V is the growth rate, and indices designate the direction and phase, respectively [45].

Isotropic model	$V_{x\text{-Ni}}$	$V_{y\text{-Ni}}$	$V_{z\text{-Ni}}$	$V_{x\text{-YSZ}}$	$V_{y\text{-YSZ}}$	$V_{z\text{-YSZ}}$	DF	TPBL
The best model	0.050	0.050	0.050	0.050	0.050	0.050	0.010	0.0222
Average model	0.1	0.1	0.1	0.005	0.005	0.005	0.010	0.0145
The worst model	0.001	0.001	0.001	0.001	0.001	0.001	0.010	0.0072

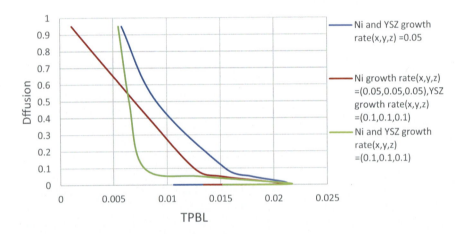

Figure 3.3 TPBL versus growth rate of Ni clusters in the Z-direction (growth rate in x- and y-directions = 0.05) [45].

illustrated in Fig. 3.3. According to this figure the largest TPBL occurs for diffusion factor values between 0.005 and 0.01, and thus diffusion is selected in this range for other samples in order to decrease the number of virtual reconstructions and increase the accuracy of tests [45].

Fig. 3.4 shows the best and worst 3D isotropic microstructures for the anode based on their TPBL.

In the second step, over 350 samples are generated to cover most possible states for anisotropic microstructures (models). Table 3.2 lists the properties of some anisometric models.

In the third step the Taguchi method [48] for the design of the experiment is employed to generate 32 samples to enhance confidence in results, and consequently, most of the possible states are considered. Table 3.3 summarizes several models used in the current study based on the Taguchi theory. Fig. 3.5 shows the

Numerical realization and characterization of random heterogeneous materials

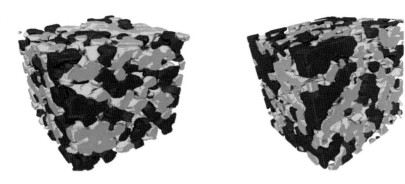

Figure 3.4 The best (left) and worst (right) 3D anode RVEs according to TPBL for isotropic models; Ni and YSZ phases are shown in different grayscale color [45].

Table 3.2 Results for anisotropic models in maximum, minimum, and average TPBL. V is the growth rate, and indices designate the direction and phase, respectively [45].

Anisotropic model	$V_{x\text{-Ni}}$	$V_{y\text{-Ni}}$	$V_{z\text{-Ni}}$	$V_{x\text{-YSZ}}$	$V_{y\text{-YSZ}}$	$V_{z\text{-YSZ}}$	DF	TPBL
The best model	0.05	0.05	0.1	0.05	0.05	0.1	0.010	0.0231
Average model	0.01	0.01	0.01	0.005	0.005	0.005	0.010	0.0144
The worst model	0.001	0.001	0.001	0.001	0.001	0.005	0.005	0.0057

Table 3.3 Summary of the results achieved based on Taguchi models with maximum, minimum, and average TPBL. V is growth rate, and indices indicate the direction and phase, respectively [45].

Taguchi model	$V_{x\text{-Ni}}$	$V_{y\text{-Ni}}$	$V_{z\text{-Ni}}$	$V_{x\text{-YSZ}}$	$V_{y\text{-YSZ}}$	$V_{z\text{-YSZ}}$	DF	TPBL
The best model	0.034	0.034	0.067	0.001	0.1	0.067	0.010	0.0217
Average model	0.001	0.034	0.034	0.034	0.034	0.034	0.005	0.0144
The worst model	0.001	0.001	0.001	0.001	0.001	0.001	0.005	0.0067

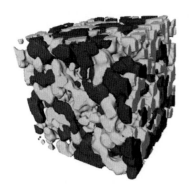 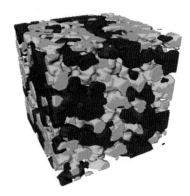

Figure 3.5 The best (left) and worst (right) 3D anode RVEs according to TPBL for anisotropic models; Ni and YSZ are shown in different grayscale color [45].

best and worst possible anisotropic 3D anode microstructures ranked based on the values of TPBL [45].

In the final step, we find a functional relationship between the growth rates, diffusion, and TPBL to discover the maximum TPBL. Since this function in general possesses a complicated nature and since it is unknown the artificial neural network (ANN) is employed to approximate it. After finding the appropriate ANN, the genetic algorithm is employed to acquire the inputs that correspond to the maximal value of TBPL. To account for the regressive nature of this problem the radial basis function (RBF) [49] network is utilized to approximate the objective function. For this purpose, about 80% of the virtual realizations in the dataset (305 samples out of all 382 samples) are chosen to train the ANN, and the remaining virtualizations are used for testing the trained network. In order to create the best possible network, networks with different numbers of nodes (from 1 to 20) are trained, and their mean square errors (MSEs) for the test datasets are calculated. The network that was trained with 10 nodes demonstrated the minimum test MSE. Therefore the RBF network that was trained with 10 nodes is used to approximate a functional form of the TBPL. Using the genetic algorithm, inputs in the range of 0.001 to 0.1 for growth rates and 0.005 to 0.01 for diffusion are generated randomly, and their corresponding outputs are computed by the above-mentioned approximation functions until the best output is achieved. Then, these inputs are simulated by our realization algorithm, and it is observed that the results of the ANN and TPBL codes are different. It is presumed that there are correlations between the samples in the dataset and the method by which they are generated. Thus only 32 data generated by the Taguchi method are utilized as the samples of the dataset. Since the number of new dataset samples is limited this time a leave-one-out method is applied to prevent the trained network from overtraining [49] and to find the number of nodes by which the most accurate RBF ANN is trained. In this technique, for each number of the nodes (from 1 to 20 nodes) the network is trained by 31 samples of the dataset, and it is tested by the remaining sample. This process is iterated 32 times until all the 32 samples

Numerical realization and characterization of random heterogeneous materials

are used as the test data one time, and the test MSE of these 32 iterations is calculated and considered as the MSE of the corresponding number of nodes. After such considerations, it is observed that the network which was trained with five nodes shows the minimum MSE. Then we used the network that was trained with five nodes for the approximation of the function to estimate TBPL for the new datasets in the genetic algorithm to gain the best inputs, but again, no improvement was achieved. Finally, it was decided to account for all 1 to 20 nodes by the RBF ANN, and all the 32 samples gained by the Taguchi method are used as their train data. Therefore, 20 approximation functions for TPBL are produced by this method. After that the same genetic algorithm is used to generate inputs randomly for each of these 20 networks in the code. For each network, this code is run 10 times, so in total, 200 TPBLs are achieved in this way. After running all these 200 data in the TPBL code, one of the obtained results through the network that was trained with 10 nodes revealed the best outcome. The results of the RBF network that was trained with 10 nodes are listed in Table 3.4, and the net results in this table indicate the predicted TPBLs from ANN [45].

These procedures led to the optimal 3D RVE of the anode with the largest TPBL that is listed as the second model in Table 3.4. Fig. 3.6 shows 3D anode RVEs with the best, average, and worst possible values of TPBL [45].

In Fig. 3.7 the volume distribution of Ni and YSZ clusters in maximum and minimum TPBL models is revealed. The unit of volume is defined as $voxel^3$.

Fig. 3.7A and B indicate that by controlling the size and the number of Ni and YSZ clusters, it is possible to improve TPBL. The volumes of the largest clusters of Ni and YSZ in the largest TPBL model that is most effective in TPBL are 877288 ($voxel^3$) and 1145023 ($voxel^3$), respectively, and those for the minimum TPBL model are 796912 ($voxel^3$) and 1095975 ($voxel^3$). This figure also illustrates that a microstructure with higher TPB has larger clusters, and this is not dictated by the growth rate and other factors such as diffusion and the direction of growth adopted in the current study. One reason can be that large clusters are connected to the electrolyte or current collector (e.g., small Ni clusters have no connection with the current collector, small pores have no connection with the gas channel, and small YSZ clusters have no connection with the electrolyte) [45].

The growth rate of Ni and YSZ crystals in diverse directions is one of the important factors that affect the TPBL. To investigate the growth rate effect on TPBL, over 150 samples with a diffusion factor of 0.005 are utilized, and the corresponding results are illustrated in Fig. 3.8 [45].

Plots in Fig. 3.8 infer that the relationship between the growth rates in x-, y-, and z-directions with TPBL is rather complex. Overall, it is possible to hypothesize that by increasing the growth rates in the z-direction for both Ni and YSZ, TPBL increases, but there is an optimal state for each procedure. In this study, all RVEs are cubic and the growth rates in x- and y-directions are the same as in the z-direction for the isotropic models, but for a real anode which is very thin the influence of an increase in growth rates in the x- and y-directions to achieve larger TPBL is more pronounced than that in the z-direction, which is the same as in the simulated anisotropic models [45].

Table 3.4 Results for ANN and genetic algorithm models in maximum, minimum, and average TPBL. V is the growth rate, and indices indicate the direction and phase, respectively [45].

$V_{x\text{-Ni}}$	$V_{y\text{-Ni}}$	$V_{z\text{-Ni}}$	$V_{x\text{-YSZ}}$	$V_{y\text{-YSZ}}$	$V_{z\text{-YSZ}}$	DF	Net result	TBPL
0.077596	0.077065	0.095834	0.085329	0.1	0.085025	0.009854	0.0224	0.0199
0.08322	0.071541	0.085261	0.094777	0.096448	0.085291	0.009986	0.02419	0.0236
0.077598	0.06299	0.079911	0.068372	0.086539	0.069128	0.009965	0.02397	0.0214
0.083376	0.091014	0.089712	0.085677	0.098154	0.093084	0.009953	0.02451	0.0214
0.078347	0.097822	0.096391	0.099945	0.095571	0.099885	0.009973	0.02508	0.0211
0.072872	0.071417	0.066629	0.069929	0.072871	0.075935	0.009993	0.02343	0.0176
0.067903	0.067936	0.065912	0.069519	0.068274	0.062508	0.007538	0.02393	0.0214
0.07031	0.057902	0.089845	0.064201	0.06886	0.0916	0.007309	0.02412	0.0208
0.061122	0.055678	0.059232	0.076735	0.080136	0.059363	0.007484	0.02417	0.0228
0.062351	0.065235	0.085964	0.085697	0.07235	0.070359	0.007458	0.02384	0.0217

Numerical realization and characterization of random heterogeneous materials 107

Figure 3.6 3D RVEs with the best (left), average (middle), and worst (right) values of TPBL. Ni and YSZ phases are shown in different colors [45].

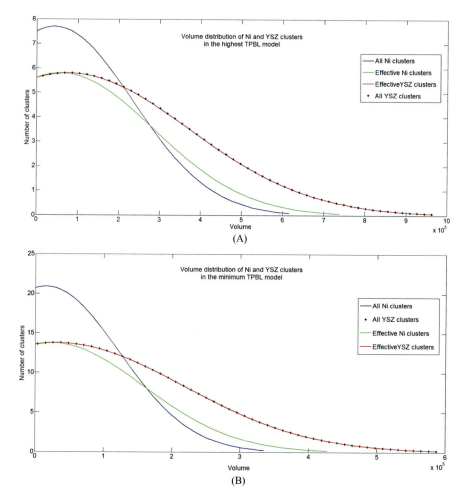

Figure 3.7 The volume distribution of Ni and YSZ clusters in maximum (A) and minimum TPBL models (B), respectively [45].

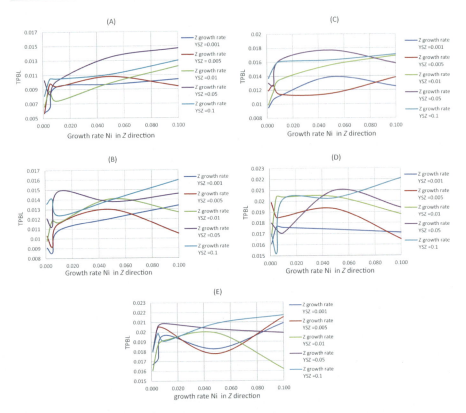

Figure 3.8 Investigation of the effects of growth rate of Ni clusters in the z-direction on TPBL, with Ni and YSZ growth rates in x- and y-directions of (A) 0.001, (B) 0.005, (C) 0.01, (D) 0.05, and (E) 0.1 [45].

3.3 Multiphase reconstruction of heterogeneous materials using a full set of TPCFs

In this section a technique for the reconstruction of a general multiphase medium is presented. The method uses limited statistical information provided from one (or at most two) cut section of the medium to reconstruct the respective 3D medium in two major steps: (1) the approximation of 3D correlation functions for all vectors in the space of the medium and (2) the reconstruction of the microstructure based on a phase-recovery algorithm that relies on the approximated correlation functions. It is shown that some of the side effects of reconstruction originating from the errors in the approximation step for a three-phase microstructure can be fully eliminated in the subsequent reconstruction step using a two-step phase-recovery algorithm, where the output of the first step is used as a constraint for the next reconstruction step. As an illustrative example the 3D reconstruction of an isotropic SOFC anode is carried out using a 2D SEM micrograph. Finally, it is observed that the proposed method accurately reconstructs an anisotropic SOFC and conserves key performance characteristics of this medium, including the three-phase boundary length and the effective conductivity and diffusivity [3].

3.3.1 Approximation of TPCFs for three-phase microstructures

3.3.1.1 N-point correlation functions and conditional probability

Correlation functions contain morphological attributes of a statistically homogeneous microstructure. Considering N points within the microstructure, randomly positioned at x_1, x_2, ...,x_N, the probability of the event in which all these points reside in the phases $i = 1 \ldots n$ is defined as the N-point correlation function as [3]

$$C_N^{ij\ldots n}(x_1.x_2.\ldots.x_N) = \left\langle p_{x_1}^i p_{x_2}^j \ldots p_{x_N}^n \right\rangle. \quad i.j.\ldots.n \in set\ of\ phases. \tag{3.3}$$

where $\langle\ldots\rangle$ is the ensemble average symbol, and p_x^i, the microstructure function, is defined as

$$p_x^i = \begin{cases} 1\ x\ in\ phase\ i \\ 0\ otherwise \end{cases}. \tag{3.4}$$

The microstructure with the characteristic function described as Eq. (3.4) is called an Eigen microstructure. It means that in every position, there is only one specified phase. TPCFs are the simplest form of these functions that contain geometrical attributes of a microstructure. These functions describe the joint probability distribution of finding starting and ending points of a predefined vector within specific phases. For a statistically homogeneous microstructure the absolute position of points is irrelevant. Therefore for such media, a base point is chosen arbitrarily, and all other points are expressed as vectors originating from this point. Denoting head and tail points of a randomly chosen vector, $\vec{r}$, by x_1 and x_2, $C_2^{ij}(x_1.x_2)$ can be expressed as [3]

$$C_2^{ij}(x_1.x_2) = C_2^{ij}(\vec{r}) = P\{(x_1 \in \varphi_i) \cap (x_2 \in \varphi_j)\}, \quad i.j = 1.2.\ldots.n. \tag{3.5}$$

Using conditional probability, higher-order correlation functions can be defined based on lower-order probability functions by [3]

$$\begin{cases} C_2^{ij}(x_1.x_2) = P\{(x_2 \in \varphi_j)|(x_1 \in \varphi_i)\}P(x_1 \in \varphi_i). \\ C_3^{ijk}(x_1.x_2.x_3) = P\{(x_3 \in \varphi_k)|((x_1 \in \varphi_i) \cap (x_2 \in \varphi_j))\}P\{(x_2 \in \varphi_j)|(x_1 \in \varphi_i)\}P(x_1 \in \varphi_i) \end{cases} \tag{3.6}$$

3.3.2 Number of independent sets of TPCFs

A set of TPCFs is an array that contains C_2^{ij} for all vectors $\vec{r}$. For an n-phase microstructure, there are n^2 sets of TPCFs; for example, for a two-phase microstructure, C_2^{11}, C_2^{22}, C_2^{12}, and C_2^{21} form all sets of TPCFs. However, there are some interdependencies among the sets. For example, for a two-phase medium, there is only one independent set of TPCFs. Due to the normality condition the following constraints must be satisfied [3,50]:

$$\sum_{i=1}^{n}\sum_{j=1}^{n} C_2^{ij}(\vec{r}) = 1 \text{ where } \vec{r} = x_2 - x_1. \tag{3.7}$$

$$\sum_{j=1}^{n} C_2^{ij}(x_1.x_2) = v_i. \tag{3.8}$$

$$C_2^{ij}(x_1.x_2) = C_2^{ji}(x_2.x_1). \tag{3.9}$$

where v_i is the volume fraction of phase i. These constraints reduce the independent sets to $n(n-1)/2$ [50]. Moreover, given the periodic microstructure, there are other interdependencies between the sets of TPCFs in the Fourier domain that can be expressed by [3]

$$P_m^{ij} = \frac{\left(P_m^{ki}\right)^* P_m^{kj}}{P_m^{kk}} \tag{3.10}$$

where P_m^{ij} and m are the Fourier transform of $C_2^{ij}(\vec{r})$ and the transform variable in Fourier space, respectively [see Eqs. (3.17) and (3.18)], and the $*$ denotes a complex conjugate [51]. These interdependencies reduce the independent sets of TPCFs to $(n-1)$ for an n-phase medium. For example, for a three-phase medium, defining only two sets is sufficient for the reconstruction of the microstructure. By solving Eqs. (3.7) through (3.10) simultaneously, it is observed that for a three-phase medium with exact TPCFs (means not approximated), at least one set must be cross-correlation (means $i \neq j$ for F_m^{ij}), and using the reconstruction procedure just once is sufficient to reconstruct the whole three-phase microstructure [3].

3.3.3 Approximation formulation

2D cross-sections only provide TPCFs of 2D vectors, so to reconstruct a 3D microstructure, TPCFs of all 3D vectors should be approximated based on the statistical information of 2D cross-sections. For every desired 3D vector $\vec{r}$, with head and tail points, x_1 and x_3, respectively, an intermediate point x_2 is considered on the xy- or yz-plane to decompose the 3D vector into two 2D vectors. Arrangement of the points is shown in Fig. 3.9. Point x_1 is chosen as the origin of the coordinate system,

Numerical realization and characterization of random heterogeneous materials

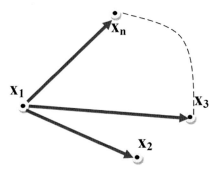

Figure 3.9 Schematic of correlation vectors of the n-point correlation functions [3].

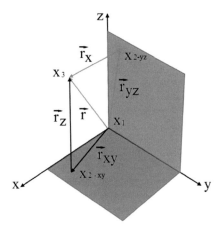

Figure 3.10 Decomposition of vector $\vec{r}$ into two perpendicular vectors on both xy- and yz-planes [3].

and point x_2 is the projection of point x_3 on the xy- or yz-plane. When there is no preference in choosing the projection plane for point x_2, it is better to project this point on both planes. Therefore $\vec{r}_z$ and $\vec{r}_x$ are the projections of $\vec{r}$ on the z- and x-axes, respectively, and $\vec{r}_{xy}$ and $\vec{r}_{yz}$ are expressed as [3]

$$\vec{r}_{xy} = \vec{r} - \vec{r}_z, \vec{r}_{yz} = \vec{r} - \vec{r}_x. \tag{3.11}$$

where $C_2^{ik}(x_1.x_3)$ is averaged (Fig. 3.10). Averaging reduces the error and leads to better results.

Using conditional probability [Eqs. (3.5) and (3.6)], it is possible to exactly express $C_2^{ik}(x_1.x_3)$ based on three-point correlation functions [10]. Then the three-point correlation functions are approximated using 2D TPCFs and some weight functions [52]. Weigh functions are omitted finally (for details, see [10]), and for an n-phase microstructure, $C_2^{ik}(x_1.x_3)$ can be expressed based on TPCFs of its two perpendicular cut sections as [3]

$$C_2^{ik}(x_1.x_3) \approx \sum_{j=1}^{n} \frac{C_2^{ij}(x_1.x_2)C_2^{jk}(x_2.x_3)}{v_j}. \tag{3.12}$$

where v_j is the volume fraction of phase j. As explained earlier, there are only two independent sets of TPCFs for identifying full sets of TPCFs for a three-phase periodic microstructure. Therefore for 3D reconstruction of a microstructure, it is required to approximate only two sets, for example, $C_2^{11}(x_1.x_3)$ and $C_2^{22}(x_1.x_3)$. Using Eq. (3.12), $C_2^{11}(x_1.x_3)$ can be expressed as [3]

$$C_2^{11}(x_1.x_3) \approx \frac{C_2^{11}(x_1.x_2)C_2^{11}(x_2.x_3)}{v_1} + \frac{C_2^{12}(x_1.x_2)C_2^{21}(x_2.x_3)}{v_2} + \frac{C_2^{13}(x_1.x_2)C_2^{31}(x_2.x_3)}{v_3}. \tag{3.13}$$

A similar relation can be written for $C_2^{22}(x_1.x_3)$.

3.3.4 Modification of the phase recovery algorithm for three-phase reconstruction

There are several methods for the realization of a microstructure based on its statistical information. Methods such as simulated annealing, gradient-based schemes, and Gaussian random fields are usually applicable for isotropic and two-phase media [1,9,23,26,28,53,54]. For multiphase and anisotropic media, most of these methods tend to be computationally expensive. The phase-recovery algorithm is an alternative method that is successfully used for the rapid and accurate reconstruction of multiphase and anisotropic media [10,27]. This method, initially proposed by Gerchberg and Saxton [55], was first proposed as a rapid solution technique to obtain the phase of the complete wave function with known intensity in the diffraction and imaging planes of an imaging system. Later, Fineup [56] proposed an algorithm for the reconstruction of an object from the modulus of its Fourier transform. This algorithm is applicable to X-ray crystallography, image processing, and microstructure reconstruction or generally with a known modulus and unknown phase [3].

The first step in the reconstruction is to digitize the media using the microstructure function, that is, applying Eq. (3.4) to all positions and phases in order to obtain p_x^i. The microstructure state, p_x^i, is the existence probability for phase i in position x, 0 or 1 for eigen microstructures. Mathematically, this condition is described by [3]

$$\sum_{i=1}^{n} p_x^i = 1. p_x^i \in \{0.1\}. \tag{3.14}$$

where n denotes the number of phases in the microstructure. One-point correlation functions can be obtained simply by [3]

$$C_1^i = \frac{1}{Z}\sum_{x=0}^{Z-1} p_x^i \tag{3.15}$$

where Z is the total number of grid points of the microstructure. Similarly, discretized TPCFs are expressed by [3]

$$C_2^{ij}(r) = \frac{1}{Z}\sum_{x=0}^{Z-1} p_x^i p_{x+r}^j \tag{3.16}$$

where superscripts i and j denote the phases of interest, and subscript r enumerates location vectors in a discretized space. To use the reconstruction process with the phase-recovery algorithm, first we apply an FFT to the microstructure function to obtain [3]

$$P_m^{ij} = \mathscr{F}\left(P_x^i\right) = \frac{1}{Z}\sum_{x=0}^{Z-1} p_x^i e^{-2\Pi km/z} = \frac{1}{Z}|P_m^i|e^{k\theta_m^i}. k = \sqrt{-1}. \tag{3.17}$$

where $|P_m^i|$ is the amplitude and θ_m^i is the phase of the Fourier transform. Assuming periodicity and using the convolution theorem an FFT can be applied to a TPCF to obtain [3]

$$P_m^{ij} = \mathscr{F}\left(C_2^{ij}(r)\right) = \frac{1}{Z}|P_m^i|e^{k\theta_m^i}|P_m^j|e^{-k\theta_m^j} = \frac{1}{Z}(P_m^i)^* P_m^j. \tag{3.18}$$

The FFT is reduced to the square of the amplitude of the FFT of the respective microstructure function [51] defined by [3]

$$P_m^{ii} = \frac{1}{Z}(P_m^i)^* P_m^i = \frac{1}{Z}|P_m^i|^2. \tag{3.19}$$

Therefore given the autocorrelation of a specified phase i (P_m^{ii} or $C_2^{ii}(r)$), the modulus of P_m^i (means $|P_m^i|$) can be obtained using Eq. (3.19) and θ_m^i can be calculated by the phase-recovery algorithm to determine the entire microstructure functions for all points residing on phase i. This reference modulus is called $|P_m^i|_{ref}$.

Following the Fineup algorithm [57], our phase-recovery method has four steps: (1) starting with an initial random microstructure the FFT is calculated using Eq. (3.17); (2) replacing the modulus with the square root of the known autocorrelation multiplied by Z or $|P_m^i|_{ref}$, while keeping the phases unchanged; (3) calculating the inverse Fourier transform [using the inverse of Eq. (3.18)]; and (4) imposing the constraints in the real space [Eq. (3.4)]. The output of step 4 is used as an input to step 1, and the iteration continues until an error measure reaches to a threshold.

The reconstruction of a multiphase microstructure, with exact TPCFs, will be done after implementing the algorithm only one time for a specified phase i. Using Eq. (3.18), given P_m^i for a specified phase i, it is possible to determine the microstructure function, P_m^j, for all phases j. Because of the approximate nature of the TPCFs used for the reconstruction here, some discrepancy is unavoidable. This means that some points belong to more than one phase.

In the current study, all examples represent three-phase media, and therefore the phase-recovery algorithm is used twice to remove the contradictions mentioned above. At first the microstructure functions of one of the phases are determined using the phase-recovery algorithm. These microstructure functions are then used as constraints for the next phase-recovery step. In every iteration, for each point residing on more than one phase the first phase is preserved to avoid any contradiction in the realization of the entire microstructure [3].

3.3.5 Isotropic reconstruction

For a statistically homogeneous and isotropic microstructure the correlation functions can be extracted from an arbitrary planar cut through 3D media [1]. Fig. 3.11 illustrates a 300×300 SEM micrograph of a three-phase anode microstructure of an SOFC, where green, red, and blue denote yttria-stabilized zirconia (YSZ), Ni, and void phases with the volume fractions of 0.49, 0.17, and 0.34, respectively. Because of the isotropy and using only one cut section, it is assumed that xy- and yz-planes are identical. As expressed before (Section 2.2), full TPCFs of a periodic three-phase microstructure are defined by two independent variables. Therefore for

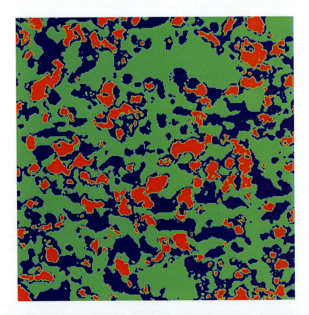

Figure 3.11 The reference cut section, used for reconstruction of a statistically isotropic medium (green: YSZ, red: Ni, and blue: void) [3].

Numerical realization and characterization of random heterogeneous materials 115

the reconstruction, it is required to calculate C_2^{11} and C_2^{22} using Eq. (3.13) as independent variables. After the calculation of C_2^{11}, the phase-recovery algorithm is utilized for phase 1. A subset of the reconstructed microstructure with a thickness of about 20 pixels is illustrated in Fig. 3.12 [3].

In an SOFC, the YSZ phase conducts ionic species, and the Ni phase is used for the electronic conduction. The YSZ phase should percolate throughout the media for the SOFC to function properly. Fig. 3.13 illustrates a subset of the YSZ phase.

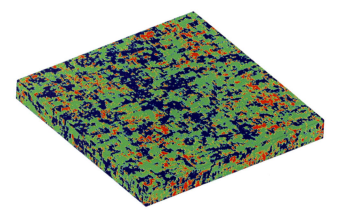

Figure 3.12 A subset of the reconstructed microstructure with a thickness of about 20 pixels (green: YSZ, red: Ni, and blue: void) [3].

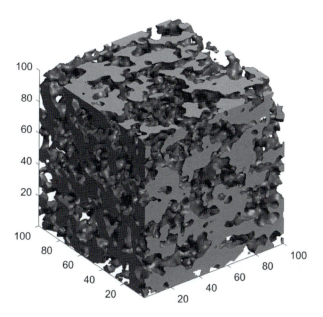

Figure 3.13 A subset of the reconstructed medium illustrating the connectivity and percolation of the YZS phase [3].

It is observed that the proposed phase-recovery reconstruction algorithm preserves important percolation characteristics of this phase.

3.3.5.1 Anisotropic reconstruction

Manufacturing processes like rolling and extrusion may result anisotropic media. These microstructures have an axis of symmetry and are referred to as transversely isotropic materials [58]. To reconstruct such anisotropic media with the proposed technique, at least two mutually perpendicular cut sections (for example, longitudinal and transverse planes) are needed to quantify the microstructure. Furthermore, given two or three mutually perpendicular cut sections, it is possible to reconstruct the microstructure with three perpendicular planes of symmetry (orthotropic materials) [3].

In order to illustrate the ability of the proposed method in the reconstruction anisotropic microstructure an SOFC anode RVE with the size of $150 \times 150 \times 150$ voxel is reconstructed and a comparison is performed between the properties of the original and reconstructed microstructures.

The three phases of the microstructure are void, YSZ, and Ni with volume fractions of 0.41, 0.33, and 0.26 shown by red, green, and blue regions, respectively. We use two cross-sections perpendicular to the x- and z-axes, shown in Fig. 3.14. Realization of the original microstructure is carried out by a hybrid stochastic methodology based on the colony and kinetic algorithms using a Monte Carlo methodology [3,11].

The reconstructed 3D anisotropic microstructure for this example is illustrated in Fig. 3.15. The original and approximated TPCFs for the diagonal direction vectors for the YSZ phase are depicted in Fig. 3.16. It is noteworthy that the autocorrelation function for nonperiodic and infinite microstructures starts from the volume fraction and approaches to square of the volume fraction [50], but the microstructure

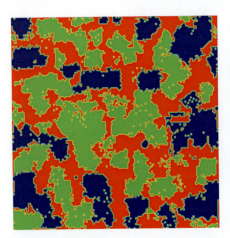 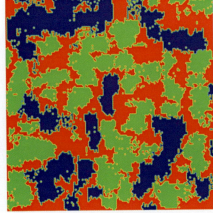

Figure 3.14 Two reference cut sections used for the approximation of TPCFs, perpendicular to the x-axis (left) and z-axis (right) (void: red, YSZ: green, and Ni: blue) [3].

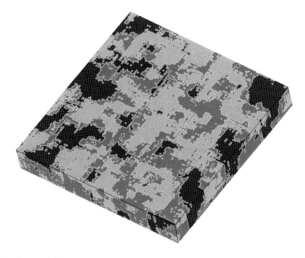

Figure 3.15 A subset of the reconstructed anisotropic microstructure based on two cut sections [3].

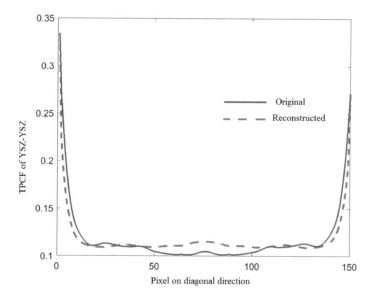

Figure 3.16 Autocorrelation function of the YSZ phase for reconstructed and original microstructures [3].

periodicity assumption in deriving Eq. (3.18) enforces the autocorrelation functions to start from the volume fraction of its related phase; then it approaches to the square of volume fraction, and ultimately, it increases to a value close to the volume fraction (Fig. 3.16). This means that the periodicity assumption changes the general form and asymptotic values of TPCFs.

3.3.6 Microstructural properties

The Ni-YSZ anodes are still commonly used in SOFCs [59]. The YSZ phase conducts oxide ions and yields an anode material with a matched coefficient of thermal expansion between the anode and electrolyte. Nickel serves as both an electrochemical catalyst and an electronic conductor. The porosities provide important percolating pathways for the transportation of the gas fuel through the anode structure. The behavior of effective transport properties, such as conductivity and diffusion, is related to the tortuosity and affected by the percolation of the respective phase. In addition the oxidation reaction of the fuel occurs only near the three-phase boundaries, where the three phases are coexisting. Therefore active TPBL and tortuosity constitute proper criterions to evaluate the performance of the proposed reconstruction procedure for SOFCs [3].

3.3.6.1 Active three-phase boundary length

TPB is the common boundary of all phases. To determine active TPBL, first the active cluster of all phases must be determined. A cluster of a phase is a subset of that phase in which every pair can be connected without crossing the other phases in the medium. A cluster is active and percolated in direction z when there is at least one point, belonging to that cluster, on each face of the domain across the z-direction. Ideally, by considering only active clusters the common boundary among them will be active too [3].

By clustering the phases of the anisotropic microstructure reconstructed in Section 4.2, using the algorithm presented in [60], the total and active clusters for all phases are determined (Table 3.5), and it is observed that there is only one dominant active cluster for each phase. Fig. 3.17 illustrates a subset of the active cluster of the Ni phase of the reconstructed microstructure [3].

After determining the active cluster for all phases the active TPB segments were counted. The results are summarized in Table 3.6, and the TPB segments of the original microstructure are illustrated in Fig. 3.18. A comparison between the original and reconstructed microstructures reveals that the percent of active cluster for YSZ and void phases are almost identical, and they differ about 9% from the original one for the Ni phase. For the TPB segments the error between the original and reconstructed microstructures increases to about 31%. This result may arise from the better distribution of the reconstructed clusters in size and number (Table 3.5) [3].

3.3.6.2 Tortuosity

The nanometer-to-micrometer pore size range restricts the gas transport mechanism in SOFC to diffusion [59]. The diffusion property of a specific phase is mathematically equivalent to the electrical/thermal conductivity problem in this case as other phases are impermeable to what passes through that specific phase [1]; therefore each phase acts as a conduit for a separate species.

Table 3.5 Clustering of all three phases (Ni, YSZ, and void) for original and reconstructed microstructures. There is only one active cluster for each phase [3].

Phase	Volume fraction	Number of clusters		Size of the active cluster/voxel		Size of the largest nonpercolated cluster/voxel		Active cluster percent	
		Original	Reconstructed	Original	Reconstructed	Original	Reconstructed	Original	Reconstructed
Ni	0.26	155	8287	779579	854681	25678	1188	88%	97%
YSZ	0.33	158	7138	1090736	1106130	14044	499	97%	99%
Void	0.41	21419	10858	1346424	1352053	53	316	98%	98%

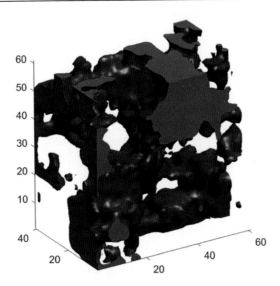

Figure 3.17 A subset of the active cluster of the Ni phase of the reconstructed microstructure [3].

A dimensionless effective conductivity and diffusivity factor can be defined as [3]

$$F = \frac{\sigma_e}{\sigma} = \frac{\mathscr{D}_e}{\mathscr{D}}. \qquad (3.20)$$

where σ_e and $\mathscr{D}_e$ are the effective conductivity and diffusion coefficient of the RVE, respectively, and σ and $\mathscr{D}$ denote the bulk conductivity and diffusion coefficient of each phase. The F factor defined in Eq. (3.20) also represents the percolation and tortuosity of each phase. Tortuosity, an important characteristic of a porous medium, is related to the F factor [1] by [3]

$$\tau = F^{-1} v. \qquad (3.21)$$

where v and τ are the volume fraction and tortuosity of a respective phase.

Using the analogy expressed in Eq. (3.20) and assuming $\sigma = 1$ the F factor can be calculated for each phase in all three directions using a thermal conductivity simulation. In these simulations, for each direction, two opposing faces normal to that direction are used with different temperature boundary conditions, and all other faces are assumed to be insulators. It is worth to mention that the F factor is a material property; therefore it is obtained independent of the selected boundary conditions. The F factor is only dependent on the geometrical arrangement of the specified phase. The F factor and values for all three phases in the three directions of the original and reconstructed anisotropic microstructures are summarized in Table 3.7 [3].

Table 3.6 Specifications of TPB segments of active and total clusters [3].

Total cell of the microstructure/voxel	Total TPB segment		Active TPB segment		Percent of active TPB		Density of active TPBL/ Segment cell $^{-1}$		Percent of error
	Original	Reconstructed	Original	Reconstructed	Original	Reconstructed	Original	Reconstructed	
3375000	114395	166179	79708	104815	69.7%	63.1%	0.0236	0.0311	31%

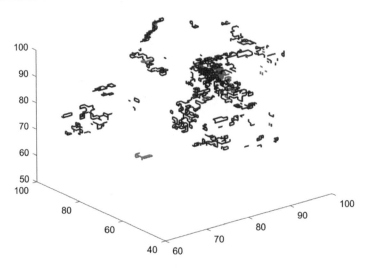

Figure 3.18 Active and total TPB segments of the original microstructure are shown in different grayscale color [3].

An error measure representing the accuracy of the reconstruction is postulated here as

$$\text{error} = \frac{1}{9}\sum \frac{|F_{\text{original}} - F_{\text{reconstructed}}|}{F_{\text{original}}}. \tag{3.22}$$

Using Eq. (3.22) the error values for the reconstructed medium remains at about 15%. It is worth to note that this error in the approximation of the properties is achieved using only two cut sections of a $150 \times 150 \times 150$ RVE [3].

3.4 Numerical characterization of thermomechanical properties: finite element method

In this work a digital 3D microstructure of the porous Ni-YSZ composite was successfully realized using a recently developed Monte Carlo technique. With a direct comparison of the TPCFs between the realized material and a real composite, the statistical equivalency was verified. An FE method was applied to study elastic behavior, thermal conductivity, and thermal expansion characteristics of the realized material as functions of temperature. All the computed room-temperature properties of the material are in the order of experimentally reported values. It suggests that at higher temperatures, where a full set of experimental values is not available, the computed properties follow the same trend as in room temperature and are very close to properties of the real materials. It is seen that in elastic response the digital material is not fully anisotropic and shows a behavior similar to that of orthotropic

Table 3.7 F factor and tortuosity for all phases in three mutually normal directions [3].

Phase	ν	X-direction				Y-direction				Z-direction			
		Original		Reconstructed		Original		Reconstructed		Original		Reconstructed	
		F	τ	F	τ	F	τ	F	τ	F	τ	F	τ
Ni	0.26	0.0221	11.7	0.0245	10.6	0.0339	7.7	0.0334	7.8	0.0283	9.2	0.0233	11.2
YSZ	0.33	0.0600	5.5	0.0819	4.0	0.0751	4.4	0.0739	4.5	0.0585	5.6	0.0765	4.36
Void	0.41	0.1393	1.9	0.1258	2.0	0.1557	1.7	0.1321	2.0	0.1530	1.7	0.1301	2.0

124 Applied Micromechanics of Complex Microstructures

materials. In addition the trends in shear and tensile behaviors are not the same. However, the thermal conductivity and thermal expansion of the same realized sample respond in a manner very close to those of the fully isotropic type. Furthermore, there are indications of the fact that thermal conductivity behavior of the digital material is mostly controlled by the Ni phase [47].

Percolation of the Ni phase promises the necessary electrical conductivity. Other practically important properties include elastic response, thermal conductivity, and thermal expansion coefficient (TEC). The studied temperature range starts from 25°C and continues to 1000°C in 100°C intervals. The FE method was applied to homogenize the digital microstructure in order to reveal the temperature dependence of the desired properties. The VCAD software package was used to transfer the digital composite into a ~ 1.2 million irregular tetrahedral element mesh readable by Simulia ABAQUS, which was the software package used to conduct FE analyses. For obtaining results the following assumptions were mandatory to be made: (1) the microstructure does not change by increasing the temperature, meaning that sintering, grain growth, and formation of new phases at a higher temperature were neglected. This assumption should not be a major source of error in simulations since all the aforementioned effects are negligible at the working temperature range of real SOFC anode materials [61,62]. (2) Aggregates of Ni and YSZ in the RVE are mechanically and thermally isotropic, meaning that the properties of aggregates in the RVE are the same along any direction. Considering the polycrystalline nature of the aggregates in real samples, this assumption sounds fair as well. No assumption was made about anisotropy of the whole RVE. In addition, it was also assumed that (3) both Ni and YSZ behave linearly elastic and (4) the RVE remains perfectly bonded while applying mechanical loads [47].

The elastic response of the composite was fully investigated by means of performing three tensile and three shear tests at each temperature interval. The total six tests stand for six independent components of the stress tensor [47].

In all the tests, 1% deformation, either tensile or shear, was introduced to the RVE to develop corresponding stress (tensile or shear) and subsequently strain. At each test, all the components of strain tensor and the component of stress tensor corresponding to the test were found by volume averaging over all the elements using the following equations [47]:

$$\varepsilon_{ij} = \frac{1}{V_{tot}} \sum_{n=1}^{N} V_n \varepsilon_{ij}^n$$

$$\sigma_{ij} = \frac{1}{V_{tot}} \sum_{n=1}^{N} V_n \sigma_{ij}^n \tag{3.23}$$

where ε_{ij} and σ_{ij} are components of the strain and stress tensor, respectively. The superscript n denotes the strain or stress component in the nth element. V_{tot} is the total volume of the RVE, and V_n is the volume of the nth element. N is the total number of elements. Knowing all the components of strain and stress tensor at each temperature interval, finding the components of compliance tensor is straightforward [47].

$$\varepsilon_{ij} = S_{ijkl}\sigma_{kl} \tag{3.24}$$

The two independent temperature-dependent elastic parameters of pure Ni and 8YSZ have been experimentally measured and reported elsewhere [47,63,64].

In this work, thermal conductivity (TC) was computed along the three mutually perpendicular directions of the RVE. At each temperature T a temperature difference equal to $T - T_0$ was applied to the RVE along each direction. T_0 is a reference temperature and for simplicity was chosen as $0°C$ in all the simulations. Like the elastic case the volume average of temperature gradient component and heat flux vector component along the desired direction was assessed using the following equations [47]:

$$\kappa_i = \frac{1}{V_{tot}}\left(\sum_{n=1}^{N} V_n q_i^n / \sum_{n=1}^{N} V_n \frac{\partial T^n}{\partial x_i}\right), \tag{3.25}$$

where κ_i is the thermal conductivity coefficient along i ($i = 1$, 2, or 3). q_i^n is the heat flux vector component along i in element n. $\frac{\partial T^n}{\partial x_i}$ is the component of temperature gradient vector in element n along direction i. The rest of the symbols have their regular meanings.

The TEC of the material was calculated by measuring the volume average thermal strain developed in each direction when the temperature difference $T - T_0$ was introduced. Eq. (3.26) further clarifies the calculations [47]:

$$\alpha_i = \frac{1}{V_{tot}}\left(\sum_{n=1}^{N} V_n \varepsilon_{ii,th}^n / T - T_0\right), \tag{3.26}$$

where α_i is the TEC along i and $\varepsilon_{ii,th}^n$ is the thermal strain component in the ith direction in the nth element. The rest of the symbols have their regular meanings. In FE simulations the previously reported experimental values of specific heat, TEC, and thermal conductivity of Ni, 8YSZ, and void (air, in the case of thermal properties) were used [47,65−68].

3.4.1 Results and discussion

3.4.1.1 Microstructural aspects

The novel model of 3D reconstruction successfully converged to build a relatively large RVE of the porous Ni-YSZ composite (Fig. 3.19). The lengths along 1, 2, and 3 axes are 20.5, 15.4, and 11.6 μm, respectively. The independent TPCFs for the real and reconstructed microstructures are shown in Fig. 3.20. There is a decent coincidence of the curves in all the cases, implying the prominent fact that the reconstructed RVE is sufficiently statistically equivalent to the real material. Although complete equivalence of a digital microstructure to a real sample needs more analyses of other morphological

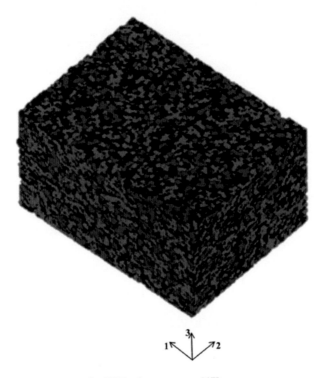

Figure 3.19 The digital RVE for YSZ microstructure [47].

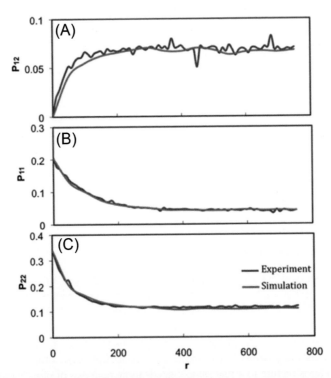

Figure 3.20 Comparison of independent TPCFs of the digital and real materials. Phases 1 and 2 stand for Ni and void, respectively [47].

Numerical realization and characterization of random heterogeneous materials 127

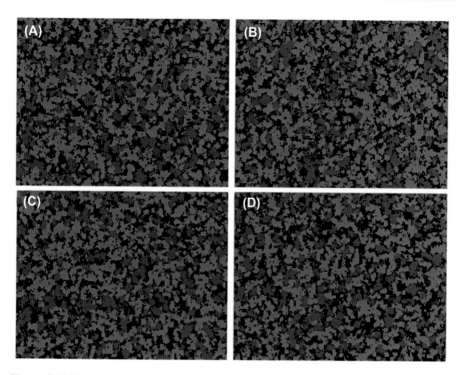

Figure 3.21 Four 2D sections of the YSZ RVE at (A) 1 mm distance from origin along axis 3; (B) at 4.5 mm distance; (C) 8.5 mm and (D) 10.5 mm [47].

and geometrical factors, similitude in TPCFs and subsequent assumption of similar properties are usually considered acceptable [54,69–73]. Fig. 3.21 illustrates four 2D sections of the 3D composite at different positions along axis 3. These 2D sections illustrate that the material is completely random in the interior of the composite and there is no gradient in average aggregate sizes of any phase [47].

The length along axis 1 is 20.5 mm, that along axis 2 is 15.4 mm, and that along axis 3 is 11.6 mm.

3.4.2 Elastic properties

The full 21-component compliance tensor at room temperature is as follows:

$$S((GPa)^{-1})$$

$$= \begin{bmatrix} 0.058613 & -0.016284 & -0.016988 & 2.63 \times 10^{-4} & -2.71 \times 10^{-4} & -5.73 \times 10^{-5} \\ & 0.05755 & -0.01664 & 2.11 \times 10^{-4} & -1.78 \times 10^{-4} & -2.78 \times 10^{-4} \\ & & 0.05653 & -9.08 \times 10^{-5} & -1.82 \times 10^{-4} & -6.06 \times 10^{-5} \\ & & & 0.21 & 1.23 \times 10^{-5} & -8.04 \times 10^{-7} \\ & & & & 0.155 & 3.82 \times 10^{-5} \\ & & & & & 0.155 \end{bmatrix}$$

(3.27)

The components shown by scientific numbers (all the off-diagonal components in the right half of the matrix) are at least 2 orders of magnitude smaller than the other components. For the sake of conciseness and practical usefulness of the results, we assume that these values are 0 in the compliance tensor. Therefore the general compliance tensor at any temperature would have the following form [47]:

$$
S = \begin{bmatrix}
\dfrac{1}{E_1} & -\dfrac{v_{12}}{E_2} & -\dfrac{v_{13}}{E_3} & 0 & 0 & 0 \\[2ex]
-\dfrac{v_{21}}{E_1} & \dfrac{1}{E_2} & -\dfrac{v_{23}}{E_3} & 0 & 0 & 0 \\[2ex]
-\dfrac{v_{31}}{E_1} & -\dfrac{v_{32}}{E_2} & \dfrac{1}{E_3} & 0 & 0 & 0 \\[2ex]
0 & 0 & 0 & \dfrac{1}{G_1} & 0 & 0 \\[2ex]
0 & 0 & 0 & 0 & \dfrac{1}{G_2} & 0 \\[2ex]
0 & 0 & 0 & 0 & 0 & \dfrac{1}{G_3}
\end{bmatrix}
\tag{3.28}
$$

This is the elastic response of an orthotropic material. The v_{ij} and E_i values are correlated to each other [47]:

$$
\frac{v_{ij}}{E_j} = \frac{v_{ji}}{E_i},
\tag{3.29}
$$

Thus orthotropic materials possess nine independent components, that is, a type of material between the two extremes cases: fully anisotropic and fully isotropic. With the assumption that the RVE can be fairly described by the orthotropic behavior the components of the compliance tensor are assessed as functions of temperature [47].

Fig. 3.22 shows the temperature dependence of E_i moduli. The room-temperature values are close to 18 GPa, which is in the order of the widely scattered values reported for real porous Ni-YSZ samples with 35% void in the microstructure [47,74,75].

E_1 and E_2 are different from each other by at most 2%, while the values of E_3 are slightly higher than the other two with at most a 4% difference. Fig. 3.23 exhibits the trend in change of G_i module versus temperature. The values start around 8 GPa at room temperature and reduce down to 5 GPa. Compared to E_i s, shear moduli behavior of the digital material along the axes is different. In the case of shear the different value is G_1 being lower than G_2 and G_3 by an average of $\sim 20\%$. This reflects the contrast in shear and tension moduli of the digital composite [47].

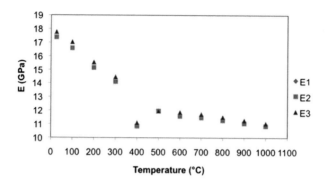

Figure 3.22 Temperature dependence of effective elastic moduli of the digital RVE at three mutually orthogonal axes. E1 and E2 fully overlap [47].

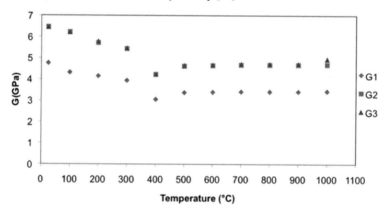

Figure 3.23 Temperature dependence of shear moduli of the digital RVE. G2 and G3 fully overlap [47].

The independent ν_{ij} components against temperature are plotted in Fig. 3.24. The average room-temperature value is 0.285, which decreases to ~ 0.23 at 1000°C. It suggests that effects of lateral strains are smaller compared to room temperature. This is a useful point in design of the fuel cells since the anode and electrolyte need to be in contact during service, and lateral strains can deteriorate the contact and should be considered in design. It is essential to notice that in such three-phase porous composites, all the properties are strictly dependent on the microstructure, in the case of elastic properties, specifically on void volume fraction and average pore size [47].

3.4.3 Thermal conductivity and thermal expansion

Fig. 3.25 shows thermal conductivity coefficients along the three orthogonal axes. The room-temperature values of the reconstructed material are $\sim 28\ W/mK$, which

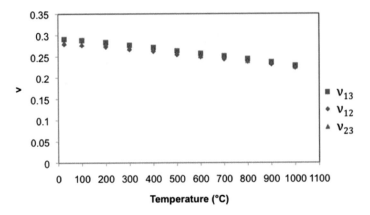

Figure 3.24 Temperature behavior of lateral strain coefficients (Poisson's ratios) versus temperature [47].

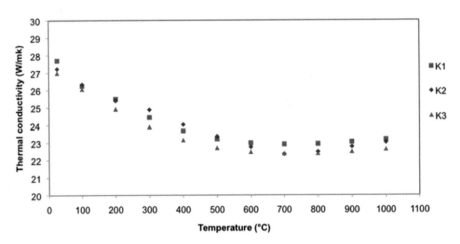

Figure 3.25 Changes of thermal conductivity coefficients as a function of temperature in the digital RVE along three orthogonal axes [47].

is in the range of experimentally reported values [76]. Thermal conductivity of the digital composite decreases by increasing T till $\sim 500°C$, which it slightly increases/remains constant. Thermal conductivity of pure Ni follows the same temperature-dependence trend [67], suggesting that Ni controls thermal conductivity behavior in the digital RVE. This is a clue of percolation of the Ni phase which promises excellent electrical conductivity if a composite with this microstructure is used in practice. Closeness in three κ_i values at all temperatures implies isotropy in thermal conductivity response of the microstructure [47].

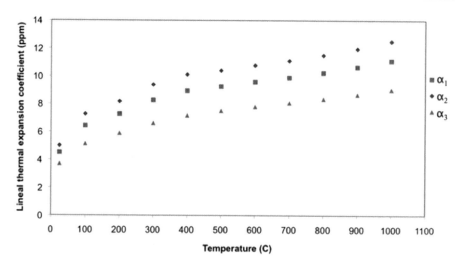

Figure 3.26 TEC of the RVE versus temperature along three orthogonal axes [47].

Fig. 3.26 exhibits TEC of the RVE versus temperature. There is a constant increase from the average $\sim 4 \times 10^{-6}$ at room temperature to an average of $\sim 10.2 \times 10^{-6}$ at 1000°C. Computed TECs are in the range of experimentally measured values for porous Ni-YSZ composites [70,76]. TEC values of the RVE at higher temperatures are distributed far apart compared to room temperature, implying an increase in anisotropy of thermal expansion as the temperature increases. At temperatures above 800°C the difference between TEC of the digital composite and the electrolyte (YSZ) [65] gets above 50%, which can in practice lead to failure of the device [47].

3.5 FFT approach

In recent decades, remarkable progress has been achieved in developing computational homogenization methods. Computational homogenization as a powerful tool can handle microstructures with complex geometry and material nonlinearity where classic analytical micromechanics models have huge difficulties.

General numerical methods such as FEM [77–79], FVM [80,81], and FDM [35] have been used frequently to address microstructure−property relationship problems. Other methods such as FFT-based methods [82–84] and methods of cells [85–87] are specifically developed for calculating effective properties. Hopefully an increase in computing power and improvement of the numerical methods alleviate the problem of a high computational cost of the numerical methods.

132 Applied Micromechanics of Complex Microstructures

Among the numerical methods, FFT-based methods received much attention due to the simplicity and robustness and became a customized method in computational homogenization. Generally speaking, it is shown that compared to other numerical methods, FFT-based methods are faster and have a lower computational cost [38]. Since the introduction 20 years ago [82] a large number of research studies have been devoted to overcoming some drawbacks of the FFT methods, such as a low convergence rate or even divergence in high phase contrast [88,89] and difficulty to impose the load direction [90,91]. Also, some challenges still exist, such as dealing with solving time-dependent equations.

In this section numerical homogenization of structural and thermal properties of composite materials were formulated using FFT methods. The developed methods rely on the accelerated Eyre–Milton FFT algorithm [92] to ensure robustness, efficiency, and ease of implementation [93].

3.5.1 Linear elastic properties

The homogenization problem for a periodic RVE under linear elastostatic conditions is described by [93]

$$
\begin{cases}
\varepsilon(\boldsymbol{x}) = \dfrac{1}{2}\left(\nabla \boldsymbol{u}(\boldsymbol{x}) + \nabla^t \boldsymbol{u}(\boldsymbol{x})\right). \\[2mm]
\sigma(\boldsymbol{x}) = \mathbb{C}(\boldsymbol{x}) : \varepsilon(\boldsymbol{x}). \\[1mm]
\nabla \cdot \sigma(\boldsymbol{x}) = \boldsymbol{0}. \\[1mm]
\boldsymbol{u}(\boldsymbol{x}) - \overline{\boldsymbol{E}} \cdot \boldsymbol{x} \quad \text{periodic.} \\[1mm]
\sigma(\boldsymbol{x}) \cdot \boldsymbol{n} \quad \text{antiperiodic.}
\end{cases}
\tag{3.30}
$$

where $\mathbb{C}(\boldsymbol{x})$ is the local elastic modulus of each phase, $\overline{\boldsymbol{E}}$ is a predefined strain tensor, and $\boldsymbol{n}$ is the normal vector for the faces of RVE. Following [92] the problem is cast into an iterative formulation such that [93]

$$
\varepsilon^{i+1}(\boldsymbol{x}) = \varepsilon^i(\boldsymbol{x}) + 2(\mathbb{C}(\boldsymbol{x}) + \mathbb{C}_0)^{-1} : \mathbb{C}_0 : \left[\Gamma^0(\boldsymbol{x}) * \left(\mathbb{C}_0 : \varepsilon^i(\boldsymbol{x}) - \sigma^i(\boldsymbol{x})\right) - \varepsilon^i(\boldsymbol{x}) + \overline{\boldsymbol{E}}\right].
\tag{3.31}
$$

where the symbol $*$ denotes the convolution product and $\mathbb{C}_0$ is the elasticity modulus of the reference medium, and its associated Green operator in the Fourier space $\hat{\Gamma}^0_{khij}$ [36] is given by [93]

$$
\hat{\Gamma}^0_{khij}(\boldsymbol{\xi}) = \frac{1}{4\mu^0 |\boldsymbol{\xi}|^2}\left(\delta_{ki}\xi_h\xi_j + \delta_{hi}\xi_k\xi_j + \delta_{kj}\xi_h\xi_i + \delta_{hj}\xi_k\xi_i\right) - \frac{\xi_i\xi_j\xi_k\xi_h}{|\boldsymbol{\xi}|^4}\frac{\lambda^0 + \mu^0}{\mu^0(\lambda^0 + 2\mu^0)}.
\tag{3.32}
$$

Numerical realization and characterization of random heterogeneous materials 133

where $\boldsymbol{\xi}$ is the frequency and λ^0 and μ^0 are the Lamé constants of the reference medium. The following is used for the reference medium, which ensures a faster convergence rate [93,94]:

$$\mu^0 = \sqrt{\mu_{\min} \times \mu_{\max}} \text{ and } K^0 = \sqrt{K_{\min} \times K_{\max}}, \tag{3.33}$$

where K is the bulk modulus ($K = \lambda + 2\mu/3$) and subscripts min and max denote the minimum and maximum of the composite constituents properties, respectively. The effective (homogenzied) elasticity tensor is then given by [93]

$$<\sigma(\boldsymbol{x})> = \mathbb{C}_{\text{eff}} : <\varepsilon(\boldsymbol{x})>. \tag{3.34}$$

The components of $\mathbb{C}_{\text{eff}}$ were obtained by applying six individual unit strains to RVE for each of the corresponding components of the strain tensor [95]. The solution procedure for the problem is further detailed, and the algorithm is as follows:

Algorithm 3.1 summarizes the procedure used to solve the elastic problem. Starting with a voxel-based representation a uniform strain is first applied to all voxels, and the objective of the method is to obtain the equilibrium state of the microstructure. In Algorithm 3.1, η_1 and η_2 are convergence criteria (in order to achieve less than 0.1% variance in homogenized properties, selected to be 10^{-6} and 10^{-3} in the current study, respectively) [93].

Algorithm 3.1: *Accelerated Eyre−Milton Elastic analysis algorithm* [93].

iteration $i = 0$:
$$\varepsilon^0(\boldsymbol{x}) = \overline{\boldsymbol{E}}, \qquad\qquad \forall \boldsymbol{x} \in \Omega_0$$
$$\sigma^0(\boldsymbol{x}) = \mathbb{C}(\boldsymbol{x}) : \varepsilon^0(\boldsymbol{x}) \qquad \forall \boldsymbol{x} \in \Omega_0$$
where $\boldsymbol{x}$ is the coordinate of each voxel, $\overline{\boldsymbol{E}}$ is a predefined strain tensor, and Ω_0 is the RVE domain.

while (true)

    1: $\tau^i(\boldsymbol{x}) = \mathbb{C}_0 : \varepsilon^i(\boldsymbol{x}) - \sigma^i(\boldsymbol{x})$

    2: $\hat{\tau}^i(\boldsymbol{\xi}) = \text{FFT}[\tau^i(\boldsymbol{x})]$

    3: $\hat{\varepsilon}^i_c(\boldsymbol{\xi}) = \hat{\Gamma}^0(\boldsymbol{\xi}) : \hat{\tau}^i(\boldsymbol{\xi}) \quad \forall \boldsymbol{\xi} \neq \boldsymbol{0}$ and $\hat{\varepsilon}^i_c(\boldsymbol{0}) = \overline{\boldsymbol{E}}$

    4: $\varepsilon^i_c(\boldsymbol{x}) = \text{FFT}^{-1}[\hat{\varepsilon}^i_c(\boldsymbol{\xi})]$

    5: $\varepsilon^{i+1}(\boldsymbol{x}) = \varepsilon^i(\boldsymbol{x}) + 2(\mathbb{C}(\boldsymbol{x}) + \mathbb{C}_0)^{-1} : \mathbb{C}_0 : (\varepsilon^i_c(\boldsymbol{x}) - \varepsilon^i(\boldsymbol{x}))$

    6: $\sigma(\boldsymbol{x})^{i+1} = \mathbb{C}(\boldsymbol{x}) : \varepsilon^{i+1}(\boldsymbol{x})$

    7: $\text{err}^i_1 = \dfrac{<\|\sigma(\boldsymbol{x})^{i+1} - \sigma(\boldsymbol{x})^i\|>}{\|\mathbb{C}_0 : \overline{\boldsymbol{E}}\|}$, $\text{err}^i_2 = \dfrac{\sqrt{<\|\boldsymbol{\xi}.\hat{\sigma}^i(\boldsymbol{\xi})\|>}}{\|\hat{\sigma}^i(\boldsymbol{0})\|}$ where $\hat{\sigma}^i(\boldsymbol{\xi}) = \text{FFT}[\sigma(\boldsymbol{x})^i]$

    8: **if** ($\text{err}^i_1 < \eta_1$ and $\text{err}^i_2 < \eta_2$)

        break

    else

        $i = i + 1$

3.5.2 Thermal properties

Similarly, for a periodic RVE the mathematical thermal homogenization problem is described by

$$\begin{cases} e(x) = -\nabla T(x). \\ q(x) = K(x)e(x), \\ \nabla \cdot q(x) = 0, \quad \text{on boundary} \\ T(x) - \bar{e} \cdot x \quad \text{periodic.} \\ q(x) \cdot n \quad \text{antiperiodic.} \end{cases} \tag{3.35}$$

where T, e, and q denote the temperature, its gradient, and the heat flux, respectively. Moreover $K(x)$ and $\bar{e}$ denote the local thermal conductivity and a uniform predefined temperature gradient tensor, respectively. Similarly, this problem can be cast into Eyre−Milton iterative formulation such that [93]

$$e^{i+1}(x) = e^i(x) + 2(K(x) + K_0)^{-1} K_0 \left[G^0(x) * \left(K_0 e^i(x) - q^i(x) \right) - e^i(x) + \bar{e} \right]. \tag{3.36}$$

In Eq. (3.36) the K_0 denotes the thermal conductivity of the reference medium, and in the case of an isotropic reference medium, $K_0 = k_0 I$, where I is the second-order identity tensor and k_0 is a scalar thermal conductivity. The $\hat{G}^0$ is the Green operator associated with thermal conductivity of the reference medium defined, in Fourier space, by [93,96]

$$\hat{G}^0_{ij}(\xi) = \frac{\xi_i \xi_j}{\sum_{m.n} K_{0mn} \xi_m \xi_n}. \tag{3.37}$$

The algorithm developed to solve Eq. (3.35) is presented in Algorithm 3.2 with guaranteed convergence [94] subject when properties for the reference medium are selected such that [93]

$$k_0 > \sqrt{k_{\min} \times k_{\max}}. \tag{3.38}$$

where $k_{\min}$ and $k_{\max}$ are the minimum and maximum thermal conductivities of the composite constituent, respectively. The effective thermal conductivity tensor is given by [93]

$$\langle q(x) \rangle = K_{\text{eff}} \langle e(x) \rangle. \tag{3.39}$$

The algorithm illustrates the procedure used for the thermal homogenization analysis. Similarly an accelerated Eyre−Milton algorithm is employed with the same convergence criteria as presented for elastic analysis. In Algorithm 3.2, η_1 and η_2 are convergence criteria (in order to achieve less than 0.1% variance in homogenized properties, selected to be 10^{-6} and 10^{-3} in the current) [93].

Algorithm 3.2: Accelerated FFT thermal analysis algorithm.

Iteration $i = 0$:
$$e^0(x) = \bar{e} \qquad\qquad \forall x \in \Omega_0$$
$$q^0(x) = K(x)e^0(x), \qquad \forall x \in \Omega_0$$
where x is the coordinate of each voxel, $\bar{e}$ is the predefined temperature gradient vector, and Ω_0 is the RVE domain

while (true)

    1: $\tau^i(x) = K_0 e^i(x) - q^i(x)$

    2: $\hat{\tau}^i(\xi) = \text{FFT}[\tau^i(x)]$

    3: $\hat{e}_c^i(\xi) = \hat{\mathbf{G}}^0(\xi)\hat{\tau}^i(\xi) \,\forall\, \xi \neq 0$ and $\hat{e}_c^i(0) = \bar{e}$

    4: $e_c^i(x) = \text{FFT}^{-1}[\hat{e}_c^i(\xi)]$

    5: $e^{i+1}(x) = e^i(x) + 2(K(x) + K_0)K_0(e_c^i(x) - e^i(x))$

    6: $q(x)^{i+1} = K(x)e^{i+1}(x)$

    7: $\text{err}_1^i = \dfrac{<\|q(x)^{i+1} - q(x)^i\|>}{\|Ke\|}$, $\text{err}_2^i = \dfrac{\sqrt{<\|\xi.\hat{q}^i(\xi)\|>}}{\|\hat{q}^i(0)\|}$ where $\hat{q}^i(\xi) = \text{FFT}[q(x)^i]$

    8: **if** $(\text{err}_1^i < \eta_1$ and $\text{err}_2^i < \eta_2)$
        break
    else
        $i = i + 1$

4−2−3)

In this section the homogenization results of elastic and thermal properties of a random heterogeneous sample are presented. Fig. 3.27 shows a two-phase random heterogeneous RVE with size 201^3 voxels, where elastic and thermal properties of each phase are shown in Table 3.8. Fig. 3.28 shows the contour plot of the strain and stress distribution after applying the shear loading as presented by Eq. (3.40):

$$\bar{\varepsilon} = \begin{bmatrix} 0 & 0 & 0 \\ 0 & 0 & 0.001 \\ 0 & 0.001 & 0 \end{bmatrix} \tag{3.40}$$

The effective stiffness tensor of the RVE ($\mathbb{C}_{\text{eff}}$) shown in Fig. 3.27 is given by Eq. (3.41)

$$\mathbb{C}_{\text{eff}} = \begin{bmatrix} 3.73 & 1.32 & 1.32 & 0.00 & 0.02 & 0.00 \\ 1.32 & 3.68 & 1.33 & 0.03 & 0.00 & 0.00 \\ 1.32 & 1.33 & 3.69 & 0.02 & 0.01 & 0.00 \\ 0.00 & 0.01 & 0.01 & 2.38 & 0.00 & 0.00 \\ 0.01 & 0.00 & 0.00 & 0.00 & 2.38 & 0.00 \\ 0.00 & 0.00 & 0.00 & 0.00 & 0.00 & 2.38 \end{bmatrix} \text{GPa} \tag{3.41}$$

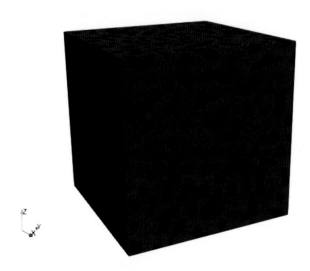

Figure 3.27 Two-phase random heterogeneous RVE with a size of 201^3 voxels.

Table 3.8 Volume fraction, elastic properties, and thermal conductivity of the blue and red phases in Fig. 3.26.

	Blue phase	Red phase
Volume fraction	0.3	0.7
Elastic modulus	50 GPa	1 GPa
Poisson's ratio	0.3	0.3
Thermal conductivity	50 W/m.K	1 W/m.K

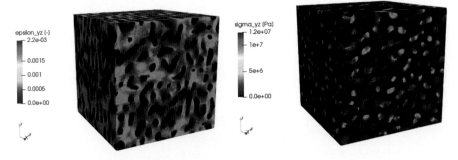

Figure 3.28 Strain (right) and stress (left) distribution in RVE shown in Fig. 3.27 after applying predefined strain shown in Eq. (3.38).

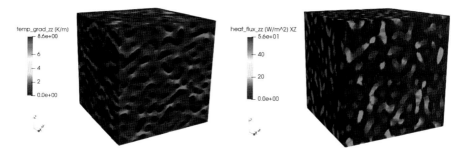

Figure 3.29 Contour plot of the heat flux and temperature gradient after applying the predefined temperature gradient equal to [0, 0, 1] in RVE shown in Fig. 3.26.

Similar to the elastic case the distribution of the temperature gradient and heat flux in the z-direction after applying the predefined temperature gradient equal to $\bar{e} = [0.0.1]$ is shown in Fig. 3.29, and the homogenized thermal conductivity tensor (K_{eff}) is given by Eq. (3.42)

$$K_{\text{eff}} = \begin{bmatrix} 5.39 & 0.02 & 0.03 \\ 0.02 & 5.32 & 0.04 \\ 0.03 & 0.02 & 5.33 \end{bmatrix} \text{W/m.K} \qquad (3.42)$$

3.6 Conclusion and remarks

An accurate realization and reconstruction of heterogeneous materials is the first step to homogenize effective properties such as mechanical and thermal properties. In this chapter, we have used two scenarios to realize and reconstruct heterogeneous materials. The first scenario used a Monte Carlo approach, and the second scenario used a phase recovery approach. A Monte Carlo approach has been used to realize various virtual 3D microstructures based on the major steps: generation, distribution, and growth of cells. The approach is powerful and useful to a model variety of natural and handmade heterogeneous materials. The Monte Carlo approach can be used to realize a variety of microstructures and reconstruction of microstructures. The reconstructed microstructures can be performed based on the target statistical correlation functions or effective properties. In the second scenario the microstructure reconstruction has been performed based on the statistical correlation function and a phase-recovery algorithm implanted to reconstruct multiphase heterogonous microstructures. Finally, to homogenize effective thermal and mechanical properties of realized microstructures, FE and FFT approaches have been used.

References

[1] Torquato S. Random heterogeneous materials: microstructure and macroscopic properties. New York: Springer-Verlag; 2002.

[2] Torquato S. Optimal design of heterogeneous materials. In: Clarke DR, Ruhle M, Zok F, editors. Annual review of materials research, 40. 2010. p. 101−29.

[3] Hasanabadi A, Baniassadi M, Abrinia K, Safdari M, Garmestani H. Efficient three-phase reconstruction of heterogeneous material from 2D cross-sections via phase-recovery algorithm. J Microsc 2016;264(3):384−93. Available from: https://doi.org/10.1111/jmi.12454.

[4] Talukdar MS, Torsaeter O, Ioannidis MA, Howard JJ. Stochastic reconstruction, 3D characterization and network modeling of chalk. J Pet Sci Eng 2002;35(1−2):1−21.

[5] Sheidaei A, Baniassadi M, Banu M, Askeland P, Pahlavanpour M, Kuuttila N, et al. 3-D microstructure reconstruction of polymer nano-composite using FIB−SEM and statistical correlation function. Compos Sci Technol 2013;80:47−54.

[6] Lu B, Torquato S. Lineal-path function for random heterogeneous materials. Phys Rev A 1992;45(2):922−9.

[7] Jiao Y, Stillinger FH, Torquato S. A superior descriptor of random textures and its predictive capacity. Proc Natl Acad Sci U S A 2009;106(42):17634−9. Available from: https://doi.org/10.1073/pnas.0905919106.

[8] Torquato S, Stell G. Microstructure of Two-Phase Random Media. I. The n-Point Probability Functions. J Chem Phys 1982;77:2071−7.

[9] Jiao Y, Stillinger FH, Torquato S. Modeling heterogeneous materials via two-point correlation functions: basic principles. Phys Rev E Stat Nonlin Soft Matter Phys 2007;76 (3 Pt 1):031110. Available from: https://doi.org/10.1103/PhysRevE.76.031110.

[10] Hasanabadi A, Baniassadi M, Abrinia K, Safdari M, Garmestani H. 3D microstructural reconstruction of heterogeneous materials from 2D cross sections: a modified phase-recovery algorithm. Computat Mater Sci 2016;111:107−15. Available from: https://doi.org/10.1016/j.commatsci.2015.09.015.

[11] Baniassadi M, Garmestani H, Li DS, Ahzi S, Khaleel M, Sun X. Three-phase solid oxide fuel cell anode microstructure realization using two-point correlation functions. Acta Mater 2011;59(1):30−43. Available from: https://doi.org/10.1016/j.actamat.2010.08.012.

[12] Adams BL, Kalidindi SR, Fulwood, David T. Microstructure-sensitive design for performance optimization. Waltham, MA 02451: Butterworth-Heinemann; 2013.

[13] Fullwood DT, Niezgoda SR, Adams BL, Kalidindi SR. Microstructure sensitive design for performance optimization. Prog Mater Sci 2010;55(6):477−562. Available from: https://doi.org/10.1016/j.pmatsci.2009.08.002.

[14] Panchal JH, Kalidindi SR, McDowell DL. Key computational modeling issues in integrated computational materials engineering. Comput Des 2013;45(1):4−25. Available from: https://doi.org/10.1016/j.cad.2012.06.006.

[15] Ruggles TJ, Rampton TM, Rose SA, Fullwood DT. Reducing the microstructure design space of 2nd order homogenization techniques using discrete Fourier Transforms. Mech Mater 2013;59:14−23. Available from: https://doi.org/10.1016/j.mechmat.2012.11.007.

[16] Safdari M, Baniassadi M, Garmestani H, Al-Haik MS. A modified strong-contrast expansion for estimating the effective thermal conductivity of multiphase heterogeneous materials. J Appl Phys 2012;112(11):114318. Available from: https://doi.org/10.1063/1.4768467.

[17] Beran MJ. Statistical continuum theories. Monographs in statistical physics and thermo-dynamics. New York: Interscience; 1968.

[18] Sheehan N, Torquato S. Generating microstructures with specified correlation function. J Appl Phys 2001;89(1):53−60.

[19] Adams BL, Gao X, Kalidindi SR. Finite approximations to the second-order properties closure in single phase polycrystals. Acta Mater 2005;53(13):3563−77. Available from: https://doi.org/10.1016/j.actamat.2005.03.052.

[20] Torquato S. Effective stiffness tensor of composite media − I. Exact series expansion. J Mech Phys Solids 1997;45(9):1421−48.

[21] Yeong CLY, Torquato S. Reconstructing random media II. Three-dimensional media from two-dimensional cuts. Phys Rev 1998;58:224−33.

[22] Cule D, Torquato S. Generating random media from limited microstructural informa-tion via stochastic optimization. J Appl Phys 1999;86(6):3428−37. Available from: https://doi.org/10.1063/1.371225.

[23] Ballani F, Stoyan D. Reconstruction of random heterogeneous media. J Microsc 2015;258(3):173−8. Available from: https://doi.org/10.1111/jmi.12234.

[24] Quiblier JA. A new three-dimensional modeling technique for studying porous media. J Colloid Interface Sci 1984;98(1):84−102.

[25] Adler PM, Jacquin CG, Quiblier JA. Flow in simulated porous media. Int J Multiph Flow 1990;16:691−712.

[26] Jiang Z, Chen W, Burkhart C. Efficient 3D porous microstructure reconstruction via Gaussian random field and hybrid optimization. J Microsc 2013;252(2):135−48. Available from: https://doi.org/10.1111/jmi.12077.

[27] Fullwood DT, Niezgoda SR, Kalidindi SR. Microstructure reconstructions from 2-point statistics using phase recovery algorithms. Acta Mater 2008;52:942−8.

[28] Lanzini A, Leone P, Asinari P. Microstructural characterization of solid oxide fuel cell electrodes by image analysis technique. J Power Sources 2009;194:408−22.

[29] Shi J, Xue X. Microstructure optimization designs for anode-supported planar solid oxide fuel cells. J Fuel Cell Sci Technol 2011;8(6):061006. Available from: https://doi.org/10.1115/1.4004642.

[30] Cronin JS, Chen-Wiegart Y-cK, Wang J, Barnett SA. Three-dimensional reconstruction and analysis of an entire solid oxide fuel cell by full-field transmission X-ray micros-copy. J Power Sources 2013;233:174−9. Available from: https://doi.org/10.1016/j.jpowsour.2013.01.060.

[31] Benveniste Y. A new approach to the application of Mori-Tanaka's theory in composite materials. Mech Mater 1987;6(2):147−57.

[32] Mori T, Tanaka K. Average stress in matrix and average elastic energy of materials with misfitting inclusions. Acta Metallurg 1973;21(5):571−4. Available from: https://doi.org/10.1016/0001-6160(73)90064-3.

[33] Bargmann S, Klusemann B, Markmann J, Schnabel JE, Schneider K, Soyarslan C, et al. Generation of 3D representative volume elements for heterogeneous materials: A review. Prog Mater Sci 2018;96:322−84.

[34] Hou TY, Wu X-H. A multiscale finite element method for elliptic problems in compos-ite materials and porous media. J Computat Phys 1997;134(1):169−89. Available from: https://doi.org/10.1006/jcph.1997.5682.

[35] Abdulle A, EW. Finite difference heterogeneous multi-scale method for homogeniza-tion problems. J Computat Phys 2003;191(1):18−39. Available from: https://doi.org/10.1016/S0021-9991(03)00303-6.

[36] Moulinec H., Suquet P. A FFT-based numerical method for computing the mechanical properties of composites from images of their microstructures. IUTAM Symposium on Microstructure-Property Interactions in Composite Materials. Dordrecht: Springer Netherlands; 1995.

[37] Michel JC, Moulinec H, Suquet P. Effective properties of composite materials with periodic microstructure: a computational approach. Comput Methods Appl Mech Eng 1999;172(1−4):109−43. Available from: https://doi.org/10.1016/s0045-7825(98)00227-8.

[38] Dunant CF, Bary B, Giorla AB, Péniguel C, Sanahuja J, Toulemonde C, et al. A critical comparison of several numerical methods for computing effective properties of highly heterogeneous materials. Adv Eng Softw 2013;58:1−12. Available from: https://doi.org/10.1016/j.advengsoft.2012.12.002.

[39] Anglin BS, Lebensohn RA, Rollett AD. Validation of a numerical method based on Fast Fourier Transforms for heterogeneous thermoelastic materials by comparison with analytical solutions. Computat Mater Sci 2014;87:209−17. Available from: https://doi.org/10.1016/j.commatsci.2014.02.027.

[40] Wang B, Fang G, Liu S, Fu M, Liang J. Progressive damage analysis of 3D braided composites using FFT-based method. Compos Struct 2018;192:255−63. Available from: https://doi.org/10.1016/j.compstruct.2018.02.040.

[41] Schneider M, Wicht D, Böhlke T. On polarization-based schemes for the FFT-based computational homogenization of inelastic materials. Computat Mech 2019;64(4):1073−95. Available from: https://doi.org/10.1007/s00466-019-01694-3.

[42] Zhu Q-Z, Yvonnet J. An incremental−iterative method for modeling damage evolution in voxel-based microstructure models. Computat Mech 2014;55(2):371−82. Available from: https://doi.org/10.1007/s00466-014-1106-1.

[43] Nosouhi Dehnavi F, Safdari M, Abrinia K, Hasanabadi A, Baniassadi M. A framework for optimal microstructural design of random heterogeneous materials. Computat Mech 2020;1−17.

[44] Hasanabadi A, Baniassadi M, Abrinia K, Safdari M, Garmestani H. Optimization of solid oxide fuel cell cathodes using two-point correlation functions. Computat Mater Sci 2016;123:268−76.

[45] Sebdani MM, Baniassadi M, Jamali J, Ahadiparast M, Abrinia K, Safdari M. Designing an optimal 3D microstructure for three-phase solid oxide fuel cell anodes with maximal active triple phase boundary length (TPBL). Int J Hydrog Energy 2015;40(45):15585−96.

[46] Garmestani H, Baniassadi M, Li D, Fathi M. Semi-inverse Monte Carlo reconstruction of two-phase heterogeneous material using two-point functions. Int J Theor Appl Multiscale Mech 2009;1(2):134−49.

[47] Tabei S, Sheidaei A, Baniassadi M, Pourboghrat F, Garmestani H. Microstructure reconstruction and homogenization of porous Ni-YSZ composites for temperature dependent properties. J Power Sources 2013;235:74−80.

[48] Taguchi G. System of experimental design: engineering methods to optimize quality and minimize costs. NY: UNIPUB/Kraus International Publications White Plains; 1987.

[49] Haykin S. Neural networks: a comprehensive foundation. Prentice Hall PTR; 1994.

[50] Gokhale AM, Tewari A, Garmestani H. Constraints on microstructural two-point correlation functions. Scr Mater 2005;53:989−93.

[51] Fullwood DT, Niezgoda SR, Kalidindi SR. Microstructure reconstructions from 2-point statistics using phase-recovery algorithms. Acta Mater 2008;56(5):942−8. Available from: https://doi.org/10.1016/j.actamat.2007.10.044.

[52] Baniassadi M, Safdari M, Garmestani H, Ahzi S, Geubelle PH, Remond Y. An optimum approximation of n-point correlation functions of random heterogeneous material systems. J Chem Phys 2014;140(7):074905. Available from: https://doi.org/10.1063/1.4865966.

[53] Fullwood DT, Kalidindi SR, Niezgoda SR, Fast A, Hampson N. Gradient-based microstructure reconstructions from distributions using fast Fourier transforms. Mater Sci Eng A 2008;494:68–72.

[54] Jiao Y, Stillinger FH, Torquato S. Modeling heterogeneous materials via two-point correlation functions. II. Algorithmic details and applications. Phys Rev E Stat Nonlin Soft Matter Phys 2008;77(3 Pt 1):031135. Available from: https://doi.org/10.1103/PhysRevE.77.031135.

[55] Gerchberg RW, Saxton WO. A practical algorithm for the determination of phase from image and diffraction plane pictures. OPTIK 1972;35(2):237–46.

[56] Fienup JR. Reconstruction of an object from the modulus of its Fourier transform. Opt Lett 1978;3(1):27–9. Available from: https://doi.org/10.1364/ol.3.000027.

[57] Fienup JR. Reconstruction of an object from the modulus of its Fourier transform. Opt Lett 1978;3. Available from: https://doi.org/10.1364/OL.3.000027.

[58] Sadd MH. Elasticity theory, applications, and numerics. Third ed. Oxford: Academic Press; 2014.

[59] He W, Lv W, Dickerson JH. Gas transport in solid oxide fuel cells. SpringerBriefs in Energy New York: Springer; 2014.

[60] Hoshen J, Kopelman R. Percolation and cluster distribution. I. Cluster multiple labeling technique and critical concentration algorithm. Phys Rev B 1976;14(8):3438–45.

[61] Zhang SHC X, Li G, Ho HK, Li J, Feng Z. A review of integration strategies for solid oxide fuel cells. J Power Sources 2010;195:685–702.

[62] Stambouli ET AB. Solid oxide fuel cells (SOFCs): a review of an environmentally clean and efficient source of energy. Renew Sustain Energy Rev 2002;6:433–55.

[63] R. Farrarao RM. Temperature dependence of the Young's modulus and shear modulus of pure nickel, platinum, and molybdenum. Metall Mater Trans A 1977; 8:1563–5.

[64] Giraud S, Canel J. Young's modulus of some SOFCs materials as a function of temperature. J Eur Ceram Soc 2008;28:77–83.

[65] Song MX X, Zhou F, Jia G, Hao X, An S. High-temperature thermal properties of yttria fully stabilized zirconia ceramics. J Rare Earths 2011;29:155–9.

[66] Kollie T. Measurement of the thermal-expansion coefficient of nickel from 300 to 1000 K and determination of the power-law constants near the Curie temperature. Phys Rev B 1977;16(11):4872–81.

[67] Desai P. Thermodynamic properties of Nickel. Int J Thermophys 1987;8:763–80.

[68] Hayashi H, Saitou T, Maruyama N, Inaba H, Kawamura K, Mori M. Thermal expansion coefficient of yttria stabilized zirconia for various yttria contents. Solid State Ion 2005;176(5–6):613–19. Available from: https://doi.org/10.1016/j.ssi.2004.08.021.

[69] Torquato S. Random heterogeneous materials. New York: Springer; 2001.

[70] Johnson JQ J. Effective modulus and coefficient of thermal expansion of Ni–YSZ porous cermets. J Power Sources 2008;181:85–92.

[71] Baniassadi HGb M, Li DS, Ahzi S, Khaleel M, Sun X. Three-phase solid oxide fuel cell anode microstructure realization using two-point correlation functions. Acta Mater 2011;59:30–43.

[72] Rechtsman MC, Torquato S. Effective dielectric tensor for electromagnetic wave propagation in random media. J Appl Phys 2008;103(8):084901–15.

[73] Pham D, Torquato S. Strong-contrast expansions and approximations for the effective conductivity of isotropic multiphase composites. J Appl Phys 2003;94(10):6591−602.

[74] Pihlatie AK M, Mogensen M. Mechanical properties of NiO/Ni−YSZ composites depending on temperature, porosity and redox cycling. J Eur Ceram Soc 2009;29:1657−64.

[75] Yu GP J, Lee S, Woo S. Microstructural effects on the electrical and mechanical properties of Ni−YSZ cermet for SOFC anode. J Power Sources 2007;163:926−32.

[76] Kawashima MH T. Thermal Proeprties of Porous Ni/YSZ Composites at High Temperatures. Mater Trans JIM 1996;37(9):1518−24.

[77] Hachi BE, Benkhechiba AE, Kired MR, Hachi D, Haboussi M. Some investigations on 3D homogenization of nano-composite/nano-porous materials with surface effect by FEM/XFEM methods combined with Level-Set technique. Comput Methods Appl Mech Eng 2020;371:113319.

[78] Sokołowski D, Kamiński M. Probabilistic homogenization of hyper-elastic particulate composites with random interface. Compos Struct 2020;241:112118. Available from: https://doi.org/10.1016/j.compstruct.2020.112118.

[79] Efendiev Y, Hou TY. Multiscale finite element methods: theory and applications. Springer Science & Business Media; 2009.

[80] Chen Q, Chen W, Wang G. Fully-coupled electro-magneto-elastic behavior of unidirectional multiphased composites via finite-volume homogenization. Mech Mater 2021;154:103553. Available from: https://doi.org/10.1016/j.mechmat.2020.103553.

[81] He Z, Pindera M-J. Finite volume based asymptotic homogenization theory for periodic materials under anti-plane shear. Eur J Mech - A/Solids 2021;85:104122. Available from: https://doi.org/10.1016/j.euromechsol.2020.104122.

[82] Moulinec H, Suquet P. A numerical method for computing the overall response of nonlinear composites with complex microstructure. Comput methods Appl Mech Eng 1998;157(1−2):69−94.

[83] Vondřejc J, Zeman J, Marek I. An FFT-based Galerkin method for homogenization of periodic media. Comput Math Appl 2014;68(3):156−73.

[84] Lucarini S, Segurado J. DBFFT: A displacement based FFT approach for non-linear homogenization of the mechanical behavior. Int J Eng Sci 2019;144:103131. Available from: https://doi.org/10.1016/j.ijengsci.2019.103131.

[85] Aboudi J. Micromechanical analysis of composites by the method of cells. 1989.

[86] Balusu K, Skinner T, Chattopadhyay A. An efficient implementation of the high-fidelity generalized method of cells for complex microstructures. Computat Mater Sci 2021;186:110004. Available from: https://doi.org/10.1016/j.commatsci.2020.110004.

[87] Meshi I, Breiman U, Aboudi J, Haj-Ali R. The cohesive parametric high-fidelity-generalized-method-of-cells micromechanical model. Int J Solids Struct 2020;206:183−97. Available from: https://doi.org/10.1016/j.ijsolstr.2020.08.024.

[88] Schneider M, Ospald F, Kabel M. Computational homogenization of elasticity on a staggered grid. Int J Numer Methods Eng 2016;105(9):693−720.

[89] Monchiet V, Bonnet G. A polarization-based FFT iterative scheme for computing the effective properties of elastic composites with arbitrary contrast. Int J Numer Methods Eng 2012;89(11):1419−36.

[90] Lucarini S, Segurado J. An algorithm for stress and mixed control in Galerkin-based FFT homogenization. Int J Numer Methods Eng 2019;119(8):797−805.

[91] Kabel M, Fliegener S, Schneider M. Mixed boundary conditions for FFT-based homogenization at finite strains. Computat Mech 2016;57(2):193−210.

[92] Eyre DJ, Milton GW. A fast numerical scheme for computing the response of composites using grid refinement. Eur Phys J Appl Phys 1999;6(1):41−7. Available from: https://doi.org/10.1051/epjap:1999150.

[93] Nosouhi Dehnavi F, Safdari M, Abrinia K, Hasanabadi A, Baniassadi M. A framework for optimal microstructural design of random heterogeneous materials. Computat Mech 2020;66(1):123−39. Available from: https://doi.org/10.1007/s00466-020-01844-y.

[94] Moulinec H, Silva F. Comparison of three accelerated FFT-based schemes for computing the mechanical response of composite materials. Int J Numer Methods Eng 2014;97 (13):960−85. Available from: https://doi.org/10.1002/nme.4614.

[95] Colabella L, Ibarra Pino AA, Ballarre J, Kowalczyk P, Cisilino AP. Calculation of cancellous bone elastic properties with the polarization-based FFT iterative scheme. Int J Numer Method Biomed Eng 2017;33(11). Available from: https://doi.org/10.1002/cnm.2879.

[96] Ghazavizadeh A, Soltani N, Baniassadi M, Addiego F, Ahzi S, Garmestani H. Composition of two-point correlation functions of subcomposites in heterogeneous materials. Mech Mater 2012;51:88−96.

Numerical characterization of tissues

4

Abstract

In this chapter a micromechanical approach is employed to generate a cerebral cortex, bone, and liver tissue representative volume element (RVE) and simulate the mechanical behavior of these types of tissues. These samples provide comprehensive knowledge for modeling of the tissue and can be applied to other biological materials.

In the first section, we use two scenarios; at first the proposed RVE includes distributed neuron cells in a matrix which both are viscoelastic, and in the second scenario, we consider a viscohyperelastic constitutive model to predict the RVE behavior more realistically.

In the second section nonlinear material models have been utilized to describe and simulate the liver biomechanical behavior in the organ level. In this chapter the mechanical homogenization of the liver tissue with a novel combined finite element (FE) method—optimization method is explained.

Finally, in the last section the development of a predictive model for the bone becomes increasingly important for medical applications such as bone surgery or bone substitutes like prosthesis. The goal of this chapter is to homogenize effective properties of the bone and the influence of the local bone microstructure distribution on the macroscopic properties of the bone. FE and reconstruction techniques are exploited to homogenize mechanical properties of the tissue.

4.1 Introduction

In recent years, biomechanical engineering has been developed to better understand the biological behavior of living organisms [1,2]. Determination of mechanical behavior and properties of organs such as the brain, bone, and veins is one of the most interesting fields for biomechanical researchers.

The brain, as the most complex of living organs, has attracted more attention. In spite of abundant performed research studies in this field, there are many lacks in the mechanical modeling of the brain tissue. Mechanical damage in this organ can lead to severe physical and mental damages or death. Thus it is very necessary to realistically know the mechanical behavior of the brain tissue [3]. Due to the recent developments in various disciplines, assessing the mechanical properties of the brain tissue is of increasing importance in various fields, for instance, automatic robotic tools, surgical robots, and helmet producing. For this reason, investigation of brain behavior under different conditions is important, which leads to experimental and computational studies. Considering the fact that experiment studies do not show distribution

---

Applied Micromechanics of Complex Microstructures. DOI: https://doi.org/10.1016/B978-0-443-18991-3.00006-4
© 2023 Elsevier Inc. All rights reserved.

of stress and injury in the microscopic scale and these tests under many existing conditions are costly, micromechanical methods are developed to obtain appropriate predictions of brain behavior. Additionally, it is found that axons undergo higher stresses; hence they are more sensitive in accidents which lead to axonal death and would cause traumatic brain injury (TBI) and diffuse axonal injury (DAI) [4,5].

The liver, a crucial organ in vertebrates, is a highly vascularized tissue mainly connected to the portal vein that brings in blood-rich digested nutrients. The human body is unable to function without the liver, and currently, there is no transplant option for patients suffering from a long-term liver failure. Therefore multiscale homogenization combined with machine learning is used to model nonlinear anisotropic behavior of the liver and integrated within real-time analyses for application in minimally invasive surgery (MIS) [6].

The bone microstructure is a complex random heterogeneous microstructure. In the chapter, using our recently developed statistical reconstruction framework, a set of "bone like" microstructures with a variety of distributions has been created to study pseudo "patient variabilities." The method provides similar effective stiffness tensors, equivalent stresses, and strain energy distributions for the original and the statistically reconstructed samples. The main outcome of this study is the correlation of similar effective mechanical properties between samples when bone remodeling depends on the local strain energy distribution as a function of each bone microstructure. It is expected that two different microstructures with equivalent bone volume fractions will lead to identical bone remodeling in a short period of time, whereas this needs to be proven for long-term evolution.

This work could be used to develop precise predictive numerical models while developing parametric studies on an infinite number of virtual samples and correlating patient dependency with more precise mechanobiological numerical models [7].

4.2 Finite element investigation of effective mechanical behavior of the cerebral cortex tissue using 3D homogenized representative volume elements

The term TBI refers to the sudden impact to the head and consequently to the brain. Annually, about 1.7 million emergency cases, hospitalizations, and deaths in America are related to this type of injury happening in different situations such as accidents, sports, and physical collisions [3,8−10]. Axonal death and disorientation can result from axon damage originating from brain impact or TBI [11]. Damage and impacts are one of the topics widely studied in mechanics. From a biomechanics point of view, brains can also be simulated and regarded as an object/material; hence it is especially important for them to be characterized. With the advances made in finite element (FE) modeling of the brain tissue in recent years, studies on TBI have increasingly made use of computer simulations. However, the precision of material properties used in these models could be highly effective in the final accuracy [4,12,13].

To achieve a better understanding mechanism of TBI, many research groups have presented various constitutive models and obtained numerical results. The key point for the development of a realistic numerical model of the brain is the understanding of the properties of this material's constituents. Thus knowing the micromechanical behavior of the tissue under different loading conditions leads to a better understanding of TBI physics. Numerous biomechanical studies have been performed to identify and evaluate the mechanical properties of the brain tissue. Since this tissue has unique complexities in a microscale, such as geometrical complexity, many research studies have been conducted in a macroscale. Accordingly the brain tissue samples have been tested in the different experiments such as tensile, compressive, shear, and high-frequency loadings (in vitro) [14,15]. Macromechanical analysis of the brain tissue provides good knowledge of mechanical response of the brain against external forces. Since macromechanical models cannot predict the distribution of stress and strain between the structural components of the brain tissue, they have no good understanding of the mechanisms of DAI. On the other hand, micromechanical models can identify the influence of each material constituent on the response of tissue. Thus these models have a good capability to demonstrate the mechanism of DAI. Micromechanical finite element methods (FEMs) present a robust tool to express the relationship between the biological and mechanical behaviors [3,16].

Recently, hyperelastic constitutive models have been employed to model the nonlinear mechanical behavior of the brain tissue in large deformations. Rashid et al. [17] conducted high-strain rate tensile tests on brain tissue samples to examine the effect of the strain rate on the viscous behavior. They presented a two terms Ogden constitutive model for the nonlinear hyperelastic behavior of the brain tissue and calibrated the model parameters using the reverse FEM and fitting the experimental data. Feng et al. [18] utilized dynamic shear tests on the white matter of the brain to specify the required material parameters in a hyperelastic model which represents the anisotropic behavior of that matter. Laksari et al. [12] used compressive stress relaxation tests to develop a constitutive model for finite compression. They considered a modified Mooney–Rivlin model for volume preserving and a two terms Ogden model for the volume changing part of deformation and calibrated the material parameters by fitting the numerical results and experimental instantaneous and equilibrium data. Also, many biomechanical research studies have been performed to investigate the association between the TBI and intensity of the stress or strain [19–21]. In these research studies the injury limit has been specified for separated and intertwined axons under different loading conditions. In this regard, Valdez and Balachandran [22] proposed a nonlinear viscohyperelastic model to examine the effect of material nonlinearity on the stress wave propagation in the brain. In the following, Chatelin et al. [23] developed an isotropic transversely viscohyperelastic constitutive model for the human brain tissue. Employing the FE simulation of unconfined compression tests and experimental data, they evaluated their model.

Arbogast and Margulies [24] presented a fiber-reinforced composite model which includes cylindrical axons surrounded by an extracellular matrix (ECM).

Their aim was to study the mechanical behavior of the brain stem under the rotational force. Also, they used the Hashin micromechanical model [25] to assess the brain stem linear viscoelastic behavior and presented an analytical solution for determining the linear viscoelastic properties of axons and ECM. Considering the viscoelastic transverse isotropic behavior of the brain stem, Ning et al. [26] proposed a unidirectional fibrous composite model for this material. They represented the instantaneous nonlinear elastic response of the brain stem as a hyperelastic transverse isotropic behavior of a tissue. Also, they used a linear isotropic viscoelastic model for both axons and the matrix to investigate the rate dependency of the brain tissue. Considering the effects of morphologic heterogeneities, Cloots et al. [27] studied the mechanical behavior of the cerebral cortex. They developed various detailed two-dimensional (2D) FE models and finally made a homogeneous cortex FE model. Also, they utilized the loading conditions presented by Brands et al. [28]. Abolfathi et al. [16] presented a micromechanical process to study the anisotropic properties of the brain stem containing viscoelastic axons surrounded by ECM. They studied a process to determine the anisotropic behavior of the brain white matter. Also, employing the experimental data of Arbogast and Margulies [24], they obtained the viscoelastic properties of axons and ECM. Karami et al. [13] proposed a micromechanical FE model to investigate the homogenized behavior of the brain tissue in large deformation. In this model the brain tissue was assumed to be a fibrous composite containing axons and the anisotropic hyperelastic matrix. Utilizing adaptive kinematic and Ogden hyperelastic constitutive models, Pan et al. [29] micromechanically developed a representative volume element (RVE) to study the mechanical responses of axons in tissues under large tensile deformations [3].

Based on what has been mentioned, many brain tissue models have been proposed by researchers. In order to overcome the high cost of the experimental tests in the microscale the micromechanical modeling is a good choice as an alternative of laboratory tests. One of the most important limitations of the micromechanical analysis on the brain tissue is the lack of material behavior of its constituents. Also, complex geometry is another important limitation of this tissue that makes difficult the micromechanical modeling.

In this chapter the cerebral cortex tissue is micromechanically investigated by presenting a realistic RVE. In this regard a 3D geometrical model is introduced with a more realistic model in comparison with the previous developed models. This model is suitable for the nonhomogeneous part of the brain that has an isotropic behavior. Assuming the perfect link between the components a homogenized FE model is developed for the cerebral cortex tissue using ABAQUS 6.14. In this regard, we present the micromechanical modeling and material constitutive modeling, creating RVE and homogenizing of the material in this chapter.

In this study the brain tissue is micromechanically studied using an RVE. At first the proposed RVE includes randomly distributed neuron cells in a matrix, both of which are viscoelastic, and second, the brain tissue includes neurons and ECM is considered as an isotropic viscohyperelastic material. A VUMAT ABAQUS subroutine with two terms of viscoelastic and one term of hyperelastic parameters is

Numerical characterization of tissues

developed to define the material properties of the RVE. In the second study the homogenization is carried out using the self-consistent method [30]. Taking advantage of the abovementioned techniques has increased the accuracy of RVE simulation significantly, leading to a more realistic representation of stress distribution between neurons and the matrix.

4.2.1 Viscoelastic homogenization of the brain tissue

In this section the material behavior has been determined employing the micromechanical homogenization method through the RVE concept. In this method the non-homogeneous composite material has been substituted with an effective fictitious homogeneous material. Many different micromechanical models have been proposed to predict the mechanical behavior of composite materials in the literature [13,16,31−36]. In this study the micromechanical characterization method, which is used to investigate composites, has been extended to simulate the cerebral cortex tissue [31]. Also, FEM is utilized in micromechanical modeling to examine the elastic and viscoelastic behavior of the cerebral cortex tissue.

Various parts of the brain have different structures. Axons have been distributed in the brain stem in the longitudinal direction, which result in transversely isotropic behavior on the macroscale. The nerve cells have been irregularly distributed in the cerebral cortex, and the tissue has an isotropic property [32,37]. In the proposed model the cerebral cortex tissue has been considered as a two-phase composite which includes neuron cells as the fillers and ECM as a basic matrix. Also, it has been assumed that the fillers have been randomly distributed in the basic matrix. Although this type of distribution has complexity in analytical and geometrical modeling, more exact results are obtained, especially in distribution of stress and strain in the tissue [3,16].

4.2.1.1 Representative volume element and finite element modeling

A 3D representative volume element for the cerebral cortex tissue includes the neurons that are randomly distributed in the ECM. Although no neurons have the same geometry, all of them have the same general structure which contains three parts (as shown in Fig. 4.1A): cell body, dendrites, and axon. Considering these parts, we developed a reference neuron cell model in ABAQUS software, which is depicted in Fig. 4.1B. In this model, Solid3, and Beam4 elements are, respectively, used for cell body, axons, and dendrites [38]. Since the ratio of axon and dendrite length to its diameter is more than 10 [39], they are modeled by the beam element type attached to the ECM with a perfect coupling. This element type selection results to reduce the number of elements in RVE, and consequently computational costs are decreased. Also, the cell body is modeled as a sphere which is connected to the axon and dendrites with a perfect coupling. Finally the cell body is attached to the axon and dendrites through the wire elements [3].

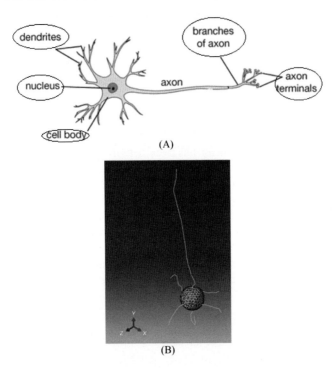

Figure 4.1 (A) Nerve cell schematic (neuron) and (B) reference neuron created in ABAQUS [3].

Creating a 3D RVE (Fig. 4.2) the random distribution of neurons in the ECM is performed by a mathematical code written in MATLAB®. This code has different inputs including the size of the cube side, the number of neurons, the radius of the cellular volume, the coordinates of axon and dendrites of a neuron, the radius of the axons and dendrites, and the minimum distance between the cellular volumes. First a cell body is assumed in the center of the cube, which is considered as the center coordinate. Then, some points, as the number of the neurons, are randomly chosen in the cube based on specified constraints including cell−cell and cell−cube collisions. By having these points, transformation matrices are derived and rotation matrices are randomly assumed as the number of those and both are applied on the axon and dendrites. Afterward, collisions that may have occurred during the distribution are identified and new rotation matrices are created and applied on collided axons and dendrites. This loop continues until no collision between two neurons is observed, and an acceptable RVE would be created. Finally a reference neuron coordinate and the transformation and rotation matrices are stored in a text file as output. To create the model in ABAQUS software a PYTHON code is developed, which generates the desired RVEs in ABAQUS employing the outputs of the MATLAB code [3].

Because the macroscopic dimensions of the brain compared to its components are so large, the use of FE simulation to study the tissue involves high

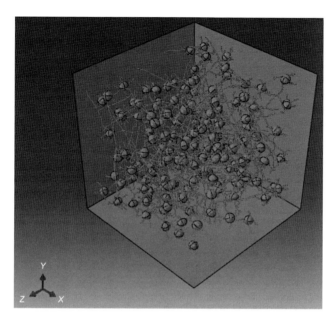

Figure 4.2 Representative volume element created in ABAQUS [3].

computational costs. That is one of the reasons due to which homogenization procedures were developed. All these procedures use a common definition of composite materials being reconstructed by repeating a volume element. RVE is the smallest volume element of a heterogeneous material that is large enough to be statistically similar to macroscale composites [39]. A 3D RVE consisting of neurons and ECM is constructed for studying the brain tissue. The optimum size of RVE is chosen by comparing the results of different sizes of RVEs. Moreover the simulated neurons are constituted of a spherical neuron cell, five undulate dendrites, and an undulate axon. It is noteworthy that the neurites have sinusoidal geometry [4].

A computational code has been developed as a MATLAB script to randomly distribute neurons and check the distance between the neurons to avoid any intersection of them into RVE. Therefore a neuron is created as the reference one; then other neurons are distributed by random translation and rotation matrixes. Neuron volume fraction (NVF), RVE size, reference neuron coordinates, and minimum distance between neuron cells are given as input values to the code, and reference neuron coordinates and translation and rotation matrixes are saved as output values. In Fig. 4.3 the distribution of neurons in RVE is shown [4].

4.2.1.2 Homogenization of the brain tissue

The brain responses in the bulk scale may vary under different conditions such as loading conditions, temperature, strain rate, and other physical parameters. To simulate these responses, different material constitutive models have been proposed, including elastic, viscoelastic, hyperelastic, and poroelastic models [40–44].

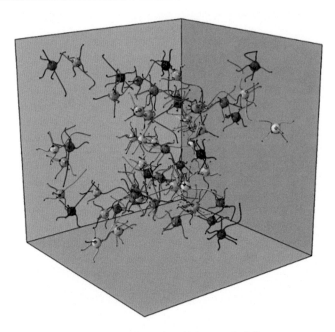

Figure 4.3 The distribution of neurons in extracellular matrix [4].

Based on the micromechanical principles, properties of a composite material can be estimated from its component properties and distribution. In this regard a viscoelastic material model is considered for material components of the cerebral cortex tissue in this study. First, both microcomponents are considered as elastic, homogeneous, and isotropic materials.

In the elastic analysis the material behavior is based on Hooke's law, which is presented as follows [3]:

$$\varepsilon_{ij} = \mathbf{S}_{ijkl}\sigma_{kl} \qquad (4.1)$$

where σ_{kl}, ε_{ij}, and $\mathbf{S}_{ijkl}$ are the stress, strain, and compliance tensors, respectively. Eq. (4.1) can be represented in the Voigt form in the following form [3]:

$$\begin{bmatrix} \varepsilon_1 \\ \varepsilon_2 \\ \varepsilon_3 \\ \varepsilon_4 \\ \varepsilon_5 \\ \varepsilon_6 \end{bmatrix} = \begin{bmatrix} S_{11} & S_{12} & S_{13} & S_{14} & S_{15} & S_{16} \\ S_{21} & S_{22} & S_{23} & S_{24} & S_{25} & S_{26} \\ S_{31} & S_{32} & S_{33} & S_{34} & S_{35} & S_{36} \\ S_{41} & S_{42} & S_{43} & S_{44} & S_{45} & S_{46} \\ S_{51} & S_{52} & S_{53} & S_{54} & S_{55} & S_{56} \\ S_{61} & S_{62} & S_{63} & S_{64} & S_{65} & S_{66} \end{bmatrix} \begin{bmatrix} \sigma_1 \\ \sigma_2 \\ \sigma_3 \\ \sigma_4 \\ \sigma_5 \\ \sigma_6 \end{bmatrix} \qquad (4.2)$$

According to the microstructural symmetry of the tissue, the number of compliance tensor coefficients is between 2 and 21. For instance the number of unknown independent coefficients of an orthotropic material is reduced to 9 [33].

Numerical characterization of tissues

To evaluate the viscoelastic behavior of the tissue a method same as that used to evaluate elastic behavior is also used. In the viscoelastic materials the relationship between stress and strain is expressed using an integral method presented in the following form:

$$\sigma_{ij}^0(t) = C_{ijkl}^t \varepsilon_{kl}^0 + \int_0^t C_{ijkl}^{t-\tau} \frac{d\varepsilon_{kl}^\tau}{dt} d\tau; \quad i,j,k,l = 1,2,3 \tag{4.3}$$

where C_{ijkl} is the stiffness matrix and t denotes the time. According to the above relation the stress in the tissue at any time depends on material behavior in history. To calculate the stiffness coefficients, mechanical properties are considered as constants in each incremental time of simulation. Thus we can discretize the integral in Eq. (4.3). In this study the relaxation test is employed to analyze where the strain is constant and stress relaxes over the time. It is noteworthy to mention that Eq. (4.3) can be represented in the Voigt form as [3]

$$
\begin{bmatrix} \sigma_{11} \\ \sigma_{22} \\ \sigma_{33} \\ \sigma_{23} \\ \sigma_{31} \\ \sigma_{12} \end{bmatrix}
=
\begin{bmatrix}
C_{11}^t & C_{12}^t & C_{13}^t & C_{14}^t & C_{15}^t & C_{16}^t \\
C_{21}^t & C_{22}^t & C_{23}^t & C_{24}^t & C_{25}^t & C_{26}^t \\
C_{31}^t & C_{32}^t & C_{33}^t & C_{34}^t & C_{35}^t & C_{36}^t \\
C_{41}^t & C_{42}^t & C_{43}^t & C_{44}^t & C_{45}^t & C_{46}^t \\
C_{51}^t & C_{52}^t & C_{53}^t & C_{54}^t & C_{55}^t & C_{56}^t \\
C_{61}^t & C_{62}^t & C_{63}^t & C_{64}^t & C_{65}^t & C_{66}^t
\end{bmatrix}
\begin{bmatrix} \varepsilon_{11}^0 \\ \varepsilon_{22}^0 \\ \varepsilon_{33}^0 \\ \varepsilon_{23}^0 \\ \varepsilon_{31}^0 \\ \varepsilon_{12}^0 \end{bmatrix}
$$

$$
+ \int_0^t
\begin{bmatrix}
C_{11}^{t-\tau} & C_{12}^{t-\tau} & C_{13}^{t-\tau} & C_{14}^{t-\tau} & C_{15}^{t-\tau} & C_{16}^{t-\tau} \\
C_{21}^{t-\tau} & C_{22}^{t-\tau} & C_{23}^{t-\tau} & C_{24}^{t-\tau} & C_{25}^{t-\tau} & C_{26}^{t-\tau} \\
C_{31}^{t-\tau} & C_{32}^{t-\tau} & C_{33}^{t-\tau} & C_{34}^{t-\tau} & C_{35}^{t-\tau} & C_{36}^{t-\tau} \\
C_{41}^{t-\tau} & C_{42}^{t-\tau} & C_{43}^{t-\tau} & C_{44}^{t-\tau} & C_{45}^{t-\tau} & C_{46}^{t-\tau} \\
C_{51}^{t-\tau} & C_{52}^{t-\tau} & C_{53}^{t-\tau} & C_{54}^{t-\tau} & C_{55}^{t-\tau} & C_{56}^{t-\tau} \\
C_{61}^{t-\tau} & C_{62}^{t-\tau} & C_{63}^{t-\tau} & C_{64}^{t-\tau} & C_{65}^{t-\tau} & C_{66}^{t-\tau}
\end{bmatrix}
\begin{bmatrix} d\varepsilon_{11}^\tau/dt \\ d\varepsilon_{22}^\tau/dt \\ d\varepsilon_{33}^\tau/dt \\ d\varepsilon_{23}^\tau/dt \\ d\varepsilon_{31}^\tau/dt \\ d\varepsilon_{12}^\tau/dt \end{bmatrix}
d\tau
\tag{4.4}
$$

Shear and compressive experiments are needed to study the viscoelastic behavior and characterization of the desired tissue. In these tests a constant strain, which should not exceed of 5% to correctly assume infinitesimal strain, is applied on the RVE [16]. In the compressive tests a compressive displacement is applied on a side of RVE and its opposite side is fixed in all directions and rotations. In addition, two nonparallel sides from the remaining four sides are fixed in the perpendicular direction (see Fig. 4.4). These boundary conditions have been introduced by Shaoning [45] which lead to appropriate results. In shear tests, all freedom degrees (directional and rotational) of a face of RVE are fixed and a shear displacement is applied on its parallel face [4].

In this study the RVE is considered as an orthotropic material whose compliance tensor has nine unknown coefficients. Nine independent equations are needed in

154 Applied Micromechanics of Complex Microstructures

Figure 4.4 Implemented boundary conditions on representative volume element [4].

order to determine these coefficients. In this regard, six kinematic loadings on RVE, including three compressive and three shear loadings, can lead to obtaining these equations. To solve these equations, applied stress and strain on the RVE should be calculated after loading. Thus a PYTHON code is developed in ABAQUS software to determine mean volumetric stress and strain values which are presented in the following form [3]:

$$\langle \sigma \rangle = \frac{1}{V} \int_V \sigma_m dv \qquad (4.5)$$

$$\langle \varepsilon \rangle = \frac{1}{V} \int_V \varepsilon_m dv \qquad (4.6)$$

where V is the RVE total volume. $\langle \sigma \rangle$ and $\langle \varepsilon \rangle$ are respectively the mean volumetric (macroscopic) stress and strain. Also, σ_m and ε_m are the local (microscopic) stress and strain, respectively. Since the reported stress and strain by the ABAQUS software are in the local coordinate, we should calculate these variables in global coordinates to use Eqs. (4.5) and (4.6). Thus Eq. (4.7) is utilized to rotate the stress tensor into global coordinates [3].

$$[\sigma] = [T][\sigma'][T^t] \qquad (4.7)$$

Numerical characterization of tissues

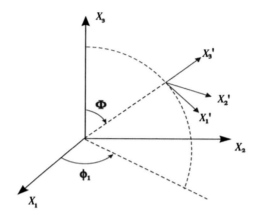

Figure 4.5 Relation between local and global coordinates [3,4].

Table 4.1 Elastic properties of the cerebral cortex tissue constituents [3,16].

	Elastic modulus (Pa)	Poisson ratio
Neuron	6160	0.4999
ECM	2040	0.4999

σ and σ' are respectively stress tensors in global and local coordinates and T^t is the transposition of rotation tensor that is presented in Eq. (4.8) in the following form [3]:

$$[T] = \begin{bmatrix} \cos\phi_1\cos\Phi & -\sin\phi_1 & \cos\phi_1\sin\Phi \\ \sin\phi_1\cos\Phi & \cos\phi_1 & \sin\phi_1\sin\Phi \\ -\sin\Phi & 0 & \cos\Phi \end{bmatrix} \quad (4.8)$$

where Φ and ϕ_1 are depicted in Fig. 4.5 as follows [3,46]:

4.2.1.3 Results and discussion

In this section, various analyses have been conducted and the results have been evaluated. Elastic properties of the ECM and neuron, as cerebral cortex tissue constituents, are listed in Table 4.1.

Viscoelastic analysis is performed utilizing ABAQUS software in which time increments are large and where stress drop is high. Although the loading time in some simulations is considered 0.1 s, many simulations are developed with various time increments to investigate the rate dependency behavior of the cerebral cortex tissue. The time domain model is used in ABAQUS software to define the viscoelastic properties. Eq. (4.9) presents the Prony series used to define the shear modulus in the viscoelastic analysis as follows [3]:

$$G(t) = G_0 \left(1 - \sum_{k=1}^{n} g_k(1 - e^{-t/\tau_k})\right) \tag{4.9}$$

where τ_k is the relaxation time and its corresponding relaxed shear modulus is $G_0 \cdot g_k$. Also, viscoelastic properties of the constituents are tabulated in Table 4.2.

In the proposed model the diameter of the neuron cell, the diameter of axons and dendrites, the length of axons, and the length of dendrites are 50, 1.5, 300, and 55 μm, respectively. It is noteworthy to mention that the size of RVE has a crucial effect on results in micromechanical analysis. Thus selecting a nonproper size and boundary conditions may influence results. Elastic analysis is employed to select the appropriate size of RVE, and the results of many RVEs with different sizes are compared by considering a fixed neuron volume fraction (NVF).

Fig. 4.6 depicts the elastic modulus of some RVEs with the NVF of 1.5% for various sizes of 300, 400, 700, 900, and 1100 μm. As observed in this figure the

Table 4.2 Viscoelastic properties of the cerebral cortex tissue constituents [3,34].

	Instantaneous modulus (Pa)	Poisson ratio	g_1	$\tau_1(sec)$	g_2	$\tau_2(sec)$
Neuron	12860	0.5	0.6039	0.60097	0.1083	0.49866
ECM	4290	0.5	0.50001	0.00623	0.25986	0.9

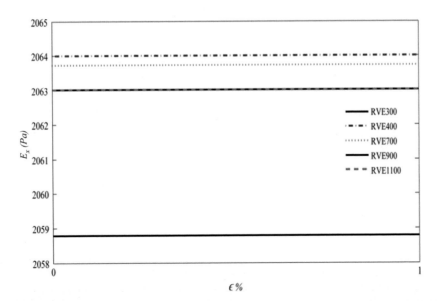

Figure 4.6 Effect of the representative volume element size with a volume fraction of 1.5% on tissue properties [3].

Table 4.3 Obtained mechanical properties of the cerebral cortex tissue with NVF of 3% [3].

Direction	Young's modulus(Pa)	Shear modulus (Pa)	Poisson ratio
X	2085.258	–	–
Y	2083.528	–	–
Z	2084.753	–	–
XY	–	698.612	0.492
XZ	–	697.736	0.495
YZ	–	697.123	0.496

RVE with a size of 900 μm is a suitable choice since by increasing the size of RVE more than that, there is not any change in elastic modulus [3].

Since the brain tissue in some areas, especially in its cerebral cortex, shows the isotropic behavior, it is necessary to investigate the isotropy of the model [32,37]. Since the random distribution is employed in the creation of the RVE, it is expected that the proposed model presents the isotropic behavior. In this regard an RVE with a dimension of 300 μm and NVF of 3% is developed. Initially, orthotropic behavior is assumed for the tissue and elastic and shear moduli are calculated (see Table 4.3). As observed in Table 4.3, there is a slight difference between the results in different directions. Thus the RVE represents an isotropic behavior which means that the mechanical properties of the model are independent of direction.

Since each of the elastic shear moduli in three directions is approximately equal, it is concluded that the behavior of isotropic elements is a representative of the volume. Also, utilizing this cellular model, there is no need to consider numerous experiments and mechanical properties of the tissue can be obtained using two types of tests.

Following the selection of an appropriate RVE the effect of NVF on the mechanical behavior of the cerebral cortex tissue is investigated. As shown in Fig. 4.7 the volume fraction is varied between 1% and 4%. Neuron cells are depicted in different colors to better see the distribution. As observed in Fig. 4.8 the elastic modulus is increased by increasing the volume fraction of neurons. Due to the low ratio of diameter to length (aspect ratio) of neuron cells, there is no significant change in elastic modulus in different volume fractions. It is noteworthy to mention that the volume ratio of axons and dendrites to volume cell is extremely low. Abolfathi et al. [16] reported that in areas of the brain such as the stem, where only axons are surrounded by the ECM, increasing the volume fraction of axons leads to enhance the elastic modulus.

Also, they reported that in areas of the brain where the volume cell has less volume and axons possess a higher volume in a neuron cell, increasing the neuron cell volume fraction can significantly increase the mechanical properties such as elastic modulus [3].

A comparison between E_x obtained from the FE results of the elastic analysis as well as the lower bound (Reuss model) and upper bound (Voigt model) [47], which are used for composite materials, is depicted in Fig. 4.9. The proposed model represents logical behavior since its results are between lower and upper bounds.

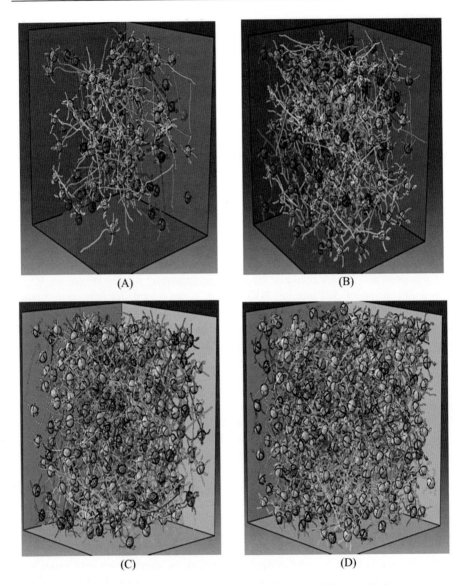

Figure 4.7 Representative volume element with a dimension of 900 μm with the neurons volume fraction of (A) 1%, (B) 2%, (C) 3%, and (D) 4% [3].

This comparison also shows that the obtained properties are closer to the lower bound due to the low aspect ratio of axons and dendrites. Due to the differences in the geometry and density of the constituents in different parts of the brain, elastic and shear moduli are different in various parts of tissues. Thus the different regions of the brain should be sampled and analyzed to obtain its mechanical properties [3].

Numerical characterization of tissues

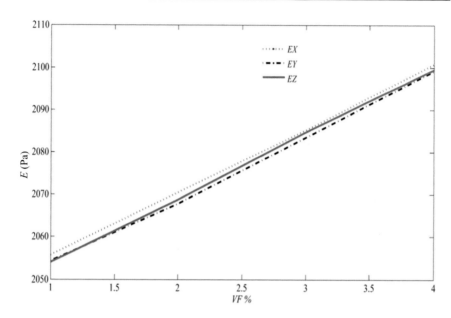

Figure 4.8 Effect of neurons volume fraction on elastic modulus of the cerebral cortex tissue [3].

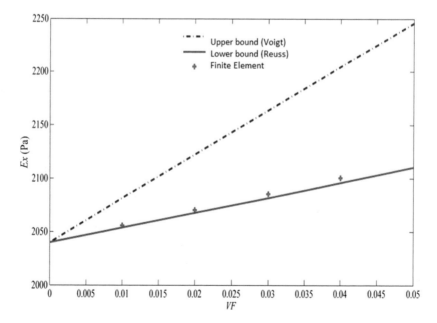

Figure 4.9 Comparison of E_x obtained from FE results and the upper bound (Voigt) and lower bound (Reuss) [3].

Table 4.4 Young's modulus for two RVEs with NVF of 2% and different random distributions [3].

Mechanical properties	1st distribution	2nd distribution
E_x(Pa)	2070.499	2069.533
E_y(Pa)	2067.772	2068.249
E_z(Pa)	2068.683	2067.231

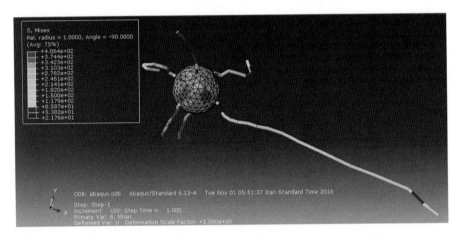

Figure 4.10 Von-Mises stress contour of a neuron after compressive load on representative volume element [3].

To investigate the effect of irregular distribution of neurons in the ECM on the mechanical properties of the cerebral cortex tissue, two RVEs with the NVF of 2% and different random distributions are developed. The simulation results in Table 4.4 demonstrate that the mechanical properties of these RVEs are close and the random distribution of neurons in the cerebral cortex tissue has no significant effect on the overall properties of the tissue.

Considering an RVE with the NVF of 3% under compressive loading, Fig. 4.10 shows the Von-Mises stress contour of a neuron.

Also, Fig. 4.11 demonstrates the stress contours of the neurons in the x-direction for an RVE with the NVF of 3%. In most models of the brain tissue the maximum Von-Mises stress in the RVE is at the junction of the volume cell and axons as well as dendrites. DAI is one of the most common brain damage where nerve cells and axons are damaged [43]. Existence of maximum stress in these regions is a confirmation of DAI [3].

After the elastic analysis the viscoelastic behavior is analyzed. In this regard a constant strain is applied on an RVE with the NVF of 2%. Fig. 4.12 shows the applied strain on RVE to simulate the relaxation test. As observed the mean strain calculated from FE results is equal to the applied strain.

Numerical characterization of tissues

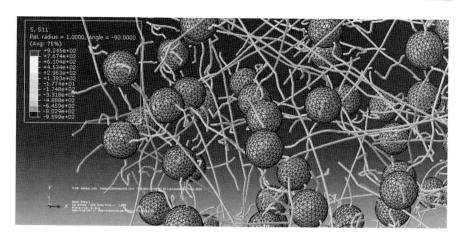

Figure 4.11 Stress contour in the x-direction in representative volume element after compression [3].

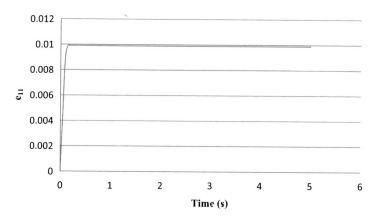

Figure 4.12 ε_{xx} of representative volume element with the neurons volume fraction of 2% under loading [3].

Fig. 4.13A and B demonstrate the stress value respect to the time for loading in x- and y-directions, respectively. As observed in these figures, increasing the NVF leads to increase the stresses. Also, stress rising in first 0.1 s is due to the applied strain. Since the relaxation test is conducted on the model, strain remains constant after applying 1% of strain. When the strain is fixed the stress starts to relax, and it is completely relaxed in 5 s. As observed, by increasing the NVF the maximum and relaxed stresses are increased and the RVE responds stiffer. This slight increasing in stiffness is due to the low aspect ratio in neurons, as discussed in the elastic modulus analysis [3].

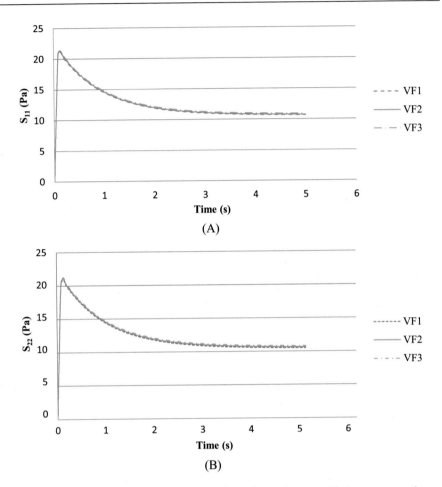

Figure 4.13 (A) σ_{xx} and (B) σ_{yy} of representative volume element with the neurons volume fraction of 1%, 2%, and 3% under loading in x- and y-directions [3].

The other item which is changed by increasing the NVF is the relaxation time in this type of loading. Assuming that neurons have less relaxation time than the ECM, it is expected that the cerebral cortex tissue relaxation time decreases by increasing the NVF. Since the neurons and ECM have almost the same relaxation time, no specific changes are seen by varying the NVF.

In the relaxation tests the rate of applied strain is one of the important parameters in loading. The strain rate should be less than the relaxation time (τ) to obtain a reliable result. Considering the results shown in previous figures the relaxation time is about 3 s. Since the time of applied strain is 0.1 s, this relaxation test is reliable in terms of loading time.

Fig. 4.14A–C, respectively, show Von-Mises, maximum tensile, and maximum compressive stresses of tissues with NVFs of 1%, 2%, and 3% under

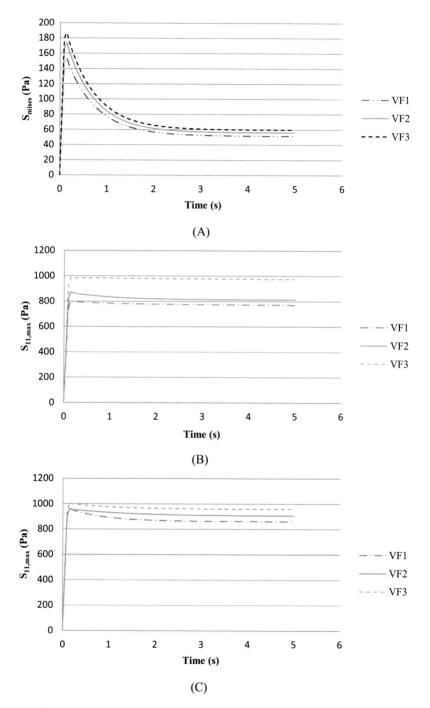

Figure 4.14 (A) Von-Mises stress and maximum (B) tensile σ_{xx} and (C) compressive σ_{xx} of tissues with neuron volume fractions of 1%, 2%, and 3% under loading in the *x*-direction [3].

loading in the x-direction. It is obvious that the maximum tensile stress in the loading direction has risen by increasing the NVF. Also the maximum tensile and compressive stresses do not have a great loss. Since the neuron distribution in the RVE with the NVF of 3% is denser compared to two other NVFs, the increase in maximum tensile and compressive stresses in NVF of 3% is larger than others [3].

To investigate the effect of strain rate on tissues an RVE with the NVF of 2% is considered. This element is loaded under four compressive loadings (1% strain) in the x-direction with loading times of 0.05, 0.1, 0.2, and 0.3 (s). The applied compressive strains for these four loadings are demonstrated in Fig. 4.15A. Also, Fig. 4.15B shows the mean volume stress in the RVE for four different strain rate loadings. It is observed that the mean volume stress increases by increasing the strain rate and the mean stress tends to a specific value in all loadings due to a constant NVF. The maximum von-Mises stress in the RVE in the first 5 s is depicted in Fig. 4.15C. Considering this figure the maximum von-Mises stresses have risen by increasing the strain rate. These stresses are relaxed to a specific value, and their relaxation times are almost equal. Also, it is obvious that the brain tissue is very vulnerable in trauma when there is a high strain rate. Fig. 4.15D and €, depict the maximum tensile and compressive stress in the x-direction, respectively. It is revealed that the compressive stress is higher than tensile stress [3].

As the last analysis the effect of neuron distribution on maximum tensile and compressive stresses in the loading direction and also Von-Mises stress is investigated. Fig. 4.16A shows the von-Mises stress in a relaxation test for three different neuron distributions where there is no reason for von-Mises stresses being equal. In this figure, Von-Mises stresses have markedly declined but relaxed to a specific value in each distribution. Fig. 4.16B and C demonstrate the maximum tensile and compressive stresses in the x-direction for three different neuron distributions, respectively. It is observed that maximum stresses are different, and the stresses have no sharp drop in both charts. Also, a great difference between two distributions in maximum tensile stress is not a reason to repeat the same behavior for the maximum compressive stress where the corresponding compressive stresses have less difference. Finally the results demonstrate that although different irregular distributions of neurons have no effect on the mechanical properties of the tissue in a constant NVF, this difference can affect the distribution of local stresses (and even the maximum stress) in the tissue [3].

4.2.2 Homogenization of the heterogeneous brain tissue under quasistatic loading: a viscohyperelastic model of a 3D representative volume element

In this section the behavior of the brain tissue is investigated under different loads with various rates and volume fractions of neurons. To this aim, RVEs with neurons are generated using MATLAB and PYTHON scripts in ABAQUS. Neurons distributed randomly in the RVEs and the interaction between neurons and ECM is considered as perfect bonding. The orthotropic elastic model is employed to study

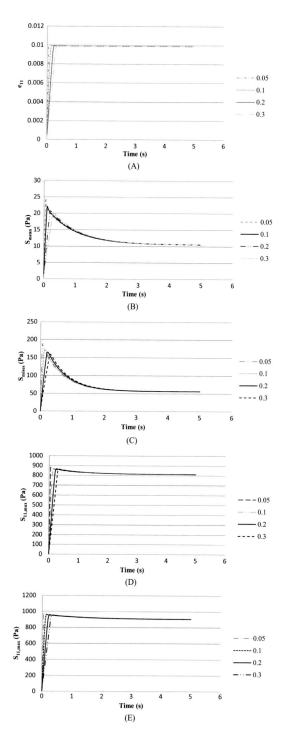

Figure 4.15 (A) Applied strain on the model for different durations of 0.05, 0.1, 0.2, and 0.3 s, (B) corresponding mean volumetric stresses in the *x*-direction, (C) corresponding Von-Mises stresses, (D) corresponding maximum tensile stresses in the *x*-direction, and (E) corresponding maximum compressive stresses in the *x*-direction [3].

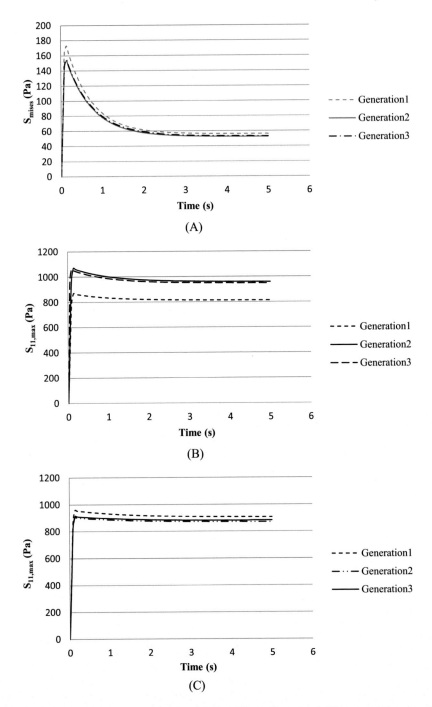

Figure 4.16 (A) Von-Mises stress and maximum (B) tensile σ_{xx} and (C) compressive σ_{xx} of tissues with the neurons volume fraction of 2% under loading in the x-direction for three different distributions of neurons in the extracellular matrix [3].

Numerical characterization of tissues

the isotropic behavior and the neuron distribution effect on the mechanical properties. The results showed that in different distributions, RVEs have the same response; also, RVEs showed isotropic behavior in all volume fractions. Additionally the optimum RVE size was evaluated using the elastic constitutive model to eliminate the boundary condition effects.

Furthermore the viscohyperelastic constitutive model is used to study loads, load rates, and NVF impacts on the mechanical response of RVEs. It was found that the maximum stress was increased with the increase of the load rate, which shows that the tissue was more vulnerable in high load rates. Moreover the load rate had a lower impact than the load itself on the maximum stress. According to the proposed result, under similar loading conditions, axons were more vulnerable than soma, which is a reason for DAI. These results could be beneficial for production of helmet, surgical robots and all tools that are employed to prevent or treat the brain injuries [4].

4.2.2.1 Constitutive model and simulation

Considering the brain structure, in some regions of the brain, neurons are distributed in ECM irregularly; hence it is expected that these regions behave in an isotropic way. In order to lower the computational costs, initially neurons and ECM are individually assumed to be elastic and isotropic materials. Utilizing the elastic model the optimum size of RVE is extracted; then its isotropy is studied. For this purpose, it is assumed that RVE follows the orthotropic model, incorporating nine independent coefficients [Eq. (4.10)] [4].

$$\varepsilon_{ij} = S_{ijkl}\sigma_{kl}$$

$$S = \begin{bmatrix} \dfrac{1}{E_1} & \dfrac{-\nu_{21}}{E_2} & \dfrac{-\nu_{31}}{E_3} & 0 & 0 & 0 \\ \dfrac{-\nu_{12}}{E_1} & \dfrac{1}{E_2} & \dfrac{-\nu_{32}}{E_3} & 0 & 0 & 0 \\ \dfrac{-\nu_{13}}{E_1} & \dfrac{-\nu_{23}}{E_2} & \dfrac{1}{E_3} & 0 & 0 & 0 \\ 0 & 0 & 0 & \dfrac{1}{G_{23}} & 0 & 0 \\ 0 & 0 & 0 & 0 & \dfrac{1}{G_{31}} & 0 \\ 0 & 0 & 0 & 0 & 0 & \dfrac{1}{G_{12}} \end{bmatrix} \tag{4.10}$$

where E_i, G_{ij}, and ν_{ij} are the Young's modulus, shear modulus, and Poisson ratios in different directions, respectively. Therefore six tests including three tensile and three shear tests in different directions are conducted. Through a comparison of the coefficients of the orthotropic model the isotropy is studied. Elastic constants for the neurons and ECM are summarized in Table 4.6.

After isotropy check and determining the optimum size of RVE, to predict a more realistic behavior of the brain a viscohyperelastic model is implemented for ECM and neurons. Assuming an isotropic nature [44] for neurons and ECM, strain energy W_e would be a function of invariants of the left or right Cauchy−Green deformation tensor [48].

$$W_e = W_e(I_1, I_2, I_3) \tag{4.11}$$

where $I_i's$ are invariants of either the left or right Green deformation tensor (**B** or **C**). Hence

$$\overline{\mathbf{B}} = J^{-2/3}\mathbf{B}, \quad \overline{\mathbf{C}} = J^{-2/3}\mathbf{C} \tag{4.12}$$

$$W_e = W_e^{iso}(\overline{I}_1, \overline{I}_2) + W_e^{vol}(I_3) \tag{4.13}$$

where $\overline{\mathbf{B}}$, $\overline{\mathbf{C}}$, J, W_e^{iso}, and W_e^{vol} are deviatoric parts of the left and right Cauchy− Green deformation tensor, the root of I_3, and deviatoric and volumetric parts of the strain energy, respectively.

In order to predict the brain's responses under different conditions, several material constitutive models have been presented, including elastic, hyperelastic, viscoelastic, and viscohyperelastic models [44,49−51]. Since the brain's responses vary under different physical conditions, such as variation of the strain rate, temperature, strain levels, and other parameters, a viscohyperelastic constitutive model is developed to consider the time-dependent behavior of the brain tissue. Moreover the stress is divided into two time-dependent and elastic parts [48].

$$\sigma = \sigma^e + \sigma^v(t) \tag{4.14}$$

$$\sigma^v(t) = \sigma^{v_1}(t) + \sigma^{v_2}(t) \tag{4.15}$$

where σ^e and σ^{v_i} are the quasistatic and viscous stresses, respectively. In addition the viscous stress is further divided into two parts including deviatoric stress and volumetric stress. Therefore the strain energy for the time-dependent part contains a deviatoric and a volumetric energy [4].

$$W_v = W_v^{iso}(\overline{I}_1, \overline{I}_2) + W_v^{vol}(I_3) \tag{4.16}$$

Based on Bergström and Boyce [52] studies, quasistatic stress equations could be used for the deviatoric part of the viscous response [4].

$$\begin{aligned} W_e^{iso} &= (c_e I_1 - 3) \\ W_{v_1}^{iso} &= (c_{v_1} I_1 - 3) \\ W_{v_2}^{iso} &= (c_{v_2} I_1 - 3) \end{aligned} \tag{4.17}$$

As for the strain energy function the Neo-Hookean strain energy model [Eq. (4.17)], which is a common material model for living tissues, is employed to include deviatoric parts of both the elastic and viscous stresses [4,6,53,54]. It is noteworthy that Neo-Hookean coefficients have different values for each of the elastic and viscous parts. Also, for the viscous part of the stress, as mentioned earlier, two relaxation time constants are assumed to consider the fast and slow loading velocities [4].

$$\sigma^{v_1}(t) = \int_0^t \frac{\partial}{\partial \tau} \left(2I_3^{-1/2} \mathbf{B} \left(\frac{\partial W_{v_1}}{\partial \mathbf{B}} \right) \right) e^{\left(-\frac{t-\tau}{\theta_1} \right)} d\tau$$

$$\sigma^{v_2}(t) = \int_0^t \frac{\partial}{\partial \tau} \left(2I_3^{-1/2} \mathbf{B} \left(\frac{\partial W_{v_2}}{\partial \mathbf{B}} \right) \right) e^{\left(-\frac{t-\tau}{\theta_2} \right)} d\tau$$

(4.18)

where θ_i is the relaxation time constant. Moreover the strain energy function related to the volumetric part of the energy is expressed with a well-known relation as [44−46].

$$W_e^{vol} = \frac{1}{2} K (J-1)^2$$

$$W_{v_1}^{vol} = \frac{1}{2} K' (J-1)^2$$

$$W_{v_2}^{vol} = \frac{1}{2} K'' (J-1)^2$$

(4.19)

where K, K', and K'' are the volumetric moduli of elastic and viscous parts, which are considered to be equal in this study. Finally, this model is numerically implemented as a VUMAT subroutine for studying the brain tissue.

4.2.2.2 Finite element simulation

To generate the geometry of the micromechanical model (RVE) in ABAQUS software a PYTHON code is developed. For this end the code reads reference neuron coordinates, rotation, and translation matrixes from a MATLAB script output. This code creates the neurons and ECM as solid parts while assigns material properties to the RVE. Additionally, it can identify contacts between the neurons and ECM and assign perfect bounding between them. Moreover, C3D4 elements are employed to geometrical discretization of both the neurons and ECM.

The presented loading method by Shaoning [45] is used to apply load on the RVE. The symmetry boundary condition is employed in tension. In this way, while one of the faces of the RVE is subjected to uniaxial displacement the opposite side is fixed in the loading direction. Two other perpendicular faces from four remained faces are fixed in their normal direction. The boundary conditions have been demonstrated in Table 4.5 using notations given in Fig. 4.17. Subsequently, each of RVEs is subjected to five quasistatic loads in two different strain rates. In order to implement the shear test, shear displacement load is applied to one of the faces, while the opposite face is fixed in a fully constrained boundary condition.

Homogenized variables at the macroscopic scale are determined using the mean volume of variables at the microscopic scale in the RVE. The mean macroscopic stress σ and strain ε could be calculated from equation [4].

$$\langle \sigma \rangle = \frac{1}{|\Omega_m|} \int_{\Omega_m} \sigma_m d\Omega$$
$$\langle \varepsilon \rangle = \frac{1}{|\Omega_m|} \int_{\Omega_m} \varepsilon_m d\Omega \tag{4.20}$$

where Ω_m, σ_m, and ε_m are the total volume of RVE and the local microscopic stress and strain, respectively.

Considering geometrical complexity in the RVE the elevation of NVE leads to an increase in computational costs. On the other hand, an increase in NVF makes the neurons keep a small distance from each other, which results in a higher number of elements in the RVE. Therefore a self-consistent procedure is necessary to be

Table 4.5 Zero boundary conditions in tension (see Fig. 4.17) [4].

	$U_1 = 0$	$U_2 = 0$	$U_3 = 0$	$UR_1 = 0$	$UR_2 = 0$	$UR_3 = 0$
Tension test Shear test	Back Back	Up Back	Right Back	Up right Back	Back right Back	Back up Back

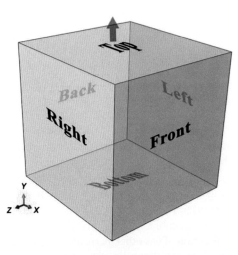

Figure 4.17 Notation of the faces for identifying load cases [4].

Numerical characterization of tissues

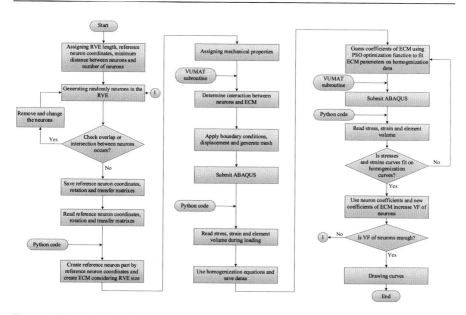

Figure 4.18 Flowchart of modeling and analysis of the brain tissue [4].

employed to resolve this problem. Considering the self-consistent method the heterogeneous composite material could be simulated as a homogeneous material, having average mechanical constants [30,55,56]. In order to implement self-consistent approximation, the particle swarm optimization (PSO) function [57,58] is implemented in MATLAB programming. Homogenized normal and shear stresses versus time curves of the heterogeneous composite are considered as input data. Utilizing the optimization function a homogeneous matrix is fitted on these curves and a new RVE is generated with a higher volume fraction. Hence MATLAB script runs a PYTHON code and the PYTHON program initiates ABAQUS software analysis; the VUMAT subroutine is also employed by ABAQUS software as the constitutive model. Finally, the PYTHON program extracts the stress−time curves when the analysis is completed. The MATLAB program, then, compares the homogeneous matrix curves with heterogeneous composite ones and keeps the loop running until the area between these curves reaches a desirable tolerance. Considering an isotropic state for the used RVE the consistency between these two curves would be enough for computing the mechanical properties. Fig. 4.18 demonstrates the procedure of the analysis in brief [4].

4.2.2.3 Results and discussion

4.2.2.3.1 Employing the elastic model to investigate isotropy, RVE size, and irregular distribution

It is known that RVE size is one of the important parameters in predicting effective properties of heterogeneous materials. Moreover, mechanical properties must be

Table 4.6 Material properties of ECM and neurons [4].

	Elastic modulus (Pa)	Poisson ratio
ECM	2040	0.5
Neuron	6160	0.5

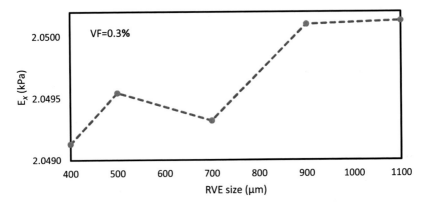

Figure 4.19 Elastic modulus (E_x) variation with representative volume element size for VF = 0.3% [4].

independent of the RVE size. Considering boundary effects on the results the calculated outcomes from a small size of RVE could not be valid. Larger sizes of RVE, on the other hand, increase the computational costs by a great deal [52]. Therefore five RVEs in different sizes (300, 400, 700, 900, and 1100 μm) are created with the same NVF (0.3%). Material properties of the components of RVEs are summarized in Table 4.6. RVEs are subsequently analyzed by the elastic model, and an optimum size for these RVEs is found considering their elastic modulus. In Fig. 4.19 the elastic modulus of different RVE sizes is plotted for the same volume fraction [4].

Fig. 4.19 depicts that an RVE size with a dimension of 900 μm would be a suitable choice since no significant variation in the results is observed for the larger sizes. Considering an elastic model for RVEs, it is obvious that the results must be independent of the RVE size. Therefore the computational cost is optimum and the results are valid when RVE dimensions are over than 900 μm.

Since random distribution of the neurons is assumed for the creation of RVEs a thorough investigation of RVEs' behavior seems essential. Therefore the elastic orthotropic behavior is initially assumed for the RVEs with a dimension of 900 μm. The results are summarized in Table 4.7 [4].

Regarding the observed similar behavior of the RVE in different perpendicular directions, it is concluded that the RVE mechanical behavior is isotropic. The same result has been demonstrated by Koser and Moeendarbary [59] for the white matter under infinitesimal tension. Due to the isotropic behavior of the RVE the characterization and

Table 4.7 Material coefficients of orthotropic assumption for the RVE [4].

Direction of properties	Elastic modulus (Pa)	Shear modulus (Pa)	Poisson's ratio
X	2050.084	–	–
Y	2050.079	–	–
Z	2050.117	–	–
XY	–	687.956	0.489982
XZ	–	687.960	0.489972
YZ	–	687.959	0.489973

Table 4.8 Elastic modulus for different distributions of neurons [4].

Elastic modulus (Pa)	Distribution 1	Distribution 2	Distribution 3
E_x	2050.084	2050.084	2050.080
E_y	2050.079	2050.092	2050.087
E_z	2050.117	2050.111	2050.100

Table 4.9 Homogenized elastic modulus in different volume fractions for the orthotropic model [4].

Elastic modulus (Pa)	0.3	0.6	0.9	1.2	1.5
E_x	2050.084	2060.080	2070.075	2080.071	2090.066
E_y	2050.079	2060.075	2070.071	2080.066	2090.062
E_z	2050.117	2060.113	2070.108	2080.104	2090.099

mechanical properties could be easily predicted by performing only two sets of tests. This significantly lowers the computational costs.

However, there are still two concerns remaining for reassurance of the isotropy, the effect of various irregular distributions and different NVFs. As for the effects of various irregular distributions of neurons on the mechanical properties, three different RVEs are created with NVF of 0.3% enjoying a random distribution. The results show that different distributions yield approximately equal mechanical properties (see Table 4.8). Hence it is deduced that the effective general mechanical properties are independent from distribution of the neurons in irregular regions of the brain [4].

To investigate the impact of NVF variation on the mechanical properties, five RVEs are generated with 0.3%, 0.6%, 0.9%, 1.2%, and 1.5% neurons and 1% tension strain load is applied on them. They are analyzed under static loads, while the self-consistent approximation is employed to increase the volume fraction. It is found that the elastic modulus rises at higher values of NVF. However, due to the low increase of NVF, the observed variation of the elastic modulus is not significant. In Table 4.9, elastic moduli of different NVFs are presented. Abolfathi and Naik [16] showed that

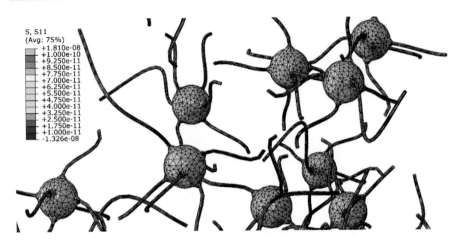

Figure 4.20 Tension stress distribution in neurons for elastic modeling (the results should be multiplied by 1e12 Pa) [4].

in some regions of the brain such as the stem where only axons are simulated, increasing the axon volume fraction results in the enhancement of elastic modulus; in addition, these regions show an anisotropic behavior. However, elastic moduli of different volume fractions are approximately equal in perpendicular directions, once more revealing an isotropic behavior in the RVEs [4].

In Figs. 4.20 1% strain load as a tension static load is applied on the RVE. It is found that the stresses in neurons are higher than those in the matrix. Moreover, Fig. 4.20 shows that in neurons the experienced stress in dendrites and axons is higher than that in soma. Therefore the neuron plays a key role in investigating DAI compared to ECM. Distribution of the stress is demonstrated in Fig. 4.20. As one may observe the micromechanical investigation on the stress distribution is one of the most important matters in TBI study. Under elastic assumption in this study the importance of micromechanical analysis is shown. Therefore to arrive at more realistic results a viscohyperelastic model is developed in this study [4].

4.2.2.3.2 Developing a viscohyperelastic model to have a more realistic study on the brain tissue

Javid and Rezaei [34] conducted an experiment on the brain tissue and presented the mechanical properties of axons and ECM separately. In this study the outcomes of that experiment are employed to find the necessary viscohyperelastic coefficients. Table 4.10 indicates these coefficients for the RVE components having an NVF of 3% in various loadings. In order to study the behavior of RVE in higher volume fractions, self-consistent approximation is also implemented to increase NVF [4].

The RVE size obtained in the elastic region (900 μm); hereon, it would be considered for the analysis. Various loads in different rates are applied, and the tissue behavior is evaluated with respect to the time. RVEs are subjected to various loads

Table 4.10 Viscohyperelastic coefficients in 3% NVF [4].

	Strain	Time of maximum load	C_e(Pa)	C_{v_1}(Pa)	C_{v_2}(Pa)	K(kPa)	θ_1(s)	θ_2(s)
ECM	1%	0.2	216.6	184.7	104.7	135.98	0.074	1.27
		0.03	216.6	184.7	104.7	135.98	0.074	1.27
	2%	0.2	213.9	183.6	95.8	145.92	0.081	1.32
		0.03	213.9	183.6	95.8	145.92	0.081	1.32
	3%	0.2	220.6	198.4	94.2	108.87	0.051	1.08
		0.03	220.6	198.4	94.2	108.87	0.051	1.08
Neuron	1%	0.2	631.9	1226.2	1552.9	80.39	1e−7	0.57
		0.03	631.9	1226.2	1552.9	80.39	1e−7	0.57
	2%	0.2	619.1	70.5	1554.1	73.22	2e−5	0.57
		0.03	619.1	70.5	1554.1	73.22	2e−5	0.57
	3%	0.2	616.5	72.6	1543.4	70.91	1e−7	0.57
		0.03	616.5	72.6	1543.4	70.91	1e−7	0.57

including 1%, 2%, 3%, 10%, and 15% tension and shear strains in a quasistatic manner; each of them is applied at 0.2 and 0.03 s. In Fig. 4.21 the strain variations versus time are plotted for different strains in various rates [4].

It should be considered that micromechanical analysis is time-consuming, especially in quasistatic and dynamic loadings in which the employed constitutive model has a time-dependent behavior. It is noteworthy that in this study, since the stress variations are not sensitive after 4 s, to avoid unnecessary computational costs the analysis is terminated at 4 s. In Fig. 4.22, variations of the mean volumetric stress in RVE in the loading direction are demonstrated under 15% tension loading for 3.9% NVF [4].

In Fig. 4.23A and B, variations of mean volumetric stress versus time are plotted with an NVF of 1.2%. As shown the stress rises until 0.2 and 0.03 s for different strain rates. Considering viscous behavior of RVE, when the strain becomes constant a reduction in the stress could be observed. Due to rate-dependent behavior of the brain tissue [60], enlarging the strain rate would lead to a stiffer response in the tissue, outcomes of Rashid and Destrade [61] also confirm the stiffer behavior of the tissue at higher rates. Thus because strain rate is usually high in the accidents the brain tissue is more vulnerable. These figures show that the developed model is suitable to study the rate dependency of the brain tissue. Additionally, Fig. 4.24 demonstrates that enlarging NVF would lead to an increase in the stress throughout the analysis [4].

One of the important points in time-dependent materials is the loading rate; in Fig. 4.25 variations of the mean volumetric stress for 2.1% NVF are plotted in two strain rates under 3% tension. Although the same values are expected at 0.2 s the presence of load history results in a nonsimilarity in the extracted values for different strain rates. The difference in the peaks of the two curves results from the strain rate dependency of RVE. Since the magnitudes of NVFs and loads are equal, the mean volumetric stresses reach an equal value in both strain rates [4].

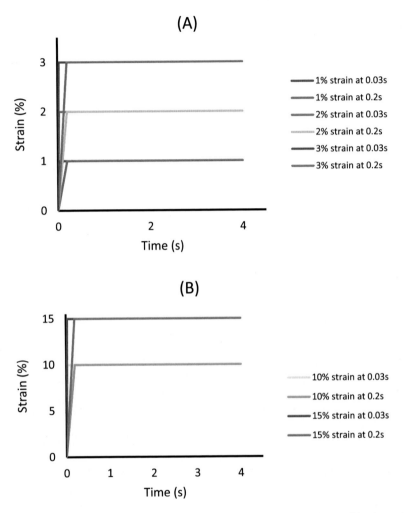

Figure 4.21 Different strain sizes in various rates: (A) 1%, 2%, and 3% logarithmic strain and (B) 10% and 15% logarithmic strain [4].

In Fig. 4.26 variations of mean volumetric stress are shown in two strain rates and various NVFs. This figure demonstrates that the stress grows with the rise of NVF, confirming that although the increase is not significant, the tissue has become stiffer [13]. It is found that in higher rates, NVF has a lower impact on the maximum stress, while it has a more severe effect in lower strain rates. Comparing the results obtained for the NVF, load and strain rates effects on the brain tissue. Fig. 4.26 shows that the brain tissue is more sensitive to the strain rate regarding NVF effects, and according to Fig. 4.24 and Figs. 4.26 and 4.27, it is also more sensitive to the magnitude of the load with respect to the other two parameters. As shown, enhancing the strain rate up to about 7 times of its initial value the maximum stress becomes about 1.18 times for both the strains. Meanwhile, tripling the

Numerical characterization of tissues

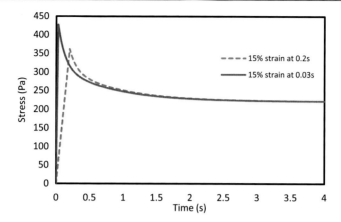

Figure 4.22 Stress variation in 3.9% VF under 15% strain [4].

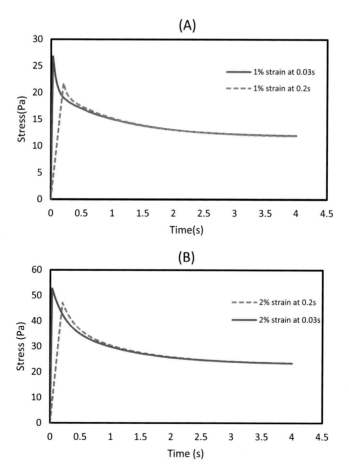

Figure 4.23 The mean volumetric stress of the RVE with 1.2% VF of neurons under different quasistatic loads in various rates: (A) 1% strain and (B) 2% strain [4].

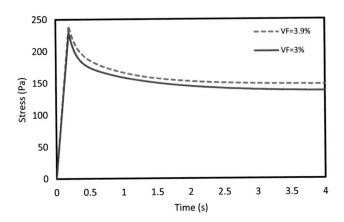

Figure 4.24 The mean volumetric stress in 10% strain applied 0.2 s in 3% VF and 3.9% VF of the RVE [4].

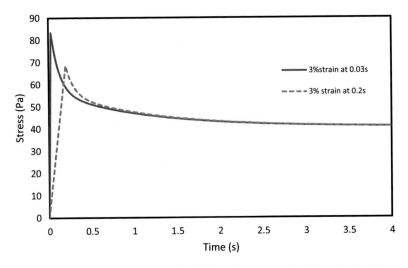

Figure 4.25 Mean volumetric stress for 2.1% VF of neurons under 3% strain load in various rates [4].

strain leads to a rise of stress by 2.87 times for two different rates. As shown in Fig. 4.28 the enhancement of mean volumetric shear stress could be reported by almost 1.2 and 2.9 times for increasing strain rates and the strain, respectively [4].

Considering an NVF of 3.9% for the RVE, Fig. 4.29 illustrates the stress contour in the loading direction while 3% strain at 0.2 s is applied. According to the contour the maximum stress in a neuron is experienced by the dendrites and axons. Therefore micromechanical analysis shows dendrites and axons having a more sensitive role in brain injury [4].

Numerical characterization of tissues 179

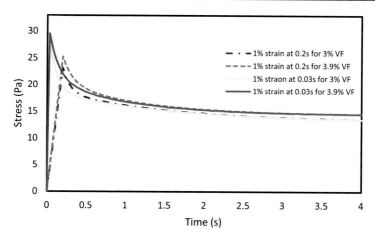

Figure 4.26 Mean volumetric stresses for different volume fractions under equal loads [4].

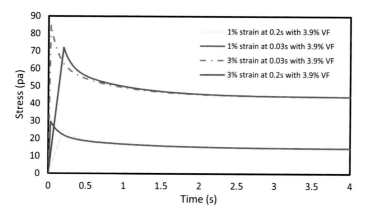

Figure 4.27 RVE with equal volume fractions under different loads in various rates [4].

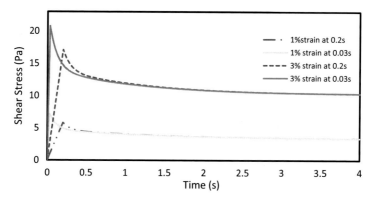

Figure 4.28 Variations of mean volumetric shear stress during the analysis having a 3% NVF under two different strains in two various rates [4].

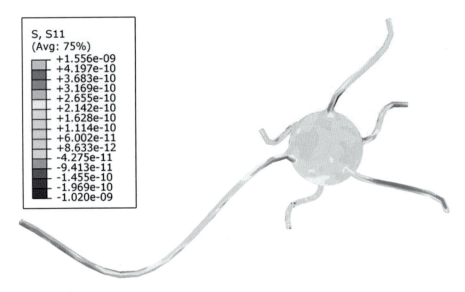

Figure 4.29 The stress distribution in the loading direction of a neuron for 3% strain applied on the RVE (the results should be multiplied by 1e12 Pa) [4].

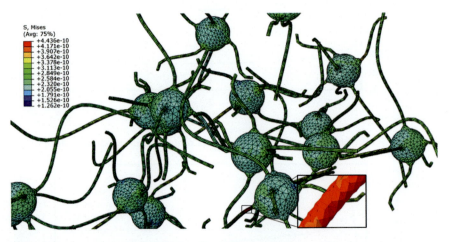

Figure 4.30 Distribution of the stress in neurons of an RVE when 3% strain is applied at 0.2 s on the RVE having 3.9% volume fraction of neurons (the results should be multiplied by 1e12 Pa) [4].

Figs. 4.30 and 4.31 depict the distribution of Von-Mises stress in the neurons of RVE at 0.2 s and homogenized stress of the RVE having an NVF of 3.9%, respectively. RVE is subjected to 3% strain, which is applied at 0.2 s. The maximum mean volume of the stress in Fig. 4.31 is about 71.89 Pa; meanwhile, according to Fig. 4.30 the experienced stress by neurons is between 126.2 and 443.6 Pa at the same time.

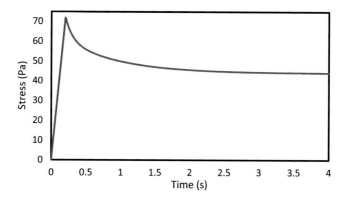

Figure 4.31 The mean volumetric stress of an RVE with 3.9% volume fraction of neurons; 3% strain is applied at 0.2 s [4].

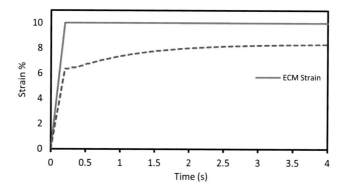

Figure 4.32 Variations of computed strain for neurons and ECM separately [4].

It is obvious that the stresses in ECM are smaller, confirming the previous outcome that in the case of loading on the brain tissue the neurons experience higher stresses compared to the ECM and are more vulnerable in brain injuries. Javid, Rezaei and Karami [34] and Karimi et al. [13] also draw the same conclusion. Moreover, in all experiments the experienced stress in neurons is higher than that in ECM [16], while the experienced strain by neurons is lower than that in ECM [16]. Fig. 4.32 shows the variation of the ECM and neurons' strain, demonstrating that the mean volume strain of the ECM is higher than that of the neurons [4].

4.3 Characterization of the liver tissue

The liver, a crucial organ in vertebrates, is a highly vascularized tissue mainly connected to the portal vein that brings in blood-rich digested nutrients. The blood exchange is made at the cellular level through the liver vascularization between the portal vein to the vena cava [62]. The human body is unable to function without the liver, and currently, there is no transplant option for patients suffering from a long-term liver failure. Minimally invasive surgery (MIS) has provided a breakthrough in surgery by increasing the success rate and patients' comfort while decreasing surgery complications. However, the loss of mechanical feedback in MIS makes the surgeon's work more challenging. A solution is to reintegrate patient-dependent 3D augmented reality with real physical behavior into the surgery room [63,64]. However, the challenge in developing such tools is the need for real-time precise surgical information, typically of the order of a tenth of a second with a precision error below 1 mm. To build such a numerical tool, multiscale and multiphysical numerical models must be developed to account for the nonlinear anisotropic behavior of the liver and integrated within real-time analyses. The exact patient's liver geometry must be integrated using medical imaging techniques such as computed tomography (CT) scan or magnetic resonance imaging (MRI), including a precise quantification of the liver mechanical properties [65−67] and accounting for the internal vascular distributions and possible tumors using noninvasive experiments for in vivo characterization. Then a multilayer macroscopic homogenization technique should be used to account for the effects of major-size vessels in the liver and obtain the corresponding effective mechanical behavior of the liver model [68−71]. Numerical models such as FEM provide accurate results through sophisticated constitutive laws but are incompatible with real-time applications in the surgery room. One way to overcome this challenge is using proper generalized decomposition (PGD) methods to calculate a set of parametric solutions (reduced-order model) with respect to the physical requirements [72]. These methods have been previously used for medical applications [72,73] and provide an FE shape parameterized model with a limited set of parameters allowing image segmentation [74,75]. This type of model has already been implemented within an augmented reality framework at IRCAD (Research Institute for Digestive Apparatus Cancer in Strasbourg, France) in the objective of real medical applications [67]. Although the PGD approach provides a real improvement of the 3D augmented reality framework in real time as a multiscale/multiphysical model through the homogenization of the heterogeneous liver tissue, it considers each constituent as an elastic material. This is a real limitation of the PGD for medical applications and more particularly soft biological tissues having a strongly nonlinear behavior [6].

In this section, we focus on the mechanical homogenization of the liver tissue based on previous published works [74−76] with a novel combined FE−optimization method. Due to the existence of major-size vessels and anisotropy of the surrounding tissue, a hyperelastic FE model of the liver tissue with different sections is computationally expensive and is not practical for surgery. With the new proposed method a homogenized, anisotropic, and hyperelastic model with the nearest response to the real heterogeneous model is created. Then homogenization is done on many samples of heterogeneous models to account for the various vascularization orientations, positions, and sizes. Finally, artificial neural networks (ANNs) as machine learning tools have been trained on this database. Then a sample of heterogeneous material was mapped to its homogenized material parameters [6].

In this study a novel mechanical homogenization method for nonlinear heterogeneous materials has been proposed and tested. It presented accurate results (below 2% relative errors) by comparing the effective overall responses of different heterogeneous and homogenized material segments. Additionally the proposed method is significantly cost-efficient as the homogenized FE processing time is a tiny fraction of the heterogeneous one. Generalizing the proved algorithm of the combined FE−optimization technique required a special application of the ANN as a machine learning method. The training has shown acceptable performance of 0.01% as the relative error. The whole process of homogenization has also been presented as a graphical user interface (GUI) with parallel processed optimization to accelerate the process and facilitate its usage for the medical team operating on the irregular heterogeneous liver tissue. Moreover the established framework of homogenization is readily applicable to other nonlinear heterogeneous materials [6].

4.3.1 Materials and method

The heterogeneity of the liver tissue influences the real-time biomechanical response in augmented reality surgery. The geometrical meshes of the liver and its vascularization, built on organ segmentations, are nonconcordant and required to be merged in order to obtain a full 3D anisotropic structure (see Fig. 4.33). Currently, there is no automated tool to merge nonconcordant traversing meshes. Hence in-house software needs to be developed [70,76]. FE models enable precise mechanical simulations but are incompatible with the real-time application. Hence there is a

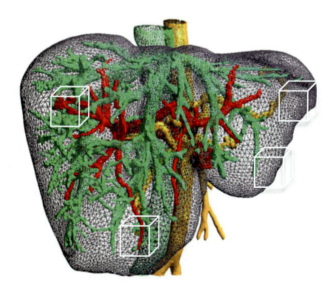

Figure 4.33 FE model of liver geometry including its vascularization (each cube represents a heterogeneous segment) [6,70,76].

need for an "intelligent" homogenized model for patient-dependent geometry and anisotropic mechanical properties to be integrated within a simpler mechanical constitutive law [6].

In this study, we assume that the problem is quasistatic, so no inertial effect is considered. Porosity was not modeled separately, but its effect has been considered within the nonlinear effective mechanical behavior. Blood vessels are assumed to be closed at their ends as blood pressure is assumed to be almost constant within the vascularization and with a negligible effect on the mechanical response of the tissue to external forces. The liver tissue is assumed to be an incompressible material [77,78]. The first objective is to define a suitable homogenization method for the heterogeneous vascularized liver tissue using a nonlinear hyperelastic material constitutive law. The hyperelastic material properties of vessels and their surrounding tissue were extracted from pig liver in the literature [79,80]. Hybrid elements within the ABAQUS FE software were used in FE simulations to satisfy the incompressibility condition in the implicit method. As the PGD method is currently restricted to simple linear elastic behavior, the second objective was replacing the costly homogenization method with a time-efficient machine learning tool to accommodate various geometrical models of heterogeneous tissues such as the liver including its vascularization and trained on many cases to accurately predict the effective mechanical properties for any heterogeneous model [6].

4.3.2 *Homogenization with the combined FE−optimization method*

The expected application lies in minimally invasive surgery, where the known experimental data are the current positions of the surgery tools and organs. Next the FE numerical simulations, especially with the implicit method, provide better and faster convergence when the boundary conditions are displacements. Consequently the mechanical simulations are controlled by the imposed displacement, and outputs are stress components. The homogenization method imposes the same boundary conditions on both heterogeneous and homogenized FE models, and the response stress values of the analyses are being stored. The homogenized properties, used as inputs of an objective function, are determined by an optimization loop aiming for the minimum differences between the average stress components of the homogenized and real heterogeneous models. These average responses were obtained from Eq. (4.21), where is the average operator sign, e is the element number, and *evol* is the deformed volume of the element; thereby the denominator is the total volume of the model [6].

$$\sigma_{ij} = \frac{\sum_{e=1}^{n} \left\{ \sigma_{ij}^{(e)} * evol(e) \right\}}{\sum_{e=1}^{n} evol(e)} \qquad (4.21)$$

The flowchart of this method is shown in Fig. 4.34.

Numerical characterization of tissues

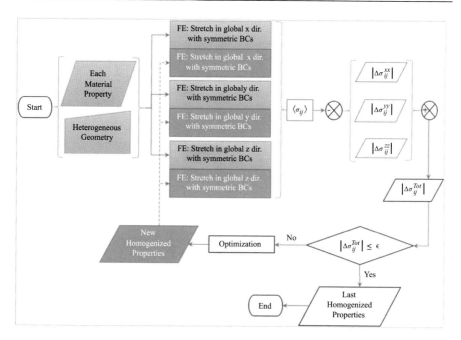

Figure 4.34 Flowchart diagram of the proposed combined FE−optimization method for material homogenization [6].

The optimization method was chosen as the gradient-based multiobjective algorithm (using the MATLAB function "fgoalattain") with the proposed method by [81]. To find the "best" starting point of the optimization loop, trial and error tests were done to reach favorable starting values of homogenized parameters by calculating the objective function of different orders of magnitudes for the input values. Stopping criteria of optimization were defined as relative differences of 2% or less in the normal stress components between homogenized and non-homogenized cases. The reported error for each stress component is the sum of average component errors for three different normal stretch tests simulated on the homogenized and heterogeneous models to minimize the error in all global directions at the same time. For each test the average value is calculated by the numerical trapezoidal integral of errors for different stretch values (from zero to 20%) divided by the stretch span of the tests.

The error is calculated at each analysis time increment by the vector $tt = \{0.0\ 0.1\ 0.2\ 0.3\ 0.4\ 0.5\ 0.6\ 0.7\ 0.8\ 0.9\ 1.0\}$, being a representation of the fraction of stretch in this study. The error calculation of each normal stress component is given in Eq. (4.22). σ_{ii}^{xx} represents the normal stress components for the heterogeneous stretch in the x-direction, while σ_{11}^{xx-sim} with the added superscript of "sim" is the same parameter for the homogeneous model [6].

$$\Delta\sigma_{ii} = \left|\sigma_{ii}^{xx}(tt) - \sigma_{ii}^{xx-sim}(tt)\right| + \left|\sigma_{ii}^{yy}(tt) - \sigma_{ii}^{yy-sim}(tt)\right|$$
$$+ \left|\sigma_{ii}^{zz}(tt) - \sigma_{ii}^{zz-sim}(tt)\right|; \quad \textit{"i" is a fixed index of } 1. 2.\textit{ or } 3 \tag{4.22}$$

Finally, in Eq. (4.23) the average error in the whole stretch interval (unit time) is calculated with the numerical trapezoidal integration on the error values of Eq. (4.22) divided by the respective maximum stress components of the heterogeneous stretch. In Eq. (4.23) the *trapz* function has two arguments: The first one is the domain of integration, and the second one is the integrand [6].

$$Diff(\%) = \left[\frac{trapz(tt.\Delta\sigma_{11})}{\max\{\sigma_{11}^{xx}(tt)\}} \cdot \frac{trapz(tt.\Delta\sigma_{22})}{\max\{\sigma_{22}^{xx}(tt)\}} \cdot \frac{trapz(tt.\Delta\sigma_{33})}{\max\{\sigma_{33}^{xx}(tt)\}} \right] * 100\% \qquad (4.23)$$

4.3.3 A heterogeneous model including vessels and the surrounding tissue

The determination of the homogenized mechanical properties of an arbitrary heterogeneous segment using the combined method of FE−optimization was done on a simplified geometry of a cube representing the surrounding tissue and partitioned with a tubular vessel. Vascularization has been reported to have a size range between 3 to 6 mm [79]. The size of the surrounding cube was defined so that the heterogeneity (tubular vascularization) has a minimum effect on the macroscopic equivalent behavior and with a minimum mesh size. In the first model the vessel was parallel to the global Y-axis passing through the center of the cube (see Fig. 4.35). The diameter varies between 3 and 6 mm within a $50 \times 50 \times 50$ mm cube. According to experimental studies [79,80] on the material characterization of the liver tissue, both the vessel and surrounding tissue materials were assumed to be isotropic, incompressible, and hyperplastic. Also the vessel and its surrounding tissue have been modeled as the first-order Ogden material and first-order

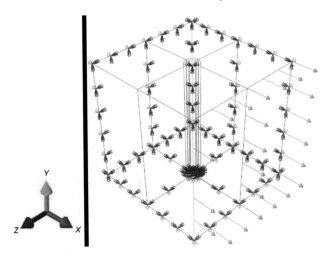

Figure 4.35 Symmetric boundary conditions for stretch simulations on the heterogeneous tissue in the global X-direction [6].

Arruda–Boyce hyperelastic material, respectively [79,80]. The loading in each of global directions was 20% stretch in the global direction of simulation while having symmetric boundary conditions on the three corner planes of the cube for better simulation of real uniaxial testing conditions as illustrated by Fig. 4.35.

An element type of C3D8H (8-node linear, brick, and hybrid element) has been chosen. Mesh sensitivity analysis has been performed, and as a result, 1 mm mesh size has been chosen as an appropriate mesh size (Fig. 4.36). These simulations have been performed on a computer with an Intel Core i7−8550U CPU and 16 GB RAM, resulting in a CPU runtime of 2.7 s for the homogenized model and 12398 s for the vascularized model.

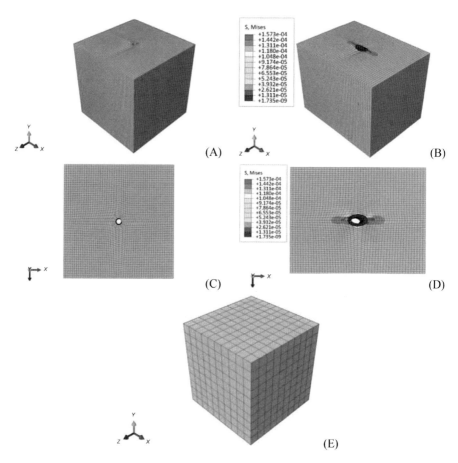

Figure 4.36 (A), (C) Mesh from isometric and XZ-plane viewpoints, respectively, in model 1. (B), (D) Von-Mises stress distribution produced by simulated stretch in the global X-direction in model 1 (number of elements: 121550 C3D8H). (E) Mesh size of 5 mm for the homogenized model (number of elements: 1000 C3D8H) [6].

4.3.4 Homogenized model

The homogenized model is a $50 \times 50 \times 50$ mm^3 cube of one material. A coarse mesh of 5 mm length or a total number of 1000 elements have been chosen due to geometric and BC simplicity as well as mesh independency analysis as shown by Fig. 4.36. Other analysis characteristics are obviously the same as the heterogeneous model to compare their simulation results and finally try to minimize the difference between their responses. The Holzapfel−Gasser−Ogden model for anisotropic hyperelastic materials defined in the material library of ABAQUS software was used to account for the fiber direction (at most three directions are accounted for in their model) [82,83].

This material model was first developed to describe the mechanical response of arterial tissues without certain mechanical, mathematical, and computational shortcomings of phenomenological models. Eq. (4.5) describes the form of the strain energy based on their model for arterial layers with distributed collagen fiber orientations [6].

$$U = C_{10}(\bar{I}_1 - 3) + \frac{1}{D}\left(\frac{(J^{el})^2 - 1}{2} - \ln J^{el}\right) + \frac{k_1}{2k_2}\sum_{\alpha=1}^{N}\left\{\exp\left[k_2\bar{E}_\alpha{}^2\right] - 1\right\}(N \leq 3);$$

$$\bar{E}_\alpha \overset{\text{def}}{=} \kappa(\bar{I}_1 - 3) + (1 - 3\kappa)(\bar{I}_{4(\alpha\alpha)} - 3); \kappa = \frac{1}{4}\int_0^\pi \rho(\theta)\sin^3\theta d\theta$$

$$(4.24)$$

$U, C_{10}, D, k_1, k_2, N, J^{el}, \bar{I}_1,$ and $\bar{I}_{4(\alpha\alpha)}$ are the strain energy per unit of reference volume, four material parameters, the number of fiber families considered for the model, the elastic volume ratio related to thermal expansion, the first deviatoric strain invariant, and a tensor invariant equal to the squared stretch in the direction of the α family of fibers, respectively. The first term of energy form accounts for the incompressible isotropic hyperelasticity by the simple neo-Hookean model, the second term is correlated to possible compressibility (absent in our study based on previous biomechanical models of the liver tissue), and the third one is associated with transversely isotropic free-energy function of one, two, or three families of collagen fibers which cannot support compressive load due to buckling. Moreover, it is assumed that the directions of the collagen fibers within each family are dispersed about a mean preferred direction. Thus κ was defined as the level of distribution in the fiber directions with $\rho(\theta)$ as the orientation density function. $\kappa = 0$ shows perfect alignment (no dispersion), while $\kappa = \frac{1}{3}$ represents an isotropic material (total random distribution).

Since the vessel in the heterogeneous tissue segments plays the role of fibers in the Holzapfel−Gasser−Ogden model, only one fiber direction is considered for our equivalent material model. For modeling convenience the preferred direction for different samples is assumed to be the (0.1.0) global direction. Since the analyses are standard/implicit, the D coefficient, associated with the material compressibility, is set to zero. As a result, inputs of the objective function are naturally the four parameters of $C_{10}, k_1, k_2,$ and κ, which should be positive, non-negative, positive, and non-negative values below one third, respectively, to have physical meaning.

At last an average error of below 2% for each of normal stress components was achieved by the combined method [6].

4.3.5 Generalization of limited homogenized samples with artificial neural networks

The combined method presented in the previous section needs a computationally expensive procedure to yield the homogenized material parameters of specified heterogeneous samples, while the available time for preprocessing of a simulated surgery is limited. Therefore ANNs have been utilized to train networks simulating the pattern of material homogenization for any arbitrary heterogeneous sample with the database of previous homogenization results of the combined method. Consequently an unknown heterogeneous sample with heterogeneity parameters would be mapped to its respective homogenized material parameters by the trained networks. The more homogenized samples the network has, the more accurate its prediction will be for the new unknown heterogeneous samples. Due to randomness in some neuron parameters of ANNs the trained network for a fixed database may slightly change. In this research a two-layer (middle layers, not including the input and output layers) feed-forward network with sigmoid neurons for the first layer and linear ones for the second layer has been used. The hidden layer consists of 10 neurons. The training method is the Levenberg–Marquardt backpropagation algorithm. The inputs of networks are the crossing vertices of the vessel with its surrounding tissue of a cubic shape, $(x1, y1, z1)$ and $(x2, y2, z2)$, and the internal and external diameters of vessel, $d1$ and $d2$, respectively. To have more generalized prediction networks, it is possible to use normalized heterogeneous parameters with respect to the cube dimension. The outputs are clearly the homogenized material parameters or Holzapfel–Gasser–Ogden model parameters of C_{10}, k_1, k_2, and κ. The schematic of the ANNs is shown in Fig. 4.37 [6].

4.3.6 Results and discussion

4.3.6.1 Generating a database of homogenized samples

Forty cases of heterogeneous materials with different geometries (different vessel orientations, positions, and sizes) were simulated using the aforementioned material properties of the vessel and its surrounding tissue to get the resultant effective stress response. For each of these models the runtime ratio of the heterogeneous FE simulations compared to the homogenous ones is about 250 (or higher), showing a

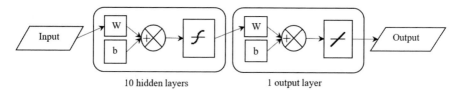

Figure 4.37 A schematic diagram of an ANN with one hidden layer [6].

significant decrease in the processing time of homogenization. In general the vessel direction would be determined by a random unit vector with polar and azimuth angles in spherical coordinates such as that done by Li et al. [84]. According to their study, if the angular spherical coordinates, polar (θ) and azimuth (φ) angles, conform to Eq. (4.25), the respective cylindrical inhomogeneities are distributed randomly [6].

$$\theta = 2\pi\alpha.\varphi = \sin^{-1}(2\beta - 1) \tag{4.25}$$

Models 10 to 15 (6 models) have been constructed with random vessel directions. Models 16–40 (25 models) have the same vessel orientations of models 1–15, although their sizes, that is, internal and external diameters of the vessel, are different in order to extract the size effect influence.

In Fig. 4.38, models 1–15 (15 models) are depicted in the ABAQUS CAE software. It also shows different vessel orientations used to generate an initial database of homogenization as the training samples of the ANNs [6].

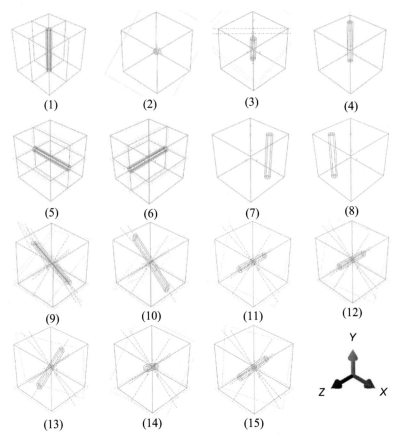

Figure 4.38 Vessel orientation with respect to the cube of heterogeneous models. The number under each picture is the number of associated heterogeneous models [6].

4.3.6.2 Results of homogenization by the combined FE–optimization method

The relative errors of homogenized material responses compared with the heterogeneous ones are below 2%. The input data of this study are the experimental results of previous studies on the material characterization of the liver tissue and vessels, separately [79,80]. Consequently the simulated responses of the heterogeneous tissue models have shown little difference in the tension tests of three global axis directions, which means that the anisotropy of the reference cases is diminutive based on those inputs and a small level of stretch (20%) imposed on the models. Moreover, it can be inferred from Fig. 4.39 showing reaction force-engineering strain curves of a cube consisting of only the liver vessel and one consisting of its surrounding tissue under simulated 1D tension tests in ABAQUS FE software as explained in previous sections. The calculated stress difference between the homogenized model and reference case increases by increasing the vessel to cube volume ratio because the increase of the inhomogeneity volume fraction increases the change in the effective material behavior to match the inclusion material properties. Nevertheless, the efficiency and strength of the proposed novel FE–optimization method of homogenization is clearly proved by the results of less than 2% difference in stress responses.

In Fig. 4.40 the effect of vessel volume fraction on the homogenized material parameters of C_{10} and k_1 are respectively shown along with fitted linear curves. The curve equations with low slope values imply that the level of anisotropy slowly increases with volume fraction of the vessel as C_{10} and k_1 are the material parameters controlling the isotropic and anisotropic terms of the strain energy, respectively, as explained in Section 2.3. k_2 and κ have not shown significant variation throughout all homogenized models. As mentioned in Section 2.3, only one fiber direction is considered for all cases. Therefore the summation in the strain energy can be simplified to one term influenced by three material parameters of k_1, k_2, and κ, which have ultimately enabled the optimization process to reach the final goal by tuning just the most influential parameter, k_1, as the multiplier of anisotropic term of energy.

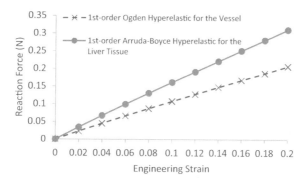

Figure 4.39 Mechanical responses of a cube consisting of only the liver vessel and another one consisting of its surrounding tissue with the material characteristics used in this study resulting from FE-simulated 1D tension tests [6].

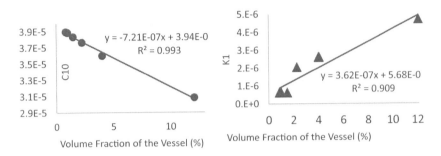

Figure 4.40 Effect of the vessel volume fraction on the homogenized material parameters. Left: C_{10}; Right: k_1 [6].

Nevertheless, the purpose of this study is the introduction of a novel method to define precise effective mechanical properties of heterogeneous living tissues using homogenization of nonlinear heterogeneous materials and ANNs for better accuracy and improved predictability. Therefore it is not necessary to achieve and use these curves for the final application of our study [6].

4.3.6.3 Artificial neural networks

The ANNs were trained by the input—output data gathered from the combined FE—optimization method. Inputs of the networks are the heterogeneous material specifications quantified by the two crossing points of the vessel and the cubic tissue, ($x1.y1.z1$) and ($x2.y2.z2$), as well as the vessel size, internal diameter d1, and external diameter d2. The outputs are the homogenized material parameters C_{10}, k_1, k_2, and κ. A separate network was used for each output parameter to avoid interconnection between all outputs. The outputs were normalized with respect to their greatest ones as unit quantity in the homogenized samples to decrease the resultant relative error. The decrease could be explained by the proportionality of the error with data values which is not considered when normalization is not done. The 40 homogenized samples were partitioned for validation and test purposes in 70% of them for training the ANNs, 15% for validation, and 15% for testing. The performance of the trained networks for normalized C_{10} and k_1 parameters are presented in Fig. 4.41. The vertical axes are the mean square error (MSE) quantities, and the horizontal ones show the epochs passed for training the ANNs. As shown in Fig. 4.41 the mean square errors at the best validation performance for the normalized C_{10} and normalized k_1 are 0.0001 in 2 epochs of training and 0.0001 in 1 epoch training, respectively, which represent acceptable errors of 0.01% with respect to the greatest parameter values. For the other parameters, k_2 and κ, the training stopped at the beginning because the calculated results were almost similar among the homogenized samples [6].

4.3.6.4 Transferring the results into the application

To train ANNs with the homogenized samples a database was generated with various locations, directions, and sizes for the vessel in the cubic heterogeneous segments.

Numerical characterization of tissues

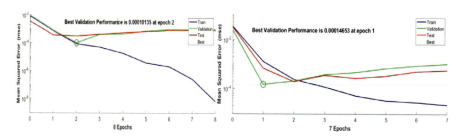

Figure 4.41 Performance curves of training for (left) C_{10} and (right) k_1 homogenized material parameters as the outputs of their ANNs [6].

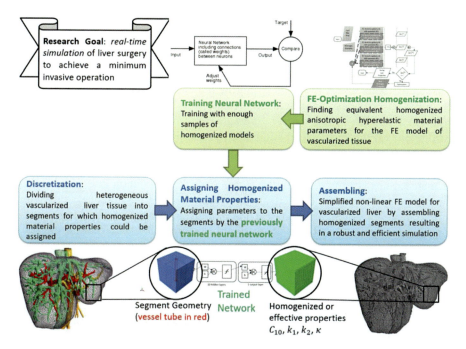

Figure 4.42 The application of our study in the greater research plan for real-time simulation of the liver surgery is shown by green color [6].

The combined method of FE–optimization takes a much greater computational cost than the ANN training and mapping. After training, assigning homogenized material parameters to new heterogeneous segments takes a negligible cost. The green parts of Fig. 4.42 show this study contribution to the greater research plan of real-time simulation of the liver surgery. As shown in the bottom of the figure, our trained network delivers a fast computational tool to homogenize the vascularized heterogeneous liver tissue. A GUI program has also been developed for the end user of our computational framework of homogenization to conveniently homogenize mechanical response of an input sample, optimize homogenized material parameters, map a heterogeneous segment to its effective homogenized material parameters, and train ANNs with a user-defined database of homogenized samples. A screenshot of the program is illustrated in Fig. 4.43 [6].

4.4 Statistical reconstruction and mechanical characterization of the bone microstructure

It is well established in the literature that the interaction between mechanics and biology plays a crucial role in the interpretation of growing tissue behavior [85−87]. This mechanobiology depends on the natural tendency of the biological tissue to develop optimizing behavior to gain proper mechanical strength with the minimum amount of material. For bone adaptation and remodeling, this optimization problem is at the origin of many studies since the early work by Wolff [88], followed by many others [89−93], exploring the internal architecture and external conformation of bones changes in accordance with mathematical laws. With the evolution of the bone microstructure being dependent on cells and microstructure distribution, various mechanical stimuli have been proposed to trigger bone adaptation [81,94−96]. While continuum models, for the description of natural bone regeneration, are widely spread in the scientific literature, scale-related effects of the material are not often integrated in the structural evolution due to its highly heterogeneous nature at the small scales. At these scales the classical continuum theory does not allow to describe the correct material behavior where the

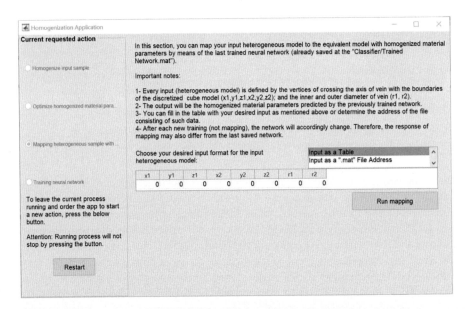

Figure 4.43 A screenshot of our developed GUI application for homogenization [6].

prediction made by Cauchy continuum theory starts being far away from the experimental evidence [63,69,97−99]. Hence adequate homogenization methods [31,100] need to be developed in order to integrate all these local phenomena at the global scale [7].

Historically, bone mechanical behavior is dealt with using classical optimization processes based mainly on the developed mechanical strain energy density through the material [75,76,101−108]. However, to be able to obtain a good predictive numerical model the biological phenomena [109] need to be linked with the surrounding physics [66,110−113] and account for the growing capillaries and nutrient supply [114−117] Lately a new form of mechanobiological stimulus for bone remodeling has emerged integrating coupled biological and mechanical effects accounting for cell migration and the nutrient supply chain [118−120]. Although this allows accounting for the couplings between the mechanics and biological phenomena, it remains a macroscopic approach and requires more insight about the locally occurring mechanobiology [7,121].

Bone microstructure distribution is optimized in order to sustain the externally applied mechanical loads and its evolution changes over the long term as a function of local biological information. However, on a macroscopic scale, where the bone density is characterized only as a homogeneous medium, the influence of this microstructure over the long-term remodeling kinematics is little investigated. Although some numerical models tried to predict the bone remodeling kinematics from known trabecular bone distributions [122−126], these do not provide insight about long-term evolutions. In addition the patient-dependent bone microstructure is very variable [127,128]; this could directly impact the numerical model predictions. Lately, some numerical models have investigated the correlations between the bone stiffness and simplified microstructure distribution [129−133] through multiscale homogenization procedures. However, the bone remodeling processes accounting for these microstructure variabilities at the macroscopic scale are not studied since these works are mostly related to the structural mechanics homogenization process [7].

The goal of this section is to study the influence of the local bone microstructure distribution on the macroscopic bone density evolution over long periods of time. The first step of this study requires an adequate method for bone microstructure reconstruction in order to study large sets of specimens as it is not possible to extract that amount of real bone samples and show that the proposed reconstruction method is able to provide the corresponding results with different structural samples as equivalent effective stresses, strain energy, and stiffness tensor. For this we present here a methodology to study the influence of the bone microstructure distribution on the developed mechanical energy field. Real bone microstructures are extracted from CT scan data and segmented/reconstructed to obtain the corresponding FE numerical model. A statistical method is used to reconstruct several equivalent bone microstructures with identical homogeneous stiffness. FE models are then constructed from these statistical microstructures. The same mechanical loads are applied on the different FE models to extract the variabilities of the local strain energy density that will be at the origin of the bone remodeling kinetics [7].

4.4.1 Specimen preparation

The real bone specimen was extracted from a woman adult's femurs of 85 years age. The adult bone was harvested from the corpse at the Timone Hospital (Marseille, France). The specimen was cut into a parallelepiped rectangle of axial, circumferential, and radial dimensions of 9.88, 6.13, and 2.29 mm, respectively, using a water-cooled low-speed diamond saw (Buehler Isomet 4000, Buehler, Lake Bluff, IL, USA). The length of each specimen was measured with a digital caliper (Absolute digimatik solar, Mitutoyo, Kanagawa, Japan, a measurement error of 0.03 mm). The specimen was stored under vacuum at $-20°C$.

The specimen was imaged at 5.5 μm voxel size using a micro-CT system (Phoenix Nanotom 180 S, General Electric, Germany). The scanning was done with a hydrated bone specimen, immersed in salt water inside a 25 mm inner diameter plastic tube. The axial specimen main axis (i.e., Haversian canal's principal orientation and medullar canal orientation) was aligned with the rotational sample holder axis. Scans were done with a field of view of 2284×2304 pixels, with a maximum voltage source of 90 kV, a current of 120 μA, and a rotation of 360°, providing 2400 projections per scan. The exposure time was 500 Ms with a total scan time of 85 min. The bone density was measured at 92% of the overall specimen volume [7].

4.4.2 Statistical reconstruction of the bone microstructure

The statistical reconstruction approach enables to obtain various microstructures with identical effective stiffness tensors by exploiting a single cut section image through the original bone sample. Bone images can be extracted from CT scan or other experimental techniques.

Following our previous works [134,135], several bone microstructures were generated with identical effective stiffness tensors to evaluate the effect of the geometrical distribution of the bone microstructure on local energy. This enabled to study many different scenarios (different virtual samples) using the full spectrum of two-point correlation function (TPCF) for Eigen microstructures, without the need of extracting real bone samples [7].

The characteristic function (CF) of the heterogeneous microstructure can be written [56] as

$$
\kappa_s^n = \begin{cases} 1 \ if \ s \in phase \ n \\ \ 0 \ otherwise \end{cases}
\tag{4.26}
$$

where n is a specific phase and the s index enumerates the voxel number of the sample.

This method is based on the fact that the material is not a functionally graded material (FGM) and that the statistical information of parallel cut sections is identical.

To reconstruct a 3D RVE based on given 2D cut sections, we introduce TPCFs, being statistical functions that give some information about the

Numerical characterization of tissues

volume fraction, shape, and distribution of the microstructure phases and are correlated with its various features such as effective thermal and mechanical properties [56].

In this research, we use FFT (fast Fourier transform) and the phase recovery algorithm to reconstruct the microstructure [136]. The TPCF of the phases n and n', $C_2^{nn'}$ for an arbitrary vector t, can be defined as

$$C_2^{nn'}(t) = \frac{1}{S} \sum_{s=1}^{S} \kappa_s^n \kappa_{s+t}^{n'} \tag{4.27}$$

where S is the number of all voxels in the RVE and t is an arbitrary vector selected from the RVE. The Fourier transform (FT) of Eq. (4.27) leads to a more convenient form [137] for computation as

$$\hat{K}_m^{nn'} = \mathcal{F}(C_s^{nn'}) = \frac{1}{S} \sum_{t=1}^{S} \sum_{s=1}^{S} \kappa_s^n \kappa_{s+t}^{n'} e^{-2\pi i (t.m)/S} \tag{4.28}$$

using the variable exchange $(t = z - s)$ and assuming a periodic extension of the RVE in all directions, Eq. (4.28) can be written as:

$$\hat{K}_m^{nn'} = \mathcal{F}(C_s^{nn'}) = \frac{1}{S} \sum_{s=1}^{S} \kappa_s^n e^{-2\pi i s.m/S} \sum_{z=1}^{S} \kappa_s^{n'} e^{2\pi i z.m/S} = \frac{1}{S} |\hat{K}_m^n| e^{-i\theta_m^n} |\hat{K}_m^{n'}| e^{i\theta_m^{n'}} \tag{4.29}$$

where $|\hat{K}_m^n|$ and θ_m^n are the module and phase of the FT for the CF, with

$$\hat{K}_m^n = \mathcal{F}(\kappa_s^n) = \sum_{s=1}^{S} \kappa_s^n e^{2\pi i s.m/S} = |\hat{K}_m^n| e^{i\theta_m^n} \tag{4.30}$$

It should be noted that the FT, expressed as $\mathcal{F}(.)$, is calculated over all axes (i.e., 3D FT for $(s_1.s_2.s_3)$).

In Eq. (4.28), if $n = n'$, the TPCF is also called an autocorrelation function, and its FT, using Eq. (4.29), can be obtained as

$$\hat{K}_m^{nn} = \frac{1}{S} |\hat{K}_m^n|^2 \tag{4.31}$$

Eq. (4.31) shows that the FT of the autocorrelation function is related only to the FT module of the CF. If the FT of the autocorrelation functions exists, it is then possible to calculate its modules. Hence for a two-phase microstructure, if the autocorrelation functions are to be known at first, it is possible to reconstruct the microstructure if the phase of the microstructure (i.e., θ_m^n) is determined [see Eq. (4.30)]. The reconstruction of an object based just on its FT modules, called phase recovery [138], is a common method in astronomy and microscopy.

The application of this procedure for microstructure reconstruction was first introduced by Fullwood et al. [137]. The reconstruction procedure using the phase recovery algorithm is as follows [7]:

1. Based on the known autocorrelation functions the module of the CF is calculated using Eq. (4.31). It is called the original amplitude. From this first step a random microstructure is generated.
2. Once the FT of the randomly generated microstructure calculated, its amplitude is replaced by the original amplitude and then the inverse FT is calculated.
3. Constraints of the generated microstructure are applied. If the difference between the FT amplitude of the obtained CF and the original amplitude is below a given threshold, the reconstruction procedure is considered completed and acceptable. Else the obtained CF is defined inadequate for this generated microstructure and the procedure is reset until the difference between the two amplitudes falls under the given threshold.

An improvement of this method was recently proposed by Hasanabadi et al. [134,135] on a new approximation for the calculation of the TPCFs for a 3D RVE. We assume that the TPCFs of an identical vector in a given RVE in two parallel planes are approximately equal. Hence only two perpendicular cut sections (planes) in the RVE are necessary for the TPCF definition and reconstruction process. This assumption is valid for statistically homogenous media [136]. As shown in Fig. 4.44 an arbitrary vector r in space can be decomposed into r_z and r_{xy}. If the TPCFs of two planes, xy and yz, are known, the full set of TPCFs for 3D RVE can be approximated [135] based on Eq. (4.32):

$$C_2^{11}(r) \approx \frac{C_2^{11}(r_{xy})C_2^{11}(r_z)}{v_1} + \frac{(v_1 - C_2^{11}(r_{xy}))(v_1 - C_2^{11}(r_z))}{(1 - v_1)} \qquad (4.32)$$

where v_1 is the volume fraction of phase 1.

The flowchart of the reconstruction procedure is presented in Fig. 4.45. More details of this method can be found in [137].

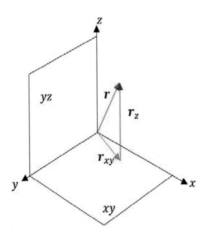

Figure 4.44 Decomposition of the vector r to r_z and r_{xy} [7].

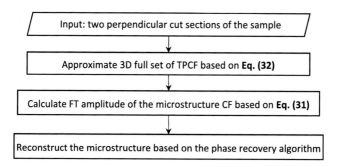

Figure 4.45 Flowchart of the reconstruction procedure [7].

In this approach, TPCFs were extracted using CT images and used to reconstruct microstructures converging to the real ones. The main reasons to choose this reconstruction technique are as follows [7]:

1. A strong contrast expansion shows that the stiffness tensor is linked to the TPCF of the microstructure; therefore the TPCF is a very good selection of the statistical distribution descriptor.
2. In the phase-recovery algorithm, different sets of TPCFs can be employed for the reconstruction procedure, therefore enabling to reconstruct a variety of microstructures with few cut-section images.

An improvement of the reconstruction process is envisaged in future works using TPCFs, but the mathematical framework for this needs to be developed.

4.4.3 Finite element homogenization

The original CT scanned images were cropped into size 351×319 pixels and were imported into the VCAT software (released by V-CAD Program, RIKEN, Japan). The images were then assembled into stacks of 150 images after thresholding and made a 3D RVE of the bone microstructure. Fig. 4.46 shows the 3D RVE for the real bone specimen and the statistical one possessing similar realistic features (size, shape, and distribution) as the real one, being suitable for the calculation of its mechanical properties. Fig. 4.46 also shows the connected network of voids inside the cortical bone. Effective properties were extracted directly from the results of the FE analysis on statistical and real bone microstructres. A direct comparison was made between the real bone sample and statistical reconstructed ones to validate the reconstruction process [7].

In this section the primary goal is to focus on the ability of the statistical reconstruction method to provide "bone like" microstructures and check whether these reconstructed microstructures show similar mechanical characteristics as the real bone. Once the statistical reconstruction method is validated on a number of samples, it is then possible to create virtually an infinite number of microstructures showing different microstructure distributions and bone densities, hence enabling to study an infinite number of "bone samples" without the need of specimen extraction. Although the computation time is an important aspect, this is not the primary

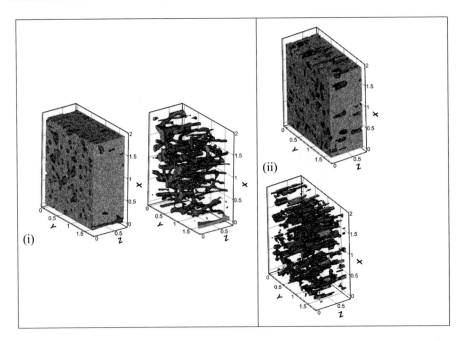

Figure 4.46 Finite element mesh for cortical bone microstructures (Left: Bone phase, Right: void phase): (i) real specimen and (ii) statistically reconstructed specimen from cut (Section 4.2) [7].

focus here as it can be addressed in future studies but rather crucial to validate the quality of the reconstruction method in the first place [7].

The V-CAT software was used with its built-in mesh generator to obtain a high-quality tetrahedral FE mesh, accounting for highly irregular and high-aspect ratio solid and void clusters. This capability enabled to preserve a good quality of the void volume fraction during the meshing process. The mesh simplifier software has been used to reduce the number of elements in the bone part without sacrificing the volume fraction of the void. The total number of elements in the bone part was around 3.5 million tetrahedral elements for all RVEs. The volume fraction of the void phase has been calculated around 8%, which is very close to the original CT scanned data. Six simulations of the mechanical tests (three tensile tests and three shear tests) were carried out on each RVE to find the modulus of elasticity and Poisson's ratios of the RVEs in the X-, Y-, and Z-directions. More details about creating the 3D model, mesh generation, and homogenization process for real and reconstructed microstructures are provided in [7,139].

The internal bone microstructure, either cortical or trabecular, is subjected to the combination of normal and shear loads defined by the external bone geometry and load conditions (muscles, body weight, bone type). However, it is rather difficult to extract quantitative data from such conditions. Hence in order to be able to validate the statistical reconstruction method by comparing developed stresses and elastic strain energy on real and statistical bone microstructures, we conduct a series of simple tensile and shear tests on the real and statistical RVEs using the FE software ABAQUS. Since the bone is a highly heterogeneous material [140,141] and this affects its overall mechanical behavior [142,143], it is necessary to compare the calculated results at the effective macroscopic scale. Based on these tests, stiffness tensors of all RVEs were calculated and influences of the microstructure distributions were quantified [7].

4.4.4 Results and discussion

Five statistical bone microstructures were reconstructed (see the example in Fig. 4.46). From these different cases, including the real bone sample, homogenized macroscopic stiffness coefficients (i.e., Young's modulus and Poisson's ratio) were extracted for the validation of the reconstruction process, assuming that the material (bone solid phase) mechanical behavior was linear elastic with Young's modulus $E = 15$ GPa (average value for the cortical bone) and Poisson's ratio $= 0.3$. The results are presented in Table 4.11. The real bone sample presented values of Young's modulus varying between 11.17 and 12.89 GPa and Poisson's coefficients between 0.228 and 0.301 in the different directions. It is observed that the Young's modulus E_x and E_y are the lowest and E_z is the highest, which is due to the fact that the cortical bone microstructure is orientated along the Haversian canal. Comparing these values with the obtained homogenized coefficients of all statistically reconstructed cases, it was observed that for all of them the obtained Young's moduli are very close to the minimum value obtained for E_y of section 6 at 10.76 GPa and the highest value for E_z of Section 4.2 at 13.326 GPa. The errors developed from the microstructure's variations for the Young's modulus are for the minimum Young's modulus value equal to 3.1%, for the maximum Young's modulus value equal to 3.38%, and for the range min-max equal to 4.39%. For the Poisson's ratio, their variability is checked between 0 and 0.06, which is a very small range. In addition, it was observed that the real sample mechanical properties are mostly situated in the middle of all reconstructed cases which show a good correlation with the different scenarios [7].

From the original bone sample, it was possible to reconstruct different bone microstructures corresponding to different case scenarios (different samples/different virtual patients) with a particularly good accuracy. The reconstruction method allows obtaining an infinite number of virtual cases (patients) from the real bone sample to study their mechanical behavior individually and compare them with the real bone sample. It is also possible to extract tendencies with given scenarios of specific microstructure distributions [7].

Once the statistical reconstruction of different microstructures was validated, both tensile and shear mechanical loads were applied on each of them to compare their mechanical behavior as well as the stresses and energy/energy density distribution fields developed. Each model is constituted of about 3.5 million tetrahedral FE elements in the solid phase. The applied load corresponds to an average standing person body weight (about 750 N) downscaled to the size of the sample surfaces. This leads to small deformation in the sample for the given mechanical properties; hence linear elastic analysis is adequate here. For the sake of lightness of this section, we present

Table 4.11 Modulus of elasticity and Poisson's ratio in different directions for the real and statistical bone microstructures [7].

	E_x (GPa)	E_y (GPa)	E_z (GPa)	xy	xz	yz	yx	zx	zy
Real bone	11.723	11.172	12.894	0.228	0.263	0.262	0.275	0.289	0.301
Cut section 2	12.506	12.064	13.326	0.291	0.272	0.274	0.281	0.29	0.301
Cut section 3	11.456	11.310	12.686	0.281	0.260	0.264	0.277	0.288	0.298
Cut section 4	11.456	11.310	12.713	0.281	0.260	0.264	0.277	0.289	0.297
Cut section 5	11.556	11.229	12.838	0.288	0.259	0.263	0.279	0.288	0.297
Cut section 6	11.185	10.765	12.370	0.282	0.259	0.258	0.271	0.287	0.298

only the results of the real bone sample with two other statistical microstructures built from cut Section 3 and cut Section 5 on the real bone sample [7].

For the tensile test the stresses are plotted in the corresponding loading directions (x) and (z), together with the corresponding strain energy density developed within the microstructures. They are presented in Figs. 4.47 and 4.48, respectively. For the shear load scenario, stresses in the (xy) direction together with strain energy density are presented in Fig. 4.49. Although the effective equivalent mechanical properties are similar for all cases (see Table 4.11), the stresses and strain energy density (SED), presented in Fig. 4.47 to Fig. 4.49, show different distributions for both tensile and shear applied loads. Despite this anisotropic distribution, the amount of developed energy within the samples remains similar (see below) [7].

The interpretations of these results tend to show that [7]

1. The statistical reconstruction method provides a practical method to study any virtual sample from the mechanical behavior point of view as homogenized mechanical variables (stiffness coefficients, stresses, and strain energy) remain very similar. It provides a viable method to extract local microstructure interpretations. More specifically, it shows the equivalent correlation between homogenized bone density measurements (as in MRI information, for example, for the gray scale of the Hue value) and different types of microstructure distributions.
2. The variabilities of the mechanobiological response for each reconstructed case can be evaluated as a function of the distributed strain energy density (SED) and the different microstructure distributions. This in turn can provide long-term variations in the bone density evolution and can show the influence of each of the different microstructures on the bone density kinetics.

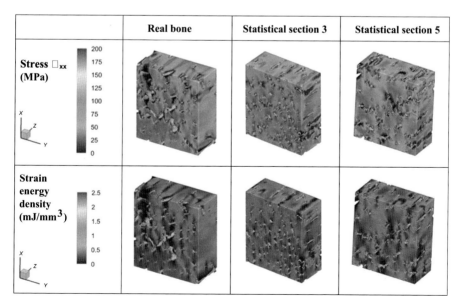

Figure 4.47 Developed stresses and strain energy density in the real bone and statistical samples (sections 3 and 5) under tensile load applied in the x-direction [7].

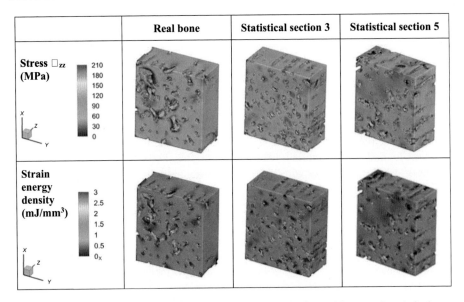

Figure 4.48 Developed stresses and strain energy density in the real bone and statistical samples (sections 3 and 5) under tensile load applied in the z-direction [7].

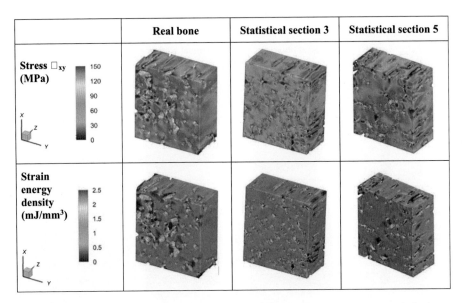

Figure 4.49 Developed stresses and strain energy density in the real bone and statistical samples (sections 3 and 5) under shear load applied in the xy-direction [7].

To evaluate the potential bone density evolution as a function of the applied mechanical loads the total developed strain energy inside each sample for the different cases (real sample and five reconstructed cases) under tensile and shear load conditions is presented in Fig. 4.50.

Small variations of the developed strain energy are observed for each of the different load scenarios. For tensile loads the strain energy varies between a minimum of 1.485 mJ/mm^3 up to 1.786 mJ/mm^3. For shear loads, it varies from a minimum of 2.284 mJ/mm^3 up to 2.622 mJ/mm^3. For all cases the real bone sample is always around the mid-value of the statistical cases in the same way as was observed for the stiffness coefficients. The strain energy variation for all cases is within 3% to 6% compared to the real bone sample showing the reliability of the statistical reconstruction method. The developed strain energy in the z-direction is slightly higher than in the x- and y-directions for tensile load conditions. This is in agreement with the higher stiffness of the z-direction reported in Table 4.11. However, interestingly, shear load cases show significantly higher strain energy than simple tensile load conditions. The average energy for shear load has increased by 53% compared to tensile load conditions. This needs to be investigated in future work with a more exhaustive screening of different bone microstructures in order to extract the coupling effects between shear and tensile load conditions [7].

Nevertheless, as bone cell activation will depend on the developed mechanical energy within the structure, we can assume that for different types of bone microstructures, this coupling effect (between shear and tensile loads) will have an impact on the bone remodeling process. If shear stresses show to be predominant, it could be correlated to the fact that cells are more sensitive to shear load conditions than hydrostatic pressure [144], hence favoring shear loads for optimized bone reconstruction [7].

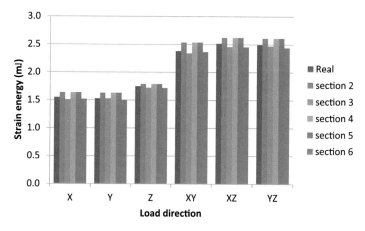

Figure 4.50 Total developed strain energy for the different bone microstructures (X, Y, and Z indicate tensile tests, and XY, XZ, and YZ indicate shear tests) [7].

To correlate the viability of the reconstructed microstructures to the real bone sample, probability distribution functions (PDFs) were extracted. It represents the number of elements in the sample having a specific strain energy and is plotted as an element count versus strain energy divided by the total number of elements in the model. Fig. 4.51 shows the average PDF value of the statistically reconstructed cases with the real bone PDF distribution. Statistical variations are small, and very good agreement is observed between the reconstructed case mechanical response and the real sample [7].

Since mechanical load type, load direction, and microstructure distribution impact directly the bone density evolution, the strain energy density for all the elements in each RVE were extracted. The PDFs of the strain energy density for each load case and microstructure are presented in Fig. 4.52.

For each load case the PDF of the real bone is plotted in red. We observe that not only the overall mechanical response of the statistically reconstructed microstructures is similar to the one of the real bone but also the distributions of strain energy densities are similar. The variabilities observed in these results are based on the fact that a 3D stochastic reconstruction technique has been used for reconstruction of the bone microstructure. The proposed method includes two steps: (1) A full set of TPCFs is approximated using correlation functions being extracted from two perpendicular images, and (2) the full set of approximated TPCFs are used to reconstruct microstructures using a phase recovery algorithm. Hence the accuracy of the reconstructing technique can directly be related to the approximation of the full set of TPCF. In our previous study [145], it was shown that for anisotropic microstructures the error of the estimated TPCF and consequently the error of the reconstruction procedure increased in the direction with high anisotropic distribution of microstructures. In addition the reference image used for the reconstruction is the source of errors as it depends on the approximated TPCF in the direction perpendicular to the microstructure distribution of the reference image [145]. Nevertheless, accounting for the anisotropic complexity of the trabecular bone microstructures the differences

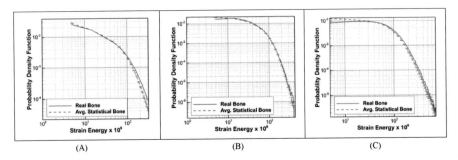

Figure 4.51 Probability density function for the *YY* (left) and *ZZ* (middle) tensile cases and *XY* (right) shear case as a function of the developed local strain energy (mJ/mm^3) [7].

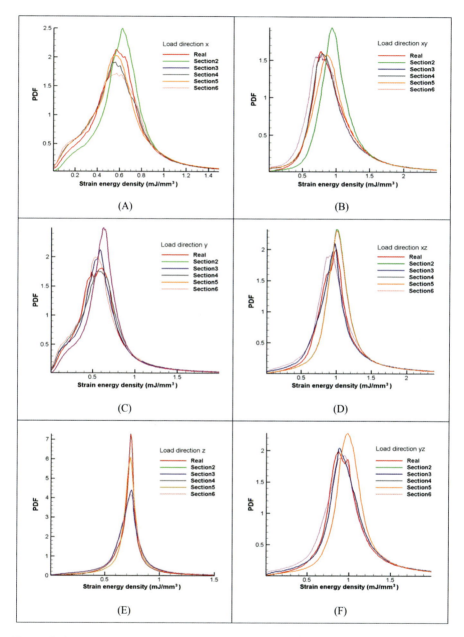

Figure 4.52 Normalized probability distribution function of strain energy density for the different microstructures with different mechanical load directions [7]. (A) Loading x direction, (B) loading xy direction, (C) loading y direction, (D) loading xz direction, (E) loading z direction, (F) loading yz direction.

obtained from the reconstruction method remain small, as presented in Table 4.12. Comparison of these results with those in Fig. 4.52 shows that [7]

1. All strain energies over the whole sample for the statistical cases remain lower than 10% variations compared with the real bone sample. For simple tensile load cases, variations were between 0.2% to 4.5%, showing the influence of the anisotropic microstructure distribution over the reconstruction method [145]. For shear load cases the differences were up to 9.4% in the yz-direction, showing in addition to the statistical error the sensitivity of the shear load scenario on the microstructure distribution, as presented in Fig. 4.50.
2. For the standard deviation variations, they are not correlated to the strain energy variations as the mean strain energy does not account for the individual microstructure distributions. However, we can still observe that for tensile cases, the z-direction shows a higher difference due to the stiffness difference with other directions and reconstruction method [145]. This does not appear in the shear load case as it is a tangential load. The full width at half maximum (FWHM) deviations of PDF for tensile load remain within the 10% variability. For the z-direction, it is almost identical.

It appears from Table 4.11, Fig. 4.50 to Fig. 4.52, and Table 4.12 that the statistical reconstruction method enables to study virtually an unlimited number of samples (bone-like microstructures) as the calculated effective mechanical properties are remarkably similar to the one of the real bone. Some improvements remain to be done as the distribution of the strain energy shows variabilities over the statistical microstructures (up to 10%) due to the abilities of the reconstruction method to best adapt to the real bone microstructure. These variations need to be evaluated with their respective impact on the bone density evolution over long periods of time in future works [7].

As bone remodeling occurs on internal surfaces of the bone where osteoblasts and osteoclasts are present, it is necessary to look at the developed strain energy densities on these surfaces. These are presented in Fig. 4.53 for three cases: (1) real bone, (2) cut section 3, and (3) cut section 5 for the mechanical load conditions of simple tension in x-, y-, and z-directions and shear xy. In addition the PDF functions versus strain energy density are plotted for the elements on the border. The tail of the PDF plots beyond a given energy is not shown in the figures as their PDF values are exceedingly small [7].

Table 4.12 Differences between the real bone sample and average of statistical cases for (1) strain energy density variation in different directions under tensile and shear load conditions (X, Y, and Z indicate tensile test, and XY, XZ, and YZ indicate shear tests) and (2) standard deviation difference at FWHM as a representative of the variation of the energy density of each element compared to the mean value [7].

	Tension X	Tension Y	Tension Z	Shear XY	Shear XZ	Shear YZ
Mean strain energy density variation (mJ/mm^3)	0.0186 (1.8%)	0.0449 (4.5%)	0.0021 (0.2%)	0.0699 (7%)	0.0138 (1.38%)	0.0942 (9.4%)
Standard deviation difference at FWHM	11%	10%	42%	6%	11%	7%

Numerical characterization of tissues

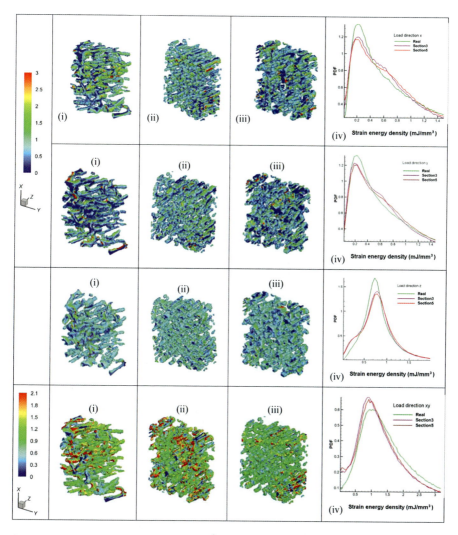

Figure 4.53 Strain energy density (mJ/mm^3) on elements surrounding the void space for mechanical load applied in the X-, Y-, Z-, and XY-directions: (i) real bone, (ii) cut section 3, (iii) cut section 5, and (iv) probability distribution function for the three cases [7].

The presented results highlight the geometrical localization where bone remodeling is more likely to occur as osteoblasts and osteoclasts (respectively responsible of construction and resorption) are located on the bone surface. Like previous results the maximum distributed strain energy density is for the shear load scenario with an average value around 1 mJ/mm^3, whereas for tensile loads the average values are around 0.2 mJ/mm^3 for x and y tensile loads and 0.7 mJ/mm^3 for z tensile load. It appears again that tensile loads provide similar results compared to the stiffness coefficient proportions. We also observe that the real bone sample shows mostly a higher strain energy density for tensile loads but lower for shear load. However, statistical samples are very close to each other. This could be correlated with the reconstruction method as explained above [145].

However, as far as the mechanical properties are concerned a particularly good correlation is observed between the statistical cases and real bone, which therefore provides good confidence in the statistical distribution of the developed strain energy. The impact of the observed differences needs to be quantified with regard to the biological distribution of cells on the developed surface microstructures and cell activation thresholds for bone remodeling. This will be done in future works [7].

All statistically reconstructed samples had an identical effective bone density compared to the real bone sample and showed similar mechanical properties. The local stress, strain, and strain energy were calculated for all samples under simple mechanical load conditions. Local variations were observed as a function of the different reconstructed microstructures and linked to effective values. Good correlation was observed for all cases [7].

The statistical reconstruction method enables us to create virtual bones, extract the variations of the local strain energy distribution as a function of the bone microstructure, and compare with the macroscopic one. From this, bone remodeling could be incurred as a function of the local cell activation processes on the bone inner free surface. This local bone remodeling could therefore be linked to the effective bone density. The knowledge of the anisotropic stiffness tensor could be then used as a complementary indicator of the long-term evolution of bone density. This needs to be validated in future work [7].

4.5 Conclusion and remarks

In this chapter, different types of complex microstructures have been generated to calculate mechanical properties of tissues. A 3D geometrical model has been developed to realize a realistic model for the cerebral cortex tissue, and an FE approach has been implemented to homogenize effective mechanical properties.

The bone microstructure experimentally and statistically has been reconstructed to calculate effective mechanical properties. Real microstructures are extracted from CT scan data and stitched to obtain a 3D reconstructed model. A statistical approach has been used to reconstruct several bone microstructures.

In the other part of this chapter a framework has been established to homogenize nonlinear heterogeneous materials based on the ANN as a machine learning method. The homogenization approach is readily applicable to nonlinear biological materials like tissues.

References

[1] Cowin SC. Structural changes in living tissues. Meccanica 1999;34(5):379−98. Available from: https://doi.org/10.1023/a:1004777501507.

[2] Ambrosi D, Beloussov LV, Ciarletta P. Mechanobiology and morphogenesis in living matter: a survey. Meccanica 2017;52(14):3371−87. Available from: https://doi.org/10.1007/s11012-017-0627-z.

[3] Shahsavari H, Jokar H, Haghighi-yazdi M, Baghani M, Remond Y, George D, et al. Micromechanical modeling of the effective mechanical behavior of cerebral cortex tissue. Iran J Sci Technol Trans Mech Eng 2020;44(2):273−85.

[4] Kazempour M, Baniassadi M, Shahsavari H, Remond Y, Baghani M. Homogenization of heterogeneous brain tissue under quasi-static loading: a visco-hyperelastic model of a 3D RVE. Biomech Model Mechanobiol 2019;18(4):969−81. Available from: https://doi.org/10.1007/s10237-019-01124-6.

[5] Kazempour M, Kazempour A, Baniassadi M, Remond Y, Baghani M. Numerical investigation of axonal damage for regular and irregular axonal distributions. Front Mech Eng 2021;7. Available from: https://doi.org/10.3389/fmech.2021.685519.

[6] Hashemi MS, Baniassadi M, Baghani M, George D, Remond Y, Sheidaei A. A novel machine learning based computational framework for homogenization of heterogeneous soft materials: application to liver tissue. Biomech Model Mechanobiol 2020; 19(3):1131−42. Available from: https://doi.org/10.1007/s10237-019-01274-7.

[7] Sheidaei A, Kazempour M, Hasanabadi A, Nosouhi F, Pithioux M, Baniassadi M, et al. Influence of bone microstructure distribution on developed mechanical energy for bone remodeling using a statistical reconstruction method. Mathematics Mech Solids 2019;24(10):3027−41. Available from: https://doi.org/10.1177/1081286519828418.

[8] Faul M, Xu L, Wald MM, Coronado V, Dellinger AM. Traumatic brain injury in the United States: national estimates of prevalence and incidence, 2002−2006. Injury Prev 2011;16(Supplement 1):A268-A. Available from: https://doi.org/10.1136/ip.2010.029215.951.

[9] Park E, Bell JD, Baker AJ. Traumatic brain injury: can the consequences be stopped? CMAJ 2008;178(9):1163−70. Available from: https://doi.org/10.1503/cmaj.080282.

[10] Bessa MA, Glowacki P, Houlder M. Bayesian machine learning in metamaterial design: fragile becomes supercompressible. Adv Mater 2019;31(48):1904845. Available from: https://doi.org/10.1002/adma.201904845.

[11] Peter SJ, Mofrad MR. Computational modeling of axonal microtubule bundles under tension. Biophys J 2012;102(4):749−57. Available from: https://doi.org/10.1016/j.bpj.2011.11.4024.

[12] Laksari K, Shafieian M, Darvish K. Constitutive model for brain tissue under finite compression. J Biomech 2012;45(4):642−6. Available from: https://doi.org/10.1016/j.jbiomech.2011.12.023.

[13] Karami G, Grundman N, Abolfathi N, Naik A, Ziejewski M. A micromechanical hyperelastic modeling of brain white matter under large deformation. J Mech Behav Biomed Mater 2009;2(3):243−54. Available from: https://doi.org/10.1016/j.jmbbm.2008.08.003.

[14] van Dommelen JA, van der Sande TP, Hrapko M, Peters GW. Mechanical properties of brain tissue by indentation: interregional variation. J Mech Behav Biomed Mater 2010;3(2):158−66. Available from: https://doi.org/10.1016/j.jmbbm.2009.09.001.

[15] Velardi F, Fraternali F, Angelillo M. Anisotropic constitutive equations and experimental tensile behavior of brain tissue. Biomech Model Mechanobiol 2006;5(1):53−61. Available from: https://doi.org/10.1007/s10237-005-0007-9.

[16] Abolfathi N, Naik A, Sotudeh Chafi M, Karami G, Ziejewski M. A micromechanical procedure for modelling the anisotropic mechanical properties of brain white matter. Comput Methods Biomech Biomed Engin 2009;12(3):249−62. Available from: https://doi.org/10.1080/10255840903097871.

[17] Rashid B, Destrade M, Gilchrist MD. Inhomogeneous deformation of brain tissue during tension tests. Computat Mater Sci 2012;64:295−300. Available from: https://doi.org/10.1016/j.commatsci.2012.05.030.

[18] Feng Y, Okamoto RJ, Namani R, Genin GM, Bayly PV. Measurements of mechanical anisotropy in brain tissue and implications for transversely isotropic material models of white matter. J Mech Behav Biomed Mater 2013;23:117−32. Available from: https://doi.org/10.1016/j.jmbbm.2013.04.007.

[19] Bain AC, Meaney DF. Tissue-level thresholds for axonal damage in an experimental model of central nervous system white matter injury. J Biomech Eng 2000;122 (6):615−22. Available from: https://doi.org/10.1115/1.1324667.

[20] Bain AC, Shreiber DI, Meaney DF. Modeling of microstructural kinematics during simple elongation of central nervous system tissue. J Biomech Eng 2003;125(6):798−804. Available from: https://doi.org/10.1115/1.1632627.

[21] Pfister BJ, Iwata A, Taylor AG, Wolf JA, Meaney DF, Smith DH. Development of transplantable nervous tissue constructs comprised of stretch-grown axons. J Neurosci Methods 2006;153(1):95−103. Available from: https://doi.org/10.1016/j.jneumeth.2005.10.012.

[22] Valdez M, Balachandran B. Longitudinal nonlinear wave propagation through soft tissue. J Mech Behav Biomed Mater 2013;20:192−208. Available from: https://doi.org/10.1016/j.jmbbm.2013.01.002.

[23] Chatelin S, Deck C, Willinger R. An anisotropic viscous hyperelastic constitutive law for brain material finite-element modeling. J Biorheol 2012;27(1−2):26−37. Available from: https://doi.org/10.1007/s12573-012-0055-6.

[24] Arbogast KB, Margulies SS. A fiber-reinforced composite model of the viscoelastic behavior of the brainstem in shear. J Biomech 1999;32(8):865−70. Available from: https://doi.org/10.1016/s0021-9290(99)00042-1.

[25] Hashin ZVI. Viscoelastic fiber reinforced materials. AIAA J 1966;4(8):1411−17. Available from: https://doi.org/10.2514/3.3686.

[26] Ning X, Zhu Q, Lanir Y, Margulies SS. A transversely isotropic viscoelastic constitutive equation for brainstem undergoing finite deformation. J Biomech Eng 2006;128 (6):925−33. Available from: https://doi.org/10.1115/1.2354208.

[27] Cloots RJ, Gervaise HM, van Dommelen JA, Geers MG. Biomechanics of traumatic brain injury: influences of the morphologic heterogeneities of the cerebral cortex. Ann Biomed Eng 2008;36(7):1203−15. Available from: https://doi.org/10.1007/s10439-008-9510-3.

[28] Eshelby JD. The determination of the elastic field of an ellipsoidal inclusion, and related problems. Proc R Soc Lond Ser A Math Phys Sci 1957;241(1226):376−96. Available from: https://doi.org/10.2307/100095.

[29] Pan Y, Shreiber DI, Pelegri AA. A transition model for finite element simulation of kinematics of central nervous system white matter. IEEE Trans Biomed Eng 2011;58 (12):3443−6. Available from: https://doi.org/10.1109/TBME.2011.2163189.

[30] Rémond Y. Homogenization of reconstructed RVE. In: Rémond Y, Ahzi S, Baniassadi M, Garmestani H, editors. Applied RVE reconstruction and homogenization of heterogeneous materials. Hoboken: Wiley; 2016.

[31] Rémond Y, Ahzi S, Baniassadi M, Garmestani H. Applied RVE reconstruction and homogenization of heterogeneous materials. Hoboken: Wiley; 2016.

[32] Cloots RJ, van Dommelen JA, Nyberg T, Kleiven S, Geers MG. Micromechanics of diffuse axonal injury: influence of axonal orientation and anisotropy. Biomech Model

Mechanobiol 2011;10(3):413−22. Available from: https://doi.org/10.1007/s10237-010-0243-5.

[33] Garnich MR, Karami G. Finite element micromechanics for stiffness and strength of wavy fiber composites. J Compos Mater 2016;38(4):273−92. Available from: https://doi.org/10.1177/0021998304039270.

[34] Javid S, Rezaei A, Karami G. A micromechanical procedure for viscoelastic characterization of the axons and ECM of the brainstem. J Mech Behav Biomed Mater 2014;30:290−9. Available from: https://doi.org/10.1016/j.jmbbm.2013.11.010.

[35] Nemat-Nasser S, Hori M. Micromechanics: overall properties of heterogeneous materials. Amsterdam: Elsevier; 2013.

[36] Tran A, Tran H. Data-driven high-fidelity 2D microstructure reconstruction via nonlocal patch-based image inpainting. Acta Mater 2019;178:207−18. Available from: https://doi.org/10.1016/j.actamat.2019.08.007.

[37] Cloots RJ, van Dommelen JA, Kleiven S, Geers MG. Multi-scale mechanics of traumatic brain injury: predicting axonal strains from head loads. Biomech Model Mechanobiol 2013;12(1):137−50. Available from: https://doi.org/10.1007/s10237-012-0387-6.

[38] Yang M, Nagarajan A, Liang B, Soghrati S. New algorithms for virtual reconstruction of heterogeneous microstructures. Comput Methods Appl Mech Eng 2018;338:275−98. Available from: https://doi.org/10.1016/j.cma.2018.04.030.

[39] Braitenberg V, Schüz A. Cortex: statistics and geometry of neuronal connectivity. Berlin: Springer; 2013.

[40] Ferrant M, et al. Registration of 3D intraoperative MR images of the brain using a finite element biomechanical model. Medical image computing and computer-assisted intervention−MICCAI. Springer; 2000.

[41] Meaney DF. Relationship between structural modeling and hyperelastic material behavior: application to CNS white matter. Biomech Model Mechanobiol 2003;1(4):279−93. Available from: https://doi.org/10.1007/s10237-002-0020-1.

[42] Miller K, Chinzei K. Constitutive modelling of brain tissue: experiment and theory. J Biomech 1997;30(11−12):1115−21. Available from: https://doi.org/10.1016/s0021-9290(97)00092-4.

[43] Škrinjar O, et al. Steps toward a stereo-camera-guided biomechanical model for brain shift compensation. Information processing in medical imaging. Springer; 2001.

[44] Wang HC, Wineman AS. A mathematical model for the determination of viscoelastic behavior of brain in vivo—I oscillatory response. J Biomech 1972;5(5):431−46. Available from: https://doi.org/10.1016/0021-9290(72)90002-4.

[45] Shaoning S. (2014) Mechanical characterization and modeling of polymer/clay nanocomposites.

[46] Mazrouei M. Evaluating the effect of mechanical loading on the effective thermal conductivity of carbon nanotube reinforced polymers (a Monte-Carlo approach). Studies. 2014;22.

[47] Pierard O, Friebel C, Doghri I. Mean-field homogenization of multi-phase thermo-elastic composites: a general framework and its validation. Compos Sci Technol 2004;64 (10−11):1587−603. Available from: https://doi.org/10.1016/j.compscitech.2003.11.009.

[48] Holzapfel GA. Nonlinear solid mechanics: a continuum approach for engineering. New York: Wiley; 2000.

[49] Miller K. Constitutive model of brain tissue suitable for finite element analysis of surgical procedures. J Biomech 1999;32(5):531−7. Available from: https://doi.org/10.1016/s0021-9290(99)00010-x.

[50] Kyriacou SK, Mohamed A, Miller K, Neff S. Brain mechanics For neurosurgery: modeling issues. Biomech Model Mechanobiol 2002;1(2):151−64. Available from: https://doi.org/10.1007/s10237-002-0013-0.

[51] Couper Z, Albermani F. Infant brain subjected to oscillatory loading: material differentiation, properties, and interface conditions. Biomech Model Mechanobiol 2008;7 (2):105−25. Available from: https://doi.org/10.1007/s10237-007-0079-9.

[52] Bergstrom J. Constitutive modeling of the large strain time-dependent behavior of elastomers. J Mech Phys Solids 1998;46(5):931−54. Available from: https://doi.org/10.1016/s0022-5096(97)00075-6.

[53] Budday S, Sommer G, Birkl C, Langkammer C, Haybaeck J, Kohnert J, et al. Mechanical characterization of human brain tissue. Acta Biomater 2017;48:319−40. Available from: https://doi.org/10.1016/j.actbio.2016.10.036.

[54] Chavoshnejad P, More S, Razavi MJ. From surface microrelief to big wrinkles in skin: a mechanical in-silico model. Extreme Mech Lett 2020;100647.

[55] Mura T. Micromechanics of defects in solids. Berlin: Springer; 2013.

[56] Torquato S. Random heterogeneous materials: microstructure and macroscopic properties. Berlin: Springer; 2013.

[57] Dréo J. Metaheuristics for hard optimization: methods and case studies. Berlin: Springer; 2006.

[58] Poli R. An analysis of publications on particle swarm optimization applications. Essex: Department of Computer Science, University of Essex;; 2007.

[59] Koser DE, Moeendarbary E, Hanne J, Kuerten S, Franze K. CNS cell distribution and axon orientation determine local spinal cord mechanical properties. Biophys J 2015;108 (9):2137−47. Available from: https://doi.org/10.1016/j.bpj.2015.03.039.

[60] Pervin F, Chen WW. Dynamic mechanical response of bovine gray matter and white matter brain tissues under compression. J Biomech 2009;42(6):731−5. Available from: https://doi.org/10.1016/j.jbiomech.2009.01.023.

[61] Rashid B, Destrade M, Gilchrist MD. Mechanical characterization of brain tissue in tension at dynamic strain rates. J Mech Behav Biomed Mater 2014;33:43−54. Available from: https://doi.org/10.1016/j.jmbbm.2012.07.015.

[62] Bonfiglio A, Leungchavaphongse K, Repetto R, Siggers JH. Mathematical modeling of the circulation in the liver lobule. J Biomech Eng 2010;132(11):111011. Available from: https://doi.org/10.1115/1.4002563.

[63] Placidi L, Andreaus U, Della Corte A, Lekszycki T. Gedanken experiments for the determination of two-dimensional linear second gradient elasticity coefficients. Z für Angew Mathematik und Phys 2015;66(6):3699−725.

[64] Hostettler A, Nicolau SA, Soler L, Remond Y. Towards an accurate real-time simulation of internal organ motions during free breathing from skin motion tracking and an a priori knowledge of the diaphragm motion. Int J Comput Assist Radiol Surg 2007;2.

[65] Hostettler A, George D, Remond Y, Nicolau SA, Soler L, Marescaux J. Bulk modulus and volume variation measurement of the liver and the kidneys in vivo using abdominal kinetics during free breathing. Comput Methods Prog Biomed 2010;100(2):149−57. Available from: https://doi.org/10.1016/j.cmpb.2010.03.003.

[66] Lemaire T, Kaiser J, Naili S, Sansalone V. Three-scale multiphysics modeling of transport phenomena within cortical bone. Math Probl Eng 2015;2015:1−10. Available from: https://doi.org/10.1155/2015/398970.

[67] Kugler M, Hostettler A, Soler L, Borzacchiello D, Chinesta F, George D, et al. Numerical simulation and identification of macroscopic vascularised liver

behaviour: Case of indentation tests. Biomed Mater Eng 2017;28(s1):S107−11. Available from: https://doi.org/10.3233/BME-171631.

[68] Abdel Rahman R, George D, Baumgartner D, Nierenberger M, Rémond Y, Ahzi S. An asymptotic method for the prediction of the anisotropic effective elastic properties of the cortical vein: superior sagittal sinus junction embedded within a homogenized cell element. J Mech Mater Struct 2012;7(6):593−611. Available from: https://doi.org/10.2140/jomms.2012.7.593.

[69] dell'Isola F, Andreaus U, Placidi L. At the origins and in the vanguard of peridynamics, non-local and higher-gradient continuum mechanics: an underestimated and still topical contribution of Gabrio Piola. Mathematics Mech Solids 2014;20(8):887−928. Available from: https://doi.org/10.1177/1081286513509811.

[70] Kugler M, Hostettler A, Soler L, Remond Y, George D. A new algorithm for volume mesh refinement on merging geometries: application to liver and vascularisation. J Computat Appl Mathematics 2018;330:429−40. Available from: https://doi.org/10.1016/j.cam.2017.09.012.

[71] George D, Baniassadi M, Hoarau Y, Kugler M, Rémond Y. Influence of the liver vascular distribution on its overall mechanical behavior: A first approach to multiscale fluid-structure homogenization. J Cell Immunotherapy 2018;4(1):35−7. Available from: https://doi.org/10.1016/j.jocit.2018.09.008.

[72] Chinesta F, Keunings R, Leygue A. The proper generalized decomposition for advanced numerical simulations: a primer. Cham: Springer; 2013.

[73] Cueto E, Chinesta F. Real time simulation for computational surgery: a review. Adv Modeling Simul Eng Sci 2014;1(1):11. Available from: https://doi.org/10.1186/2213-7467-1-11.

[74] Lauzeral N, Borzacchiello D, Kugler M, George D, Remond Y, Hostettler A, et al. A model order reduction approach to create patient-specific mechanical models of human liver in computational medicine applications. Comput Methods Prog Biomed 2019;170:95−106. Available from: https://doi.org/10.1016/j.cmpb.2019.01.003.

[75] Spingarn C, Wagner D, Remond Y, George D. Multiphysics of bone remodeling: a 2D mesoscale activation simulation. Biomed Mater Eng 2017;28(s1):S153−8. Available from: https://doi.org/10.3233/BME-171636.

[76] Madeo A, George D, Lekszycki T, Nierenberger M, Rémond Y. A second gradient continuum model accounting for some effects of micro-structure on reconstructed bone remodelling. Comptes Rendus Mécanique 2012;340(8):575−89. Available from: https://doi.org/10.1016/j.crme.2012.05.003.

[77] Marchesseau S, Chatelin S, Delingette H. Non linear biomechanical model of the liver. In: Payan Y, Ohayon J, editors. Biomechanics of living organs. Amsterdam: Elsevier; 2017.

[78] Sparks JL. Liver tissue engineering. In: Hasan A, editor. Tissue engineering for artificial organs: regenerative medicine, smart diagnostics and personalized medicine. Weinheim: Co. KGaA; 2017.

[79] Umale S, Chatelin S, Bourdet N, Deck C, Diana M, Dhumane P, et al. Experimental in vitro mechanical characterization of porcine Glisson's capsule and hepatic veins. J Biomech 2011;44(9):1678−83. Available from: https://doi.org/10.1016/j.jbiomech.2011.03.029.

[80] Kerdok AE, Ottensmeyer MP, Howe RD. Effects of perfusion on the viscoelastic characteristics of liver. J Biomech 2006;39(12):2221−31. Available from: https://doi.org/10.1016/j.jbiomech.2005.07.005.

[81] Hegedus DH, Cowin SC. Bone remodeling II: small strain adaptive elasticity. J Elast 1976;6(4):337−52. Available from: https://doi.org/10.1007/bf00040896.

[82] Holzapfel GA, Gasser TC, Ogden RW. A new constitutive framework for arterial wall mechanics and a comparative study of material models. J Elast Phys Sci Solids 2000;61. Available from: https://doi.org/10.1016/S0022-3697(99)00252-8.

[83] Gasser TC, Ogden RW, Holzapfel GA. Hyperelastic modelling of arterial layers with distributed collagen fibre orientations. J R Soc Interface 2006;3(6):15−35. Available from: https://doi.org/10.1098/rsif.2005.0073.

[84] Li DS, Baniassadi M, Garmestani H, Ahzi S, Reda Taha MM, Ruch D. 3D reconstruction of carbon nanotube composite microstructure using correlation functions. J Comput. Theor Nanosci 2010;7(8):1462−8. Available from: https://doi.org/10.1166/jctn.2010.1504.

[85] Carter DR, Van Der Meulen MC, Beaupre GS. Mechanical factors in bone growth and development. Bone 1996;18(1 Suppl):5S−10S. Available from: https://doi.org/10.1016/8756-3282(95)00373-8.

[86] Casanova R, Moukoko D, Pithioux M, Pailler-Mattéi C, Zahouani H, Chabrand P. Temporal evolution of skeletal regenerated tissue: what can mechanical investigation add to biological? Med & Biol Eng Comput 2010;48(8):811−19.

[87] Huiskes R, Ruimerman R, van Lenthe GH, Janssen JD. Effects of mechanical forces on maintenance and adaptation of form in trabecular bone. Nature 2000;405(6787):704−6. Available from: https://doi.org/10.1038/35015116.

[88] Wolff J. Das gesetz der transformation der knochen. A Hirshwald 1892;1:1−152.

[89] Bai J, Hamon A-L, Marraud A, Jouffrey B, Zymla V. Synthesis of SWNTs and MWNTs by a molten salt (NaCl) method. Chem Phys Lett 2002;365(1):184−8.

[90] Frost HM. Bone "mass" and the "mechanostat": a proposal. Anat Rec 1987;219 (1):1−9. Available from: https://doi.org/10.1002/ar.1092190104.

[91] Lekszycki T. Modelling of bone adaptation based on an optimal response hypothesis. Meccanica 2002;37(4):343−54. Available from: https://doi.org/10.1023/A:1020831519496.

[92] Lekszycki T. Functional adaptation of bone as an optimal control problem. J Theor Appl Mech 2005;43.

[93] Pivonka P, Zimak J, Smith DW, Gardiner BS, Dunstan CR, Sims NA, et al. Model structure and control of bone remodeling: a theoretical study. Bone 2008;43(2):249−63. Available from: https://doi.org/10.1016/j.bone.2008.03.025.

[94] Andreaus U, Colloca M, Iacoviello D, Pignataro M. Optimal-tuning PID control of adaptive materials for structural efficiency. Struct Multidiscip Optim 2011;43(1):43−59.

[95] Prendergast P, Taylor D. Prediction of bone adaptation using damage accumulation. J Biomech 1994;27(8):1067−76.

[96] Doblaré M, García J. Anisotropic bone remodelling model based on a continuum damage-repair theory. J Biomech 2002;35(1):1−17.

[97] Misra A, Poorsolhjouy P. Identification of higher-order elastic constants for grain assemblies based upon granular micromechanics. Mathematics Mech Complex Syst 2015;3(3):285−308.

[98] Abali BE, Müller WH, Dell'Isola F. Theory and computation of higher gradient elasticity theories based on action principles. Arch Appl Mech 2017;87(9):1495−510. Available from: https://doi.org/10.1007/s00419-017-1266-5.

[99] dell'Isola F, Corte AD, Giorgio I. Higher-gradient continua: the legacy of Piola, Mindlin, Sedov and Toupin and some future research perspectives. Mathematics Mech Solids 2016;22(4):852−72. Available from: https://doi.org/10.1177/1081286515616034.

[100] Martin M, Lemaire T, Haiat G, Pivonka P, Sansalone V. A thermodynamically consistent model of bone rotary remodeling: a 2D study (Conference Abstract). Computer Methods Biomech Biomed Eng 2017;20(S1):127−8.

[101] Madeo A, Lekszycki T, dell'Isola F. A continuum model for the bio-mechanical interactions between living tissue and bio-resorbable graft after bone reconstructive surgery. Comptes Rendus Mécanique 2011;339(10):625−40. Available from: https://doi.org/10.1016/j.crme.2011.07.004.

[102] Lekszycki T, Dell'Isola F. A mixture model with evolving mass densities for describing synthesis and resorption phenomena in bones reconstructed with bio-resorbable materials. ZAMM-J Appl Mathematics Mech/Zeitschrift für Angew Mathematik und Mechanik 2012;92(6):426−44.

[103] Andreaus U, Giorgio I, Lekszycki T. A 2-D continuum model of a mixture of bone tissue and bio-resorbable material for simulating mass density redistribution under load slowly variable in time. ZAMM - J Appl Mathematics Mech / Z für Angew Mathematik und Mechanik 2014;94(12):978−1000. Available from: https://doi.org/10.1002/zamm.201200182.

[104] Scala I, Spingarn C, Rémond Y, Madeo A, George D. Mechanically-driven bone remodeling simulation: application to LIPUS treated rat calvarial defects. Mathematics Mech Solids 2016;22(10):1976−88. Available from: https://doi.org/10.1177/1081286516651473.

[105] Giorgio I, Andreaus U, Scerrato D, dell'Isola F. A visco-poroelastic model of functional adaptation in bones reconstructed with bio-resorbable materials. Biomech Model Mechanobiol 2016;15(5):1325−43. Available from: https://doi.org/10.1007/s10237-016-0765-6.

[106] Cuomo M. Forms of the dissipation function for a class of viscoplastic models. Mathematics Mech Complex Syst 2017;5(3):217−37.

[107] Giorgio I, Andreaus U, dell'Isola F, Lekszycki T. Viscous second gradient porous materials for bones reconstructed with bio-resorbable grafts. Extreme Mech Lett 2017;13:141−7. Available from: https://doi.org/10.1016/j.eml.2017.02.008.

[108] Chavoshnejad P, Ayati M, Abbasspour A, Karimpur M, George D, Remond Y, et al. Optimization of Taylor spatial frame half-pins diameter for bone deformity correction: application to femur. Proc Inst Mech Eng H 2018;232(7):673−81. Available from: https://doi.org/10.1177/0954411918783782.

[109] Burr D, Allen M. Basic and applied bone biology. China: Elsevier; 2014.

[110] Pivonka P, Komarova SV. Mathematical modeling in bone biology: from intracellular signaling to tissue mechanics. Bone 2010;47(2):181−9. Available from: https://doi.org/10.1016/j.bone.2010.04.601.

[111] Lemaire T, Naili S, Sansalone V. Multiphysical modelling of fluid transport through osteo-articular media. An Acad Bras Cienc 2010;82(1):127−44. Available from: https://doi.org/10.1590/s0001-37652010000100011.

[112] Lemaire T, Capiez-Lernout E, Kaiser J, Naili S, Sansalone V. What is the importance of multiphysical phenomena in bone remodelling signals expression? A multiscale perspective. J Mech Behav Biomed Mater 2011;4(6):909−20. Available from: https://doi.org/10.1016/j.jmbbm.2011.03.007.

[113] Sansalone V, Gagliardi D, Desceliers C, Haiat G, Naili S. On the uncertainty propagation in multiscale modeling of cortical bone elasticity. Comput Methods Biomech Biomed Engin 2015;18(Suppl 1):2054−5. Available from: https://doi.org/10.1080/10255842.2015.1069619.

[114] Bednarczyk E, Lekszycki T. A novel mathematical model for growth of capillaries and nutrient supply with application to prediction of osteophyte onset. Z für Angew Mathematik und Phys 2016;67(4). Available from: https://doi.org/10.1007/s00033-016-0687-2.

[115] Lu Y, Lekszycki T. A novel coupled system of non-local integro-differential equations modelling Young's modulus evolution, nutrients' supply and consumption during bone fracture healing. Z für Angew Mathematik und Phys 2016;67(5). Available from: https://doi.org/10.1007/s00033-016-0708-1.

[116] Moya A, Paquet J, Deschepper M, Larochette N, Oudina K, Denoeud C, et al. Human mesenchymal stem cell failure to adapt to glucose shortage and rapidly use intracellular energy reserves through glycolysis explains poor cell survival after implantation. Stem Cell 2018;36(3):363−76.

[117] Paquet J, Deschepper M, Moya A, Logeart-Avramoglou D, Boisson-Vidal C, Petite H. Oxygen tension regulates human mesenchymal stem cell paracrine functions. Stem Cell Transl Med 2015;4(7):809−21. Available from: https://doi.org/10.5966/sctm. 2014-0180.

[118] George D, Allena R, Remond Y. Mechanobiological stimuli for bone remodeling: mechanical energy, cell nutriments and mobility. Comput Methods Biomech Biomed Engin 2017;20(sup1):91−2. Available from: https://doi.org/10.1080/10255842.2017. 1382876.

[119] George D, Allena R, Rémond Y. A multiphysics stimulus for continuum mechanics bone remodeling. Mathematics Mech Complex Syst 2018;6(4):307−19. Available from: https://doi.org/10.2140/memocs.2018.6.307.

[120] Allena R, Maini PK. Reaction-diffusion finite element model of lateral line primordium migration to explore cell leadership. Bull Math Biol 2014;76(12):3028−50. Available from: https://doi.org/10.1007/s11538-014-0043-7.

[121] Shahmohammadi A, Famouri S, Hosseini S, Farahani MM, Baghani M, George D, et al. Prediction of bone microstructures degradation during osteoporosis with fuzzy cellular automata algorithm. Mathematics Mech Solids 2022;. Available from: https://doi.org/10.1177/10812865221088520 10812865221088520.

[122] Hollister SJ, Brennan JM, Kikuchi N. A homogenization sampling procedure for calculating trabecular bone effective stiffness and tissue level stress. J Biomech 1994;27 (4):433−44. Available from: https://doi.org/10.1016/0021-9290(94)90019-1.

[123] Tsubota K-i, Suzuki Y, Yamada T, Hojo M, Makinouchi A, Adachi T. Computer simulation of trabecular remodeling in human proximal femur using large-scale voxel FE models: Approach to understanding Wolff's law. J Biomech 2009;42(8):1088−94.

[124] Jang IG, Kim IY. Computational simulation of simultaneous cortical and trabecular bone change in human proximal femur during bone remodeling. J Biomech 2010; 43(2):294−301. Available from: https://doi.org/10.1016/j.jbiomech.2009.08.012.

[125] Marzban A, Nayeb-Hashemi H, Vaziri A. Numerical simulation of load-induced bone structural remodelling using stress-limit criterion. Computer Methods Biomech Biomed Eng 2015;18(3):259−68.

[126] Famouri S, Baghani M, Sheidaei A, George D, Farahani MM, Panahi MS, et al. Statistical prediction of bone microstructure degradation to study patient dependency in osteoporosis. Mathematics Mech Solids 2022;. Available from: https://doi.org/10.1177/10812865221098777 10812865221098777.

[127] Kersh ME, Zysset PK, Pahr DH, Wolfram U, Larsson D, Pandy MG. Measurement of structural anisotropy in femoral trabecular bone using clinical-resolution CT images. J Biomech 2013;46(15):2659−66.

[128] Lian W-D, Legrain G, Cartraud P. Image-based computational homogenization and localization: comparison between X-FEM/levelset and voxel-based approaches. Computat Mech 2013;51(3):279−93.

[129] Goda I, Assidi M, Belouettar S, Ganghoffer JF. A micropolar anisotropic constitutive model of cancellous bone from discrete homogenization. J Mech Behav Biomed Mater 2012;16:87−108. Available from: https://doi.org/10.1016/j.jmbbm.2012.07.012.

[130] Räth C, Baum T, Monetti R, Sidorenko I, Wolf P, Eckstein F, et al. Scaling relations between trabecular bone volume fraction and microstructure at different skeletal sites. Bone. 2013;57(2):377−83.

[131] Goda I, Assidi M, Ganghoffer JF. A 3D elastic micropolar model of vertebral trabecular bone from lattice homogenization of the bone microstructure. Biomech Model Mechanobiol 2014;13(1):53−83. Available from: https://doi.org/10.1007/s10237-013-0486-z.

[132] Wierszycki M, Szajek K, Łodygowski T, Nowak M. A two-scale approach for trabecular bone microstructure modeling based on computational homogenization procedure. Computat Mech 2014;54(2):287−98.

[133] Goda I, Ganghoffer J-F, Czarnecki S, Wawruch P, Lewiński T. Optimal internal architectures of femoral bone based on relaxation by homogenization and isotropic material design. Mech Res Commun 2016;76:64−71. Available from: https://doi.org/10.1016/j.mechrescom.2016.06.007.

[134] Hasanabadi A, Baniassadi M, Abrinia K, Safdari M, Garmestani H. 3D microstructural reconstruction of heterogeneous materials from 2D cross sections: a modified phase-recovery algorithm. Computat Mater Sci 2016;111:107−15. Available from: https://doi.org/10.1016/j.commatsci.2015.09.015.

[135] Hasanabadi A, Baniassadi M, Abrinia K, Safdari M, Garmestani H. Efficient three-phase reconstruction of heterogeneous material from 2D cross-sections via phase-recovery algorithm. J Microsc 2016;264(3):384−93. Available from: https://doi.org/10.1111/jmi.12454.

[136] Fullwood DT, Niezgoda SR, Adams BL, Kalidindi SR. Microstructure sensitive design for performance optimization. Prog Mater Sci 2010;55(6):477−562.

[137] Fullwood DT, Niezgoda SR, Kalidindi SR. Microstructure reconstructions from 2-point statistics using phase-recovery algorithms. Acta Mater 2008;56(5):942−8.

[138] Fienup JR. Phase retrieval algorithms: a comparison. Appl Opt 1982;21(15):2758−69. Available from: https://doi.org/10.1364/AO.21.002758.

[139] Sheidaei A, Baniassadi M, Banu M, Askeland P, Pahlavanpour M, Kuuttila N, et al. 3-D microstructure reconstruction of polymer nano-composite using FIB−SEM and statistical correlation function. Compos Sci Technol 2013;80:47−54. Available from: https://doi.org/10.1016/j.compscitech.2013.03.001.

[140] Allena R, Cluzel C. Heterogeneous directions of orthotropy in three-dimensional structures: finite element description based on diffusion equations. Mathematics Mech Complex Syst 2018;6(4):339−51. Available from: https://doi.org/10.2140/memocs.2018.6.339.

[141] Cluzel C, Allena R. A general method for the determination of the local orthotropic directions of heterogeneous materials: application to bone structures using μCT images. Mathematics Mech Complex Syst 2018;6(4):353−67.

[142] Altenbach H, Eremeyev V. On the constitutive equations of viscoelastic micropolar plates and shells of differential type. Mathematics Mech Complex Syst 2015;3(3):273−83.

[143] Eremeyev VA, Pietraszkiewicz W. Material symmetry group and constitutive equations of micropolar anisotropic elastic solids. Mathematics Mech Solids 2016; 21(2):210−21.

[144] Becquart P, Cruel M, Hoc T, Sudre L, Pernelle K, Bizios R, et al. Human mesenchymal stem cell responses to hydrostatic pressure and shear stress. Eur Cell Mater 2016;31:160−73. Available from: https://doi.org/10.22203/ecm.v031a11.

[145] Izadi H, Baniassadi M, Hormozzade F, Dehnavi FN, Hasanabadi A, Memarian H, et al. Effect of 2D image resolution on 3D stochastic reconstruction and developing petrophysical trend. Transp Porous Media 2018;125(1):41−58.

Mechanical characterization of Voronoi-based microstructures

Abstract

In this chapter the mechanical properties of Voronoi microstructures containing two different elastic materials are investigated. Representative volume elements (RVEs) containing regular and random tessellations are created using a numerical procedure.

The procedure divides the RVE surface by Voronoi tessellations, and the elastic behavior of the surface is analyzed under tensile and shear deformations using the finite element (FE) method. Components of stress tensor for each element obtained from FE analysis are used to compute the overall elastic properties of the microstructure. Percolation threshold is defined based on the instantaneous gradient of the tensile and shear modulus diagrams. The results for regular Voronoi microstructures show that the dependency of tensile and shearing properties of the microstructure on volume fraction of particles can completely have different trends in some ranges of volume fractions. Also the effects of mechanical percolation on the elastic properties of microstructures are obtained. The presented approach opens a systematic method for investigating the enhanced properties of microstructures with different shapes of tessellations.

For random Voronoi microstructures, results reveal that the percolation thresholds in tensile and shear modes for isotropic RVE are almost the same, while there is a remarkable difference between percolation thresholds for an anisotropic case. Furthermore, in the procedure performed in this study a distinct inconsistency in elastic properties of anisotropic microstructures in longitudinal and transverse directions is observed. The mentioned method presents a paradigmatic overview for generating random isotropic and anisotropic tessellations with different aspect ratios on microstructures and evaluating their overall properties and percolation limit for them.

5.1 Introduction

During the past decades, improving mechanical behavior of microstructures with reinforcing inclusions is considered as one of the areas of interest for researchers in diverse interdisciplinary fields associated with material science. The formation of the interface domain between the matrix and particles which leads to localized enhancement in matrix properties plays a pivotal role in reinforcing the process of nanocomposites [1–4]. The interface region is formed as a surface effect such that it does not depend on the volume of the inclusions that it surrounds [5,6]. Chen et al. performed a comprehensive evaluation of the effects of fiber, matrix, and interface properties on the composite fracture toughness using a completely deterministic approach [7]. They have discovered that thinner reinforcing fibers do not necessarily result in better fracture toughness on composites [8].

Applied Micromechanics of Complex Microstructures. DOI: https://doi.org/10.1016/B978-0-443-18991-3.00001-5
© 2023 Elsevier Inc. All rights reserved.

Voronoi tessellation methods are utilized as general tessellation techniques regarding both proper representativeness of elements and simple mathematical implementation. Different species of Voronoi tessellations are now employed as a mesh generation framework in the thermal, mechanical, and electrical analyses [9−13]. A combination of three-dimensional (3D) Laguerre-Voronoi tessellation along with the finite element (FE) method (FEM) is employed to consider microstructure effects on the thermal properties of high porosity open cell foams [10]. Zhu et al. derived theoretical results about the effects of regularity on the geometrical properties of a random 3D Voronoi tessellation [14]. Falco et al. proposed a method for generating numerical models of polycrystalline microstructures, founded on the Laguerre-Voronoi tessellation technique [15]. Park et al. employed a weighted method to improve the Voronoi tessellation technique employed to describe the atomic structure of amorphous materials [16]. Wang et al. studied mechanisms of the fracture in polycrystalline alumina subjected to bending loads using a full-scale 3D Voronoi cell along with FE analysis [17]. Two-dimensional micromechanical analysis of fracture behavior for multiphase brittle composites was performed using different theoretical and numerical approaches including continuum mechanics and fully discrete methods [8].

Zhao et al. offered a computational outline deliberating both grain growth and diffusion in nanocrystalline structures [18] (Zhao, Wang, Ye, and Dong). Kruglova et al. introduced an approach to model the effects of dimples on the fracture surface of Al-Si alloys using Voronoi tessellation [19]. This model can be applied to metallographic images to simulate a fracture surface appearance and would probably predict if the structure had been fractured under uniaxial tensile loading. Voronoi tessellation was applied to make a segmentation pattern of a texture [20]. The pattern was conformed to a mineral texture image to analyze the mineralogical composition of the texture [21]. Lou et al. employed a mass density function, which regulates the size and distribution of Voronoi tessellations, in order to optimize the centroidal Voronoi tessellations for mesh generation [22]. Said et al. proposed a numerical scheme for multiscale modeling of strongly heterogeneous 3D composites using spatial Voronoi tessellation [23]. Ghazlan et al. evaluated the performances of nacre-like composites under blast loading by a 3D Voronoi model of an aluminum/vinylester composite structure [8,24].

Effective properties of Voronoi microstructures are causally linked to the percolation and volume faction of tessellations. Several researchers have attempted to include more mechanics in modeling mechanical percolation. Early work included the Generalized Effective Media model [25], which interpolated between a mean field model, at low volume fractions, and percolation theory, above the percolation threshold. This model has been used to predict both electrical and mechanical percolation [26,27]. The series−parallel model included an intermediate parameter that described the volume fraction of a material that was active in the transfer of forces [28]. A limitation of both models is that a previously identified value for the percolation threshold is required as input. In some cases, there is an influence of an interface region as well as the effects of clustering using the concentric cylinder micromechanics model, but not in the context of percolation thresholds [29]. A hybrid numerical analytic model was used in to investigate polymer nanocomposites with complex microstructural configurations; the model included the effects of an interface as a third, independent phase, that is, not

Mechanical characterization of Voronoi-based microstructures 223

linked to particle placement [30]. The relative influence of the competing and compounding effects of the spatial position/distribution of the particles (microstructure) and of the composite constitution (micromechanics) is examined [31,32]. Numerical approaches to provide evaluations about percolation threshold for the experimental investigations, however, are rare due to the difficulty in analyzing the evolution of the structure in nanoscale [33,34]. In addition, applications of percolation theory in derivation of some quantities by computerized tomography (CT) scan are frequent. Navarre-Sitchler et al. utilized micro-CT images to generate a pore network model [35]. Afterward, they determined the effective diffusivity and percolation limit as a function of porosity by carrying out experiments. A modified approximation for percolation threshold using 3D X-ray microtomography along with stochastic analyses was presented by Liu et al. [36]. Ikeda et al. carried out a systematic examination about the 3D interconnection and correlation function of a granite samples [8,37].

In general the effective parameters which determine the percolation threshold of the microstructure are geometrical and orientational properties of the reinforcing particles. Some of the studies use manufacturing techniques for design and manufacturing of new reinforcing inclusions [38−41]. These studies may include the design of reinforcing elements with structural hierarchy originating from their geometric design to be directly manufactured from computer-aided design (CAD) data, employing additive manufacturing techniques based on polymeric and/or metallic materials, at different scales. Several experimental and theoretical techniques can be modified as the next steps of these studies in future [31].

In this chapter a computational framework along with a micromechanics-based approach is introduced for modeling the effects characterized as mechanical percolation in the case of isotropic and anisotropic Voronoi microstructures. In a step forward in this work an approach to produce Voronoi tessellation on a random anisotropic microstructure with a specified aspect ratio is introduced. The random structure can be considered as the input model for the FE process. In the following a differential viewpoint to determine mechanical percolation threshold is proposed, and then the threshold is evaluated by FEM. To eliminate the dependency of the results to the initial random structure, five models are used as the inputs of the FE process for isotropic and anisotropic cases. The main objective of this chapter is to derive the procedure of determining percolation threshold and stress distribution in randomly isotropic and anisotropic microstructures partitioned by Voronoi cells [8].

5.2 Computational elucidation of isotropic and anisotropic microstructures with Voronoi tessellation

In this section the effects of geometrical shape of inclusion on elastic properties of microstructures have been investigated. FEM is used to evaluate mechanical properties of the microstructure reinforced by inclusions with different shapes. FEM results show that based on shapes of inclusions, adding them to the matrix enhances the mechanical properties of the whole microstructure in different states. The more the particles added to the microstructure, the more enhanced the stretching properties in "x" and "y" directions, while this is not confirmed for shear modulus. In the shear loading mode the microstructure is weakened in some ranges of volume fractions. In the stretching mode in the "x" direction the honeycomb particles have the most enhancing effect on mechanical properties of the microstructures. However, for the stretching mode in the "y" direction the rhombus particles have the most enhancing effect on mechanical properties of microstructures. A comparison has also been made among the shear loading results. Unlike the stretching mode the ascending trend for shear modulus has not been verified when the volume fraction of particles is increased. Rhombus particles had the maximum range of shear modulus among the other shapes. The ascending trend for high volume fractions also shows the effects of percolation of stiffer particles on the overall elastic properties of the microstructure. In the next steps, this research can be continued by investigating the effects of dissimilar particles in the overall behavior of the microstructure [31].

5.2.1 Homogenization and elastic percolation of composite structures with regular tessellations of the microstructure

In this section a geometrical approximation about the effects of adding inclusions with different shapes on mechanical properties of the microstructure is investigated. As a rule of thumb, one can conclude that if the particle length in the "y" direction is longer than that in the "x" direction, adding them to the matrix has a greater effect in the "y" direction. To develop a simple geometrical approximation a nondimensional length can be defined as follows [31]:

$$D_c = \frac{L_G|_y}{L_G|_x} \tag{5.1}$$

where D_c denotes the characteristic parameter associated with the effect of inclusion shape on mechanical properties of the microstructure stretched in the "y" direction. "$L_G|_y$" is defined as the length of the particle in the "y" direction across the center of the surface. Similarly, "$L_G|_x$" is defined for the maximum length of the particle in the "x" direction. The characteristic parameter can be applied to have an estimation about the mechanical behavior of the microstructure in "x" and "y" directions when different shapes of inclusions are applied for reinforcement. Fig. 5.1 shows the parameters "$L_G|_y$" and "$L_G|_x$" for a triangular inclusion [31].

Characteristic parameters for different inclusion shapes are listed in Table 5.1. It should be mentioned that inclusion with a smaller characteristic length shows stiffer behavior in the "x" direction [31].

5.2.1.1 Finite element modeling of the microstructure

As a paradigmatic example, consider the case of a microstructure with different shapes of particles in it. In general the microstructure function is somehow with perfect bonding between the matrix and particles. Effects of four different types of particle shapes on elastic properties of the microstructure were investigated: honeycomb, rhombus, square, and triangular. In each case the matrix and particles were

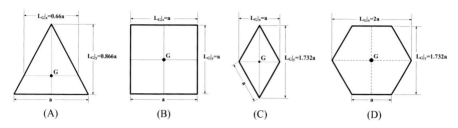

Figure 5.1 Parameters "$L_G|_y$" and "$L_G|_x$" for (A). Triangular, (B). Square, (C). Rhombus, and (D). Honeycomb inclusion [31].

considered to have isotropic behavior. To specify different values for elastic modulus of the matrix and fiber, creating partitions with different mechanical properties can be helpful. Hence in the next step, it is needed to create partitions on the microstructure based on the particle shapes. For square, triangular, and rhombus shapes the partitions were created using horizontal, vertical, and angled lines on the XY plane of the microstructure. For the honeycomb structure, at first, coordinates of the center of particles were produced using a Python code. Then coordinates of these points were read in a MATLAB® code, and finally, coordinates of the beginning and the end of the honeycomb particles edges were produced. Using these points, it is feasible to create honeycomb particles in microstructures. Fig. 5.2 shows two different partitioning states for the honeycomb and square particles [31].

The elastic modulus for inclusions is $E_i = 1000$ MPa, whereas the elastic modulus for the matrix is $E_m = 10$ MPa. The Poisson's ratio for both inclusions and the matrix is $\nu = 0.3$. The upper bound (limit) of elastic properties can be determined through the following equation [31]:

$$E_L = E_i V_f + E_m(1-V_f) \tag{5.2}$$

where V_f is the fiber volume fraction. It should be pointed out that the fiber volume fraction V_f is mathematically expressed as [31]:

$$V_f = \frac{\nu_f}{\nu_{tot.}} \tag{5.3}$$

where ν_f and ν_{tot} are the volume of fibers (reinforcing elements) and the total volume of the microstructure, respectively [31].

5.2.1.2 The protocol for generation of the random microstructure

A random number is assigned to each partition. If the random number is smaller than the volume fraction of the particles the stiffer material will be assigned to the partition. Otherwise the partition is considered to have weaker elastic properties. At first the microstructure includes 5% volume fraction of the particles. In each step the volume fraction of particles in the microstructure increases sequentially until it reaches to 95%. The FE process was performed for each step [31].

Table 5.1 Characteristic parameter for particles in the "*y*" direction [31].

Particle type	Rhombus	Triangle	Square	Honeycomb
Particle shape				
D_c	$\sqrt{3}$	$3\sqrt{3}/4$	1	$\sqrt{3}/2$

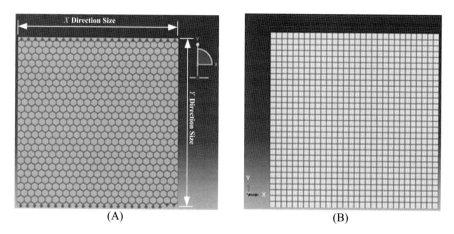

Figure 5.2 Two different partitioning states for the (A) honeycomb and (B) square particles [31].

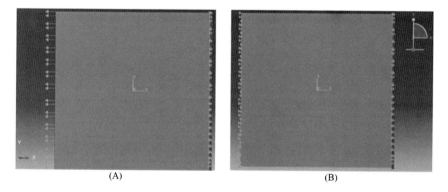

Figure 5.3 Loading conditions on microstructures: (A) in-plane stretching conditions and (B) in-plane shear loading [31].

Elastic properties of the microstructure with different shapes of particles have been evaluated using three types of loading conditions: in-plane stretching in the "x" direction, in-plane stretching in the "y" direction, and in-plane shear loading. In the case of stretching in the "x" ("y") direction a horizontal (vertical) displacement (equal to 0.001 strain) on the left (upper) edge of the microstructure was applied, and on the right (lower) edge of the microstructure a boundary condition was applied to restraint any displacement in the "x" ("y") direction. In the case of in-plane shear conditions the microstructure was subjected to an in-plane shear displacement (equal to 0.001 strain) on its left edge in the negative direction of the "y" axis and the right edge was fixed. Fig. 5.3 shows the loading conditions on the microstructure under stretching and shearing conditions [31].

ABAQUS commercial software (Abaqus/Standard; Simulia Dassault Systemes Simulia Corp. Providence, RI, United States) was used for FE modeling and computations of the microstructure deformation under the mentioned loading and

Table 5.2 Modeling characteristics for each type of microstructure [31].

	Square	Honeycomb	Triangular	Rhombus
X-direction size (Unit of length)	50	79.6743371482	50	50
Y-direction size (Unit of length)	50	78	48.497422612	48.497422612
Number of meshes	422,500	1,216,891	514,201	421,552
Cell number	2500	2464	5656	2878

boundary conditions. As was noted before, it is supposed that the microstructure has a rectangular shape in the xy plane. In Fig. 5.2 the size of the mentioned plane in 2D space was defined. The size of the plane in x- and y-directions is presented in Table 5.2 for each type of reinforcing shape. Moreover the total number of meshes and cells for each type of microstructure is indicated in Table 5.2 [31].

The edge size for each type of inclusion (parameter "A" in Fig. 5.1) is equal to 1.0 unit of length. The seed size for global elements is 0.08 unit of length. FE analyses are performed using the CPS4R element type (a four-node bilinear plane stress quadrilateral, reduced integration, hourglass control) [31].

Three static steps for each type of loading were considered. Note that in this study, we calculate the stress and volume for each element in each step using FEM. The following equations can be used as an alternative approach to calculate the elastic constants [6] [31]:

$$E_x = C_{11} = \frac{\left(\sum_{i=1}^{N} \sigma_{11}^i v^i \right)}{\varepsilon_{11}^0 V} \tag{5.4}$$

$$E_y = C_{22} = \frac{\left(\sum_{i=1}^{N} \sigma_{22}^i v^i \right)}{\varepsilon_{22}^0 V} \tag{5.5}$$

$$G_{xy} = C_{12} = \frac{\left(\sum_{i=1}^{N} \sigma_{12}^i v^i \right)}{\varepsilon_{12}^0 V} \tag{5.6}$$

where N is the total number of elements, v^i is the volume of ith element, σ^i is the stress tensor associated with ith element, and ε^0 is the external strain tensor applied on the microstructure [31].

5.2.1.3 Results and discussion

These models were analyzed based on the previous discussions and assumptions. The first goal is to find the trend of elastic modulus in stretching and shear modes for the microstructure. The comparison is performed for added particles with different geometries (shown in Figs. 5.4–5.6). For the sake of completeness the elastic modulus curves are compared with the associated lower bound and upper bound limits. Fig. 5.4 shows the elastic modulus of the microstructure in the "x" direction by considering the effect of different shapes for inclusions on the mechanical properties of the whole microstructure [31].

It can be observed from Fig. 5.4 that the trend of elastic modulus in the "x" direction for all types of inclusion shapes is ascending by an increase in volume fraction of particles. One can conclude that among the different shapes of inclusions the Honeycomb shape has the best effect on enhancement of mechanical properties of the microstructure. After the honeycomb shape the square, triangular, and rhombus shapes can enhance the mechanical properties of the microstructure, respectively. This is in a good level of consistency with the approximation pointed out in Section 3.1. Comparing different curves in Fig. 5.4, it is evident that in a microstructure with 30% volume fraction inclusions the particle shape effects do not have a remarkable effect on stretching properties of the microstructure. It can be observed that by adding particles in ranges higher than 65% volume fractions the elastic modulus in the "x" direction for all types of microstructures increases in a higher rate (Fig. 5.4). Also, Fig. 5.4 shows that for the ranges higher than 75%

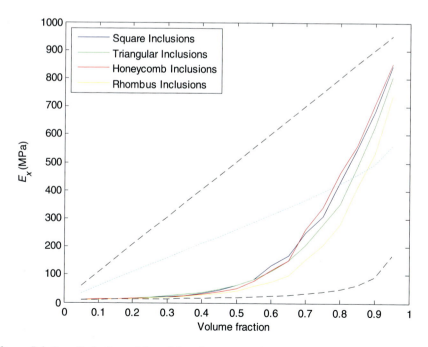

Figure 5.4 Overall elastic modulus of the microstructure in the "x" direction. Comparison of the different effects of shapes of cells on the mechanical properties of the whole microstructure. Dashed lines from top to bottom represent the upper bound, upper and lower bound average, and lower bound limits, respectively [31].

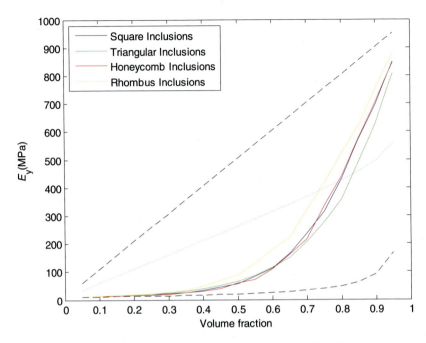

Figure 5.5 Overall elastic modulus of the microstructure in the "y" direction. Comparison of the different effects of shapes of cells on the mechanical properties of the whole microstructure. Dashed lines from top to bottom represent the upper bound, upper and lower bound average, and lower bound limits, respectively [31].

volume fraction the elastic modulus of the whole microstructure obviously converges to the upper bound. The dotted line in Fig. 5.4 shows the average of upper and lower bound limits. Taking into account the geometrical shape of particles, elastic moduli for all of the investigated structures reached to the average bound in ranges of volume fractions between 75% and 85% [31].

Fig. 5.5 shows the comparison between the elastic moduli of the microstructure in the "y" direction with different particle shapes. As in the E_x the elastic modulus in the "y" direction has an ascending trend for all types of inclusion shapes. Again, Fig. 5.5 shows a higher rate for the ascending trend of E_y in ranges higher than 65% volume fraction, and the modulus line is reached to the average of upper and lower bounds in ranges of 70%–80% volume fraction. Among the different shapes for the particles the rhombus shape has a higher range for the elastic modulus in the "y" direction, which is in good agreement with the approximation relation pointed out in Section 3–1. It should be noted that this convergence resulted in a considerable percolation of the microstructures [31].

Fig. 5.6 shows the shear modulus of the microstructure with different shapes of particles. In Fig. 5.6 (like Fig. 5.4), whenever the particle volume fraction is less than 30% the particle shape effects do not have a significant effect on shearing properties of the microstructure. Unlike the previous diagrams the shear modulus diagram does not have an additive trend for each level of particle volume fraction. In Fig. 5.6, it is seen that

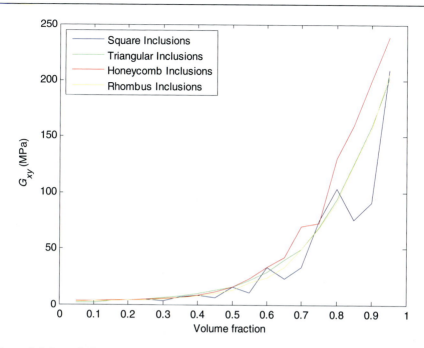

Figure 5.6 Overall shear modulus of the microstructure. Comparison of the different effects of shapes of cells on the mechanical properties of the whole microstructure [31].

the rhombus particles make the stiffest structure among other microstructures under shear loading conditions. The effect of adding square particles on overall shearing behavior of the microstructure is descending intensively when the volume fraction of particles is in a certain range ($0.7 < V_f < 0.8$). For triangular particles the descending behavior of the shear modulus was evaluated in some of intermediate values of volume fraction (0.45, 0.65, and 0.9). The intensive descending behavior in shear modulus was not observed in the case of using the rhombus particles. In this case an intensive additive trend can be observed in $V_f \approx 65\%$ instead. This effect can be observed in the case of triangular particles with $V_f \approx 65\%$ [31].

Fig. 5.7 shows the honeycomb microstructures for volume fractions from 5% to 95%. It is obvious that for the ranges from 55% to 75% of volume fraction, noticeable clusters of the stiffer material appear; however, one can conclude that the correlation of the stiffer material has improved the overall elastic properties [31].

Fig. 5.7 shows the honeycomb microstructures for volume fractions from 5% to 95%. It is obvious that for the ranges from 55% to 75% of volume fraction, noticeable clusters of the stiffer material appear; however, one can conclude that the correlation of the stiffer material has improved the overall elastic properties. Fig. 5.8 shows the rhombus microstructures for volume fractions from 5% to 95%. It should be mentioned that different shapes of particles in microstructures have different effects on percolation, and consequently, different elastic modulus values for each volume fraction are obtained. One can define a percolation threshold of elastic properties for each shape of particles, but application of these thresholds can be discussable [31].

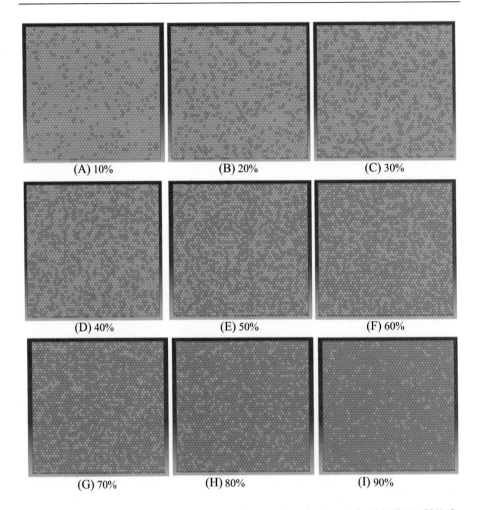

Figure 5.7 Honeycomb model overview for different volume fractions (vf), 10%vf (A), 20%vf (B), 30%vf (C), 40%vf (D), 50%vf (E), 60%vf (F), 70%vf (G), 80%vf (H), and 90%vf (I) [31].

Figs. 5.9−5.11 show the stress contour plots in the microstructure based on the shape of cells under mentioned loading and boundary conditions in a volume fraction of 25% for the stiffer material [31].

5.2.2 Homogenization and elastic percolation threshold in isotropic and anisotropic microstructures with random Voronoi tessellation

In this section the elastic properties of isotropic and anisotropic randomly shaped microstructures with different elastic materials are investigated using the FE approach. Initially, some models that can accurately represent the randomly isotropic and

Mechanical characterization of Voronoi-based microstructures 233

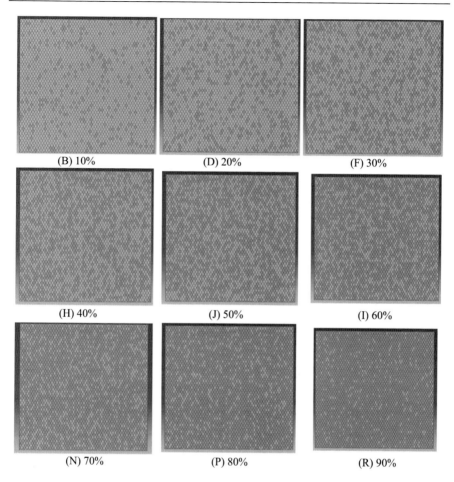

Figure 5.8 Rhombus model overview for different volume fractions (vf), 10%vf (B), 20%vf (D), 30%vf (F), 40%vf (H), 50%vf (J), 60%vf (I), 70%vf (N), 80%vf (P), and 90%vf (R) [31].

anisotropic behaviors are created. Furthermore, in this study a distinctive definition of the mechanical percolation threshold is proposed. Increasing the volume fraction of inclusions enhanced the mechanical behavior of the microstructure from the lower bound to the upper bound limit. In case of the randomly isotropic representative volume element (RVE) the proximity of the tensile and shear percolation thresholds made it possible to design a reinforced microstructure with enhanced tensile and shear properties, simultaneously. However, results of the FE analysis for randomly anisotropic microstructures revealed remarkable differences for the percolation threshold in "X" and "Y" directions. As a result the defined mechanical percolation threshold is helpful in improving the elastic properties of the anisotropic microstructures in longitudinal or transverse directions separately. It is noteworthy to mention that for the anisotropic microstructure in the shear mode the relative difference between "XY" and "YX" components of the shear modulus was less than in the tensile mode [8].

Figure 5.9 Stress contour plots for microstructures with 25% volume fraction of the stiffer material. The left column shows S_{11}, the middle column shows S_{22}, and the right column shows S_{12}. The first, second, third, and fourth rows are associated with the honeycomb, rhombus, square, and triangular microstructures, respectively [31].

There are several advantages for specifying percolation threshold for microstructures. In low percolation limits using less conductive/reinforcing inclusions the desired electrical/mechanical properties are accomplished, which ultimately can diminish the cost of fabrication. Furthermore, in a high volume fraction of the reinforcing inclusions, some alterations to the mechanical properties of the matrix can be made drastically, which manifest itself by increasing the stiffness of the microstructure. As a result a low percolation threshold can maintain the initial properties of the matrix to a great extent. For example, in the wearable electronic field, it is highly required to have structures with high electrical features and mechanical flexibility simultaneously. Evaluation of in-plane elastic properties in this way can be considered as a new method for future studies of the overall elastic properties of random microstructures. As the future work, this study can be followed by

Mechanical characterization of Voronoi-based microstructures

Figure 5.10 Stress contour plots for microstructures with 50% volume fraction of the stiffer material. The left column shows S_{11}, the middle column shows S_{22}, and the right column shows S_{12}. The first, second, third, and fourth rows are associated with the honeycomb, rhombus, square, and triangular microstructures, respectively [31].

investigating the effects of different aspect ratios of the cells for generating randomly anisotropic tessellations on the overall behavior of the microstructure [8].

5.2.2.1 Tessellation method and finite element modeling

To bring in the effects of random inclusions on overall elastic properties of the microstructure a special method for generating tessellations is utilized. In this section a general description about creating tessellation and the approach used to attribute elastic properties on inclusions are provided. For this purpose a planar model consisting of two different isotropic materials with different elastic properties is considered as the system model. The elastic modulus for inclusions is typically

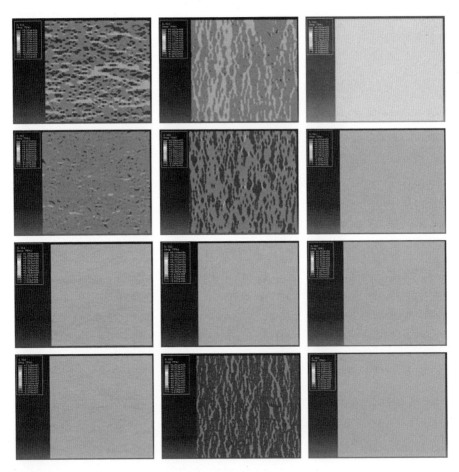

Figure 5.11 Stress contour plots for microstructures with 75% volume fraction of the stiffer material. The left column shows S_{11}, the middle column shows S_{22}, and the right column shows S_{12}. The first, second, third, and fourth rows are associated with the honeycomb, rhombus, square, and triangular microstructures, respectively [31].

considered about 100 times larger than the matrix elastic modulus. The Poisson's ratios for both the inclusion and matrix are considered to be the same. Table 5.3 shows materials parameters for the matrix and inclusion [8].

The microstructure includes 5% volume fraction of the particles initially. The volume fraction of particles in the microstructure increases gradually until it reaches 95%. Although the percolation theory deals with the connectivity of elements (e.g., particles, sites, or bonds) in a random system the elastic percolation is defined through the effect of this connectivity, which is a perceptible raise in the elastic properties of a random system. Then the elastic percolation threshold for the system would be defined as the point at which this raise commences. Therefore there is a major difference between a conductance percolation and a mechanical

Table 5.3 Materials characteristics for components of the microstructure [8].

Characteristic	Material 1 (matrix)	Material 2 (inclusion)
Elastic modulus [MPa]	10	10^3
Poisson's ratio	0.3	0.3

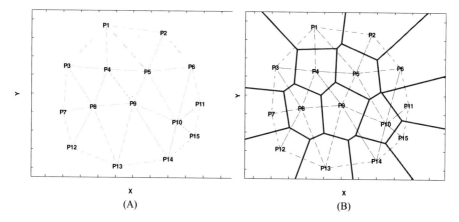

Figure 5.12 Partitioning a plane by Voronoi tessellation: (A) displaying the coordinates of Voronoi points and their connecting lines (Delaunay tessellation) and (B) drawing cell borders using half-space lines (Voronoi tessellation) [8].

percolation problem. A continuous enhancement in mechanical properties of the microstructure is observed during a gradual increase in the volume fraction of the inclusion, while electrical conductivity shows a zero-one behavior.

In this section, Voronoi tessellation is employed to generate random structures. In this method the surface is partitioned by convex polygons such that each polygon is formed by a generating point. Each point in the desired polygon is nearer to the generating point of the polygon than to any other generating points. Several points are needed to generate Voronoi tessellation on a surface. In this case, each site P_i is simply a point, and its corresponding Voronoi cell (R_i) consists of each point in the Cartesian space whose distance to P_i is less than or equal to its distance to the other points in this space. Based on this premise, each such cell is obtained from the intersection of half-spaces, and hence it is a convex polygon. Fig. 5.12 illustrates the main steps for creating partitions on a plane using Voronoi tessellation [8].

ABAQUS commercial software (ABAQUS/Standard; Simulia Dassault Systemes Simulia Corp. Providence, RI, United States) is utilized for FE modeling. Moreover the static general solver is employed to calculate the deformation of microstructures under the loading and boundary conditions, which allows to apply displacement or loads directly to the nodes/elements of the model.

To avoid the size dependency of the models, all of the RVEs are prepared in sufficiently large sizes compared with the element size. For the sake of generating a random isotropic structure a 50×50 2D planar model has been generated in ABAQUS. The mentioned surface is divided into 1×1 cells, and a Voronoi point is selected randomly in each cell. Coordinates of these points are imported into a MATLAB code, where the code makes Voronoi partitions using Voronoi points. Finally, coordinates of the vertex points in Voronoi polygons are imported into ABAQUS, and the model is fully partitioned using Voronoi tessellation. The whole surface is partitioned by Voronoi tessellation based on the preceding instructions. Five different random seeds were used to initialize the random number generator for a set of random numbers in Python, and subsequently, five primary RVE samples were generated. Moreover, to ensure that the generated structures are random, the two-point correlation function (TPCF) was calculated for all the structures, and the randomness of the structures was confirmed. Therefore this will lead to different random structures. Fig. 5.13 shows various partitioning states of the microstructure using Voronoi tessellation [8].

To generate a random anisotropic microstructure a 50×250 2D planar model is defined in ABAQUS. The surface is split into 1×5 rectangular cells, and each cell is divided into 0.2×1 rectangular regions, in each of which a random Voronoi point is selected. Based on this premise, it can be said that the aspect ratios of cells for selecting the Voronoi points in isotropic and anisotropic cases are equal to 1 and 5, respectively. After creating the Voronoi points the process of partitioning the model using the Voronoi tessellation is the same as the isotropic case. Fig. 5.14 reveals different anisotropic partitioning using Voronoi tessellation [8].

Two elastic properties for the inclusion and matrix are defined using the data in Table 5.3. Assigning the elastic properties to different partitions of the microstructure is based on generating a random number between 0 and 1 for each cell. If the specified random number of the cell is smaller than the expected volume fraction, then the stiffer material is assigned to the cell. Otherwise the softer material will be assigned. It was generally assumed that the microstructure function is mainly with perfect bonding between the matrix and inclusions. Each microstructure is analyzed considering a volume fraction of 5% to 95% for inclusions [8].

To evaluate the elastic properties of the microstructure in isotropic and anisotropic cases, different Voronoi tessellations, presented in Figs. 5.13 and 5.14, are used, respectively. Elastic properties are calculated individually for each case using the FE process. The number of Voronoi regions, seed size, and number of meshes in isotropic cases are presented in Table 5.5 [8].

Moreover the required data about the number of Voronoi regions, seed size, and number of meshes in anisotropic cases are indicated in Table 5.5 [8].

The mentioned meshing procedure is applied to the RVE model. Elastic properties of the microstructure for the foregoing cases were evaluated using different types of in-plane loading: in-plane stretching in the "X" direction, in-plane stretching in the "Y" direction, and in-plane shear loading. To stretch the RVE model in the "X" direction a simple horizontal displacement on the left border of the model is applied and a constraint to prevent any displacement in the "X" direction is assigned to the right border. The displacement causes 0.001 strain in the RVE in the "X" direction. The microstructure is subjected to the mentioned level of strain on the upper edge to model a simple stretch along the "Y" direction. In this case,

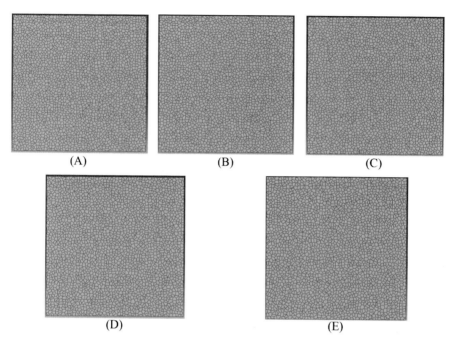

Figure 5.13 Isotropic parts partitioned by Voronoi tessellation. Tessellations were generated by different sets of random Voronoi points. Figures. (A), (B), (C), (D), and (E) correspond to (a), (b), (c), (d), and (e) cases, respectively. (See Table 5.4) [8].

Table 5.4 Meshing conditions in isotropic cases [8].

FE Model	(a)	(b)	(c)	(d)	(e)
No. of Voronoi tessellations	2627	2635	2642	2628	2639
Seed size	0.05	0.05	0.05	0.05	0.05
No. of meshes	1,136,617	1,136,528	1,129,985	1,143,654	1,136,546

boundary conditions are assigned to the lower border of the microstructure to prevent moving in the "Y" direction. Loading conditions for the isotropic and anisotropic cases are presented in Table 5.6 [8].

For the sake of modeling the RVE under shear loading the microstructure is subjected to an in-plane shear displacement (equal to 0.001 strain) on its left boundary in the negative direction of the "Y" axis, where the right boundary is fixed. Besides the preceding shear loading is applied on the upper boundary of the RVE, while the lower boundary is supposed to be fixed. Fig. 5.15 illustrates the loading and boundary conditions applied to the microstructure under stretching and shearing conditions [8].

FE processes for each case are implemented using the CPS4R element type (a four-node bilinear plane stress quadrilateral, reduced integration, hourglass control). Loading

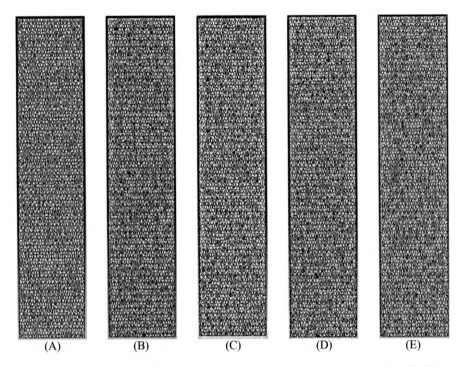

Figure 5.14 Anisotropic partitioning using Voronoi tessellation. Figures (A), (B), (C), (D), and (E) correspond to (a), (b), (c), (d), and (e) cases, respectively. (See Table 5.5) [8].

Table 5.5 Meshing conditions in the anisotropic case [8].

FE Model	(a)	(b)	(c)	(d)	(e)
No. of Voronoi tessellations	2654	2650	2652	2642	2631
Seed size	0.05	0.05	0.05	0.05	0.05
No. of meshes	5,892,168	5,906,399	5,911,129	5,934,435	5,920,546

Table 5.6 Loading conditions for isotropic and anisotropic cases [8].

Parameter	Isotropic model	Anisotropic model
Displacement in X-direction	0.05	0.05
Displacement in Y-direction	0.05	0.25

conditions for the isotropic and anisotropic cases are listed in Table 5.6. The overall elastic properties of a microstructure in "X" and "Y" directions can be calculated based on Eqs. (5.4)–(5.6).

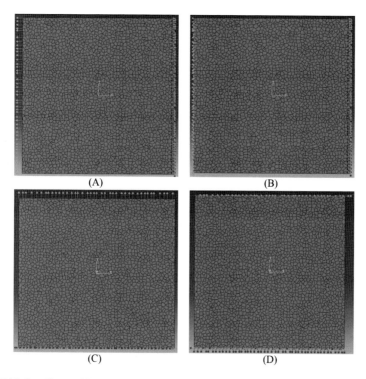

Figure 5.15 Loading and boundary conditions on the microstructure: (A) and (C) represent in-plane stretching in *X*- and *Y*-directions, respectively, and (B) and (D) represent in-plane shear loading in cases of *XY* and *YX* shear torque, respectively [8].

Employing the general form of rule of mixtures the upper bound limit for elastic modulus can be determined by the Voigt model as the following form [8]:

$$\overline{C} = \sum_{i=1}^{M} v_i c^i \tag{5.7}$$

where v_i and c^i represent the volume fraction and elastic property of the *i*th phase, respectively. *M* stands for the number of phases. Besides the lower bound for the elastic properties can be determined through the following equation (Reuss model) [8]:

$$\overline{C}^{-1} = \sum_{i=1}^{M} \frac{v_i}{c^i} \tag{5.8}$$

Components of the stress tensor and volume for each element are identified by the FEM. Hence the overall elastic properties of the microstructure can be determined using this procedure [8].

5.2.2.2 Results and discussion

The goal of this study is to investigate the effects of the elastic percolation on mechanical properties of 2D isotropic and anisotropic microstructures. Achieving a protocol to determine the percolation threshold for multiphase materials helps to generate economical structures with enhanced mechanical properties [8].

5.2.2.2.1 Isotropic microstructures

In this section the in-plane elastic properties of the RVE model are determined. Evaluation of in-plane elastic properties in this way can provide a new method for further studies on the overall properties of the isotropic and anisotropic microstructures in next steps: longitudinal Young's modulus (E_1), transverse Young's modulus (E_2), in-plane shear modulus (G_{12}), and Poisson's ratio (ν_{12}).

Fig. 5.16A shows the elastic modulus of a randomly isotropic microstructure in the "X" direction using Voronoi tessellation. Results are compared with upper and lower bond limits as well as the Mori–Tanaka method [42]. Also, Fig. 5.16B shows the elastic modulus of a randomly isotropic microstructure in the "Y" direction using Voronoi tessellation. Values displayed in Fig. 5.16A and B result from averaging the elastic properties of five isotropic models. Results are compared with the Reuss, Voigt, and Mori–Tanaka methods [8].

The elastic modulus in "X" and "Y" directions has a similar ascending trend by increasing the volume fraction under both loading conditions. Besides an increasing growth rate for both cases can be observed. A clear similarity between elastic moduli in "X" and "Y" directions is observed, which results from the isotropic behavior of the RVE model. The overall elastic modulus in stretching is determined by averaging E_x and E_y in each volume fraction. The standard deviation for the elastic modulus in a specified volume fraction is determined by the following equation [8]:

$$S = \sqrt{\frac{\sum_{i=1}^{n}(E_i - \overline{E})^2}{n-1}} \tag{5.9}$$

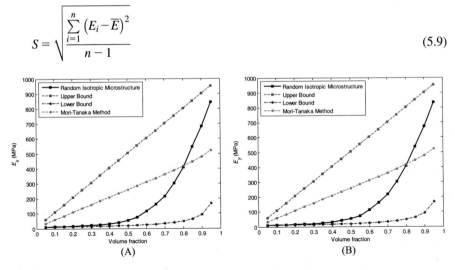

Figure 5.16 Overall elastic modulus of the isotropic microstructure: (A) elastic modulus in the X-direction and (B) elastic modulus in the Y-direction [8].

where "n" is the number of studied cases. Moreover, relative standard deviation (RSD) for a set of data is defined as the ratio of standard deviation to the average of data:

$$CV = \frac{S}{\bar{\bar{E}}} \times 100\% \tag{5.10}$$

Fig. 5.17 depicts the overall elastic modulus of the microstructure against the volume fraction. In Fig. 5.17 error bars represent standard deviation. As a paradigmatic convention employed in this study, percolation threshold is appraised as the level of the volume fraction of inclusions in which the growth rate of the elastic modulus diagram is equal to the slope of the upper bound limit. To accomplish this procedure a septic polynomial can be fitted to the set of elastic modulus data obtained from the FEM. The level of volume fraction in which the derivative of the septic polynomial is equal to the slope of the upper bound limit is introduced as the percolation threshold [8].

Based on Fig. 5.17 the overall elastic modulus for the microstructure starts with a gradual rise from the upper neighborhood of the lower bound in small volume fractions. The diagram eventually reaches to the lower neighborhood of the upper bound limit in higher volume fractions. According to Fig. 5.17, it is obvious that adding inclusions has no significant effect on the overall elastic modulus of the isotropic RVE model until it reaches more than 35% of the volume fraction. Also, it can be understood from Fig. 5.17 that in volume fractions of higher than ∼65% the elastic modulus diagram ascends with larger growth rates. According to the foregoing discussion the

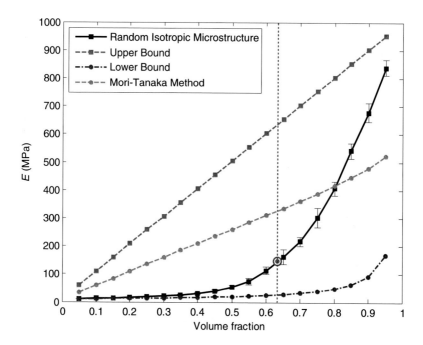

Figure 5.17 Overall elastic modulus of the isotropic microstructure compared to upper and lower bounds and Mori−Tanka method [8].

microstructure percolation limit occurs in 63.46% volume fraction of the stiffer material. In this case the overall elastic modulus is approximately 141.38 MPa. The standard deviation of the overall elastic modulus maintains an almost ascending trend by increasing the volume fraction of the inclusion. Elastic modulus behavior in terms of volume fraction is affected by mechanical percolation in the microstructure. Fig. 5.18A and B show the *XY* and *YX* components of shear modulus of the randomly isotropic microstructures, respectively. Results illustrated in Fig. 5.18A and B are obtained by averaging the elastic properties of five isotropic models. Results are compared with those of the upper and lower bound limits as well as the Mori–Tanaka method. The resemblance between Fig. 5.18A and B occurred as a result of the isotropic behavior assumption [8].

Fig. 5.19 illustrates the overall shear modulus, G, for the randomly isotropic case against the volume fraction. Error bars represent the standard deviation for the set of data in Fig. 5.19.

According to Fig. 5.19, it is obvious that adding inclusions has no significant effect on the overall shear modulus of the isotropic microstructure until it reaches more than ∼35% of the volume fraction. Also, it is inferred from Fig. 5.19 that in volume fractions of higher than ∼65% the shear modulus diagram increases at larger growth rates. According to the previously mentioned discussion the microstructure percolation limit occurs in 63.41% volume fraction of the stiffer material. In this case the overall shear modulus is approximately 55.252 MPa.

The trend for the standard deviation of the overall shear modulus remains at a partly increasing level by adding more inclusions in the microstructure. Low values of standard deviation also verify the procedure of generating isotropic models and emphasize that the size dependency and other simplifications in FE modeling do not have remarkable effects on the results. As mentioned for the elastic modulus the behavior of shear modulus also has been influenced by the mechanical percolation of the microstructure [8].

5.2.2.2.2 Anisotropic microstructures
In this section, results of the FE analysis of mechanical behavior for the anisotropic microstructure are presented. Generally as expected the mechanical behavior of the

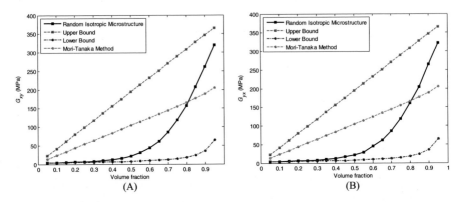

Figure 5.18 Shear modulus of the randomly isotropic microstructure: (A) *XY* component of the shear modulus and (B) *YX* component of the shear modulus [8].

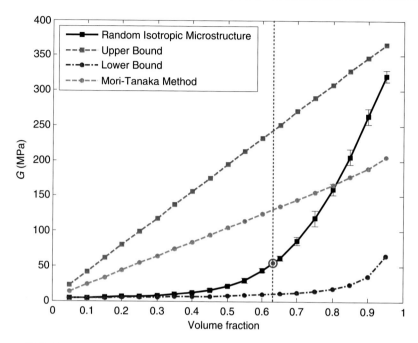

Figure 5.19 Overall shear modulus of the randomly isotropic microstructure compared to upper and lower bounds and Mori−Tanka method [8].

anisotropic microstructure along the X- and Y-axes reveals significant differences. Subsequently, anisotropic behavior of the microstructure will have a direct effect on altering the percolation threshold. This effect is taken into consideration in this section. Fig. 5.20 shows the elastic modulus in "X" and "Y" directions for the randomly anisotropic microstructure against the volume fraction [8].

Like the isotropic case the elastic modulus in small volume fractions passes through the vicinity of the lower limit and tends to the upper limit in larger volume fractions. Elastic moduli along "X" and "Y" directions have an ascending trend, but the rate of change in larger volume fractions is much higher. According to Fig. 5.20, different behaviors of the anisotropic microstructure in "X" and "Y" directions are quite distinct. Significant effects of the inclusions on the stretching behavior of the RVE model along the "X" axis are observed in ~%35 volume fraction, while the inclusions make these effects in the "Y" direction appear in ~%25 volume fraction. Mechanical percolation for the elastic modulus in the "X" direction occurs in 72.21% volume fraction, while the microstructure reaches the percolation limit in the "Y" direction in 57.96% of the volume fraction for inclusions. Fig. 5.20 illustrates that when the percolation occurs in the microstructure the elastic modulus in the "X" and "Y" directions reaches 144.93 and 144.19 MPa, respectively. As indicated in the modeling section, for the anisotropic case the "Y" direction is considered to be the longitudinal direction of the RVE, and therefore the resultant elastic properties would be greater in this direction. Based on Fig. 5.20, in the range of volume fractions between 45% and 80% the difference between the elastic moduli in "X" and "Y" directions is more than 100% [8].

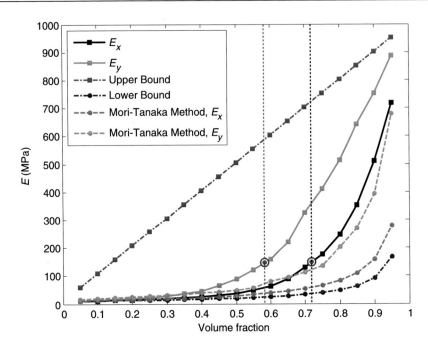

Figure 5.20 Overall elastic modulus of the anisotropic microstructure in the X and Y directions compared to upper and lower bounds and Mori–Tanka method [8].

Fig. 5.21 shows XY and YX components of the shear modulus (G_{xy} and G_{yx}) for the randomly anisotropic microstructure against the volume fraction [8].

According to Fig. 5.21 the evident difference of the XY and YX components of the shear modulus indicates the anisotropic behavior of the RVE. However, the difference in the shear mode is not as high as the relative difference in the tensile case. Remarkable effects of the reinforcing process on the shearing behavior of the RVE model are initially observed in ∼%30 volume fractions for the "YX" component, while these effects can be detected in ∼%40 volume fractions for the "XY" component. Mechanical percolation for the "XY" component of the shear modulus occurs in 67.49% volume fraction, while the microstructure reaches the percolation limit in the "Y" direction in 60.66% of volume fraction for inclusions. Fig. 5.20 illustrates that when the percolation occurs in the microstructure the elastic modulus in the "X" and "Y" directions reaches 59.37 and 54.33 MPa, respectively. For anisotropic cases the distinction between effective percolation thresholds in different directions appears for the systems with a finite size. In such systems, it seems that the tendency to spontaneous forming of a cluster system in one direction is more than the other. The difference in the percolation limits along "X" and "Y" directions in the anisotropic microstructures makes it possible to determine whether the microstructure is reinforced in the anisotropic direction or it will be reinforced entirely [8].

Fig. 5.22 shows the stress contour plots in the isotropic microstructure in 65% volume fraction for the stiffer material. Panels are separated based on different loading and boundary conditions.

Mechanical characterization of Voronoi-based microstructures 247

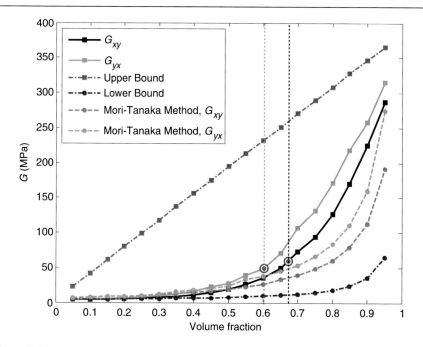

Figure 5.21 *XY* and *YX* components of the shear modulus (G_{xy} and G_{yx}) in the anisotropic microstructure compared to upper and lower bounds and Mori–Tanka method [8].

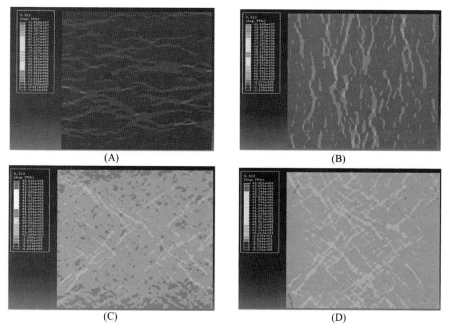

Figure 5.22 Overview of stress contour plots for the isotropic microstructures with 65% volume fraction of the stiffer material. (A) Contour plots for S_{11}, (B) contour plots for S_{22}, (C) contour plots for S_{12}, and (D) contour plots for S_{21} [8].

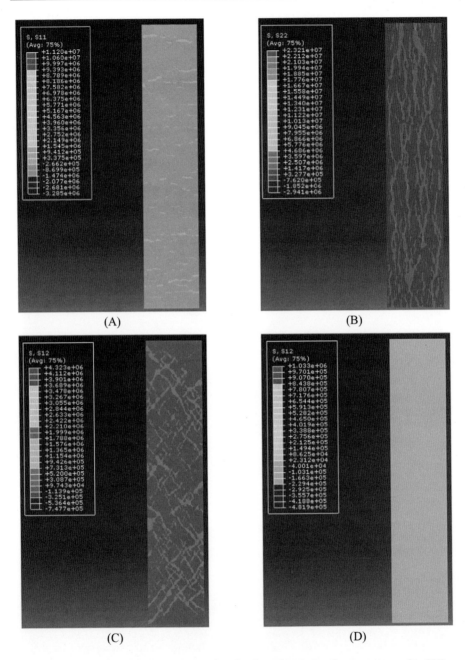

Figure 5.23 Overview of stress contour plots for the anisotropic microstructures with 65% volume fraction of the stiffer material. (A) Contour plots for S_{11}, (B) contour plots for S_{22}, (C) contour plots for S_{12}, and (D) contour plots for S_{21} [8].

Fig. 5.23 illustrates the stress contour plots in the anisotropic microstructure in 65% volume fraction for the stiffer material. Panels are separated based on different loading and boundary conditions.

As can be inferred from Figs. 5.22 and 5.23, high-stress lines in stress contour plots are aligned with the traction force applied to the boundaries of the microstructures in each case [8].

5.3 Conclusion remarks

In this chapter FEM is used to evaluate mechanical properties of the Voronoi tessellation microstructure containing two different elastic materials. RVEs containing regular and random tessellations were generated using a numerical procedure, and the elastic behavior of the microstructures was analyzed under tensile and shear deformations. The results have been used to calculate the effective elastic properties of the microstructure. Percolation threshold of mechanical properties has been defined based on the instantaneous gradient of the tensile and shear modulus diagrams. The ascending trend for high volume fractions also showed the effects of percolation of stiffer particles on the overall elastic properties of the microstructure. This research can be implemented to model tumors, tissues, and composites [8,31].

References

[1] Favier V, Chanzy H, Cavaille JY. Polymer nanocomposites reinforced by cellulose whiskers. Macromolecules 1995;28(18):6365–7.

[2] Qiao R, Brinson LC. Simulation of interphase percolation and gradients in polymer nanocomposites. Compos Sci Technol 2009;69(3):491–9.

[3] Wang HW, Zhou HW, Peng RD, Mishnaevsky L. Nanoreinforced polymer composites: 3D FEM modeling with effective interface concept. Compos Sci Technol 2011;71 (7):980–8.

[4] Kurahatti RV, Surendranathan AO, Kori SA, Singh N, Kumar AVR, Srivastava S. Defence applications of polymer nanocomposites. Def Sci J 2010;60:5.

[5] Fralick BS, Gatzke EP, Baxter SC. Three-dimensional evolution of mechanical percolation in nanocomposites with random microstructures. Probabilistic Eng Mech 2012;30 (Supplement C):1–8. Available from: https://doi.org/10.1016/j.probengmech.2012.02.002.

[6] Yi YB, Esmail K. Computational measurement of void percolation thresholds of oblate particles and thin plate composites. J Appl Phys 2012;111(12):124903.

[7] Chen Y, Wang S, Liu B, Zhang J. Effects of geometrical and mechanical properties of fiber and matrix on composite fracture toughness. Compos Struct 2015;122(Supplement C):496–506. Available from: https://doi.org/10.1016/j.compstruct.2014.12.011.

[8] Naeini VF, Zarei MH, Baniassadi M, Shirani M, Baghani M. Computational elucidation of elastic percolation threshold in isotropic and anisotropic microstructures with Voronoi tessellation. Int J Appl Mech 2019;11(03):1950029. Available from: https://doi.org/10.1142/s1758825119500297.

[9] Dotelli G. Composite materials as electrolytes for solid oxide fuel cells: simulation of microstructure and electrical properties. Solid State Ion 2002;152-153(Supplement C):509−15. Available from: https://doi.org/10.1016/s0167-2738(02)00380-6.

[10] Randrianalisoa J, Baillis D, Martin CL, Dendievel R. Microstructure effects on thermal conductivity of open-cell foams generated from the Laguerre−Voronoï tessellation method. Int J Therm Sci 2015;98(Supplement C):277−86. Available from: https://doi.org/10.1016/j.ijthermalsci.2015.07.016.

[11] Vena P, Gastaldi D. A Voronoi cell finite element model for the indentation of graded ceramic composites. Compos B Eng 2005;36(2):115−26. Available from: https://doi.org/10.1016/j.compositesb.2004.05.003.

[12] Wang H, Qin Q-H. Voronoi polygonal hybrid finite elements with boundary integrals for plane isotropic elastic problems. Int J Appl Mech 2017;9(03):1750031.

[13] Chavoshnejad P, More S, Razavi MJ. From surface microrelief to big wrinkles in skin: a mechanical in-silico model. Extreme Mech Lett 2020;100647.

[14] Zhu HX, Zhang P, Balint D, Thorpe SM, Elliott JA, Windle AH, et al. The effects of regularity on the geometrical properties of Voronoi tessellations. Phys A Stat Mech Appl 2014;406(Supplement C):42−58. Available from: https://doi.org/10.1016/j.physa.2014.03.012.

[15] Falco S, Jiang J, De Cola F, Petrinic N. Generation of 3D polycrystalline microstructures with a conditioned Laguerre-Voronoi tessellation technique. Computat Mater Sci 2017;136 (Supplement C):20−8. Available from: https://doi.org/10.1016/j.commatsci.2017.04.018.

[16] Park J, Shibutani Y. Weighted Voronoi tessellation technique for internal structure of metallic glasses. Intermetallics 2007;15(2):187−92. Available from: https://doi.org/10.1016/j.intermet.2006.05.005.

[17] Wang Z, Li P. Voronoi cell finite element modelling of the intergranular fracture mechanism in polycrystalline alumina. Ceram Int 2017;43(9):6967−75. Available from: https://doi.org/10.1016/j.ceramint.2017.02.121.

[18] Zhao J, Wang GX, Ye C, Dong Y. A numerical model coupling diffusion and grain growth in nanocrystalline materials. Computat Mater Sci 2017;136(Supplement C):243−52. Available from: https://doi.org/10.1016/j.commatsci.2017.05.010.

[19] Kruglova A, Roland M, Diebels S, Dahmen T, Slusallek P, Mücklich F. Modelling and characterization of ductile fracture surface in Al-Si alloys by means of Voronoi tessellation. Mater Charact 2017;131(Supplement C):1−11. Available from: https://doi.org/10.1016/j.matchar.2017.06.013.

[20] van der Wielen KP, Rollinson G. Texture-based analysis of liberation behaviour using Voronoi tessellations. Miner Eng 2016;89(Supplement C):93−107. Available from: https://doi.org/10.1016/j.mineng.2015.09.008.

[21] Zheng S, Li Z, Liu Z. The fast homogeneous diffusion of hydrogel under different stimuli. Int J Mech Sci 2018;137:263−70.

[22] Luo L, Jiang Z, Lu H, Wei D, Linghu K, Zhao X, et al. Optimisation of size-controllable centroidal Voronoi tessellation for FEM simulation of micro forming processes. Proc Eng 2014;81(Supplement C):2409−14. Available from: https://doi.org/10.1016/j.proeng.2014.10.342.

[23] El Said B, Ivanov D, Long AC, Hallett SR. Multi-scale modelling of strongly heterogeneous 3D composite structures using spatial Voronoi tessellation. J Mech Phys Solids 2016;88 (Supplement C):50−71. Available from: https://doi.org/10.1016/j.jmps.2015.12.024.

[24] Ghazlan A, Ngo TD, Tran P. Three-dimensional Voronoi model of a nacre-mimetic composite structure under impulsive loading. Compos Struct 2016;153(Supplement C):278−96. Available from: https://doi.org/10.1016/j.compstruct.2016.06.020.

[25] McLachlan DS, Blaszkiewicz M, Newnham RE. Electrical resistivity of composites. J Am Ceram Soc 1990;73(8):2187−203.

[26] Niklaus M, Shea H. Electrical conductivity and Young's modulus of flexible nanocomposites made by metal-ion implantation of polydimethylsiloxane: the relationship between nanostructure and macroscopic properties. Acta Mater 2011;59(2):830−40.

[27] Yves R, Ahzi S, Baniassadi M, Garmestani H. Applied RVE reconstruction and homogenization of heterogeneous materials. John Wiley & Sons; 2016.

[28] Ouali N, Cavaillé J, Perez J. Elastic, viscoelastic and plastic behavior of multiphase polymer blends. Plastics, Rubber Compos Process Appl 1991;55−60.

[29] Seidel GD, Lagoudas DC. A micromechanics model for the electrical conductivity of nanotube-polymer nanocomposites. J Compos Mater 2009;43(9):917−41.

[30] Liu H, Brinson LC. Reinforcing efficiency of nanoparticles: a simple comparison for polymer nanocomposites. Compos Sci Technol 2008;68(6):1502−12.

[31] Zarei MH, Fadaei Naeini V, Baghani M, Jamali J, Baniassadi M. Elastic percolation of composite structures with regular tessellations of microstructure. Compos Struct 2017;161:513−21. Available from: https://doi.org/10.1016/j.compstruct.2016.11.017.

[32] Baxter S.C., Burrows B.J., Fralick B.S. Mechanical percolation in nanocomposites: Microstructure and micromechanics. Probabilistic engineering mechanics. 2015.

[33] Sampaio Filho CIN, Andrade JS, Herrmann HJ, Moreira AA. Elastic backbone defines a new transition in the percolation model. Phys Rev Lett 2018;120(17):175701. Available from: https://doi.org/10.1103/PhysRevLett.120.175701.

[34] Zimmerman DT, Bell RC, Filer JA, Karli JO, Wereley NM. Elastic percolation transition in nanowire-based magnetorheological fluids. Appl Phys Lett 2009;95(1):14102.

[35] Navarre-Sitchler A, Steefel CI, Yang L, Tomutsa L, Brantley SL. Evolution of porosity and diffusivity associated with chemical weathering of a basalt clast. J Geophys Res Earth Surf 2009;114(F2).

[36] Liu J, Regenauer-Lieb K, Hines C, Liu K, Gaede O, Squelch A. Improved estimates of percolation and anisotropic permeability from 3-DX-ray microtomography using stochastic analyses and visualization. Geochem Geophys Geosyst 2009;10(5).

[37] Ikeda S, Nakano T, Nakashima Y. Three-dimensional study on the interconnection and shape of crystals in a graphic granite by X-ray CT and image analysis. De Gruyter; 2000.

[38] Boparai K, Singh R, Fabbrocino F, Fraternali F. Thermal characterization of recycled polymer for additive manufacturing applications. Compos Part B Eng 2016;106:42−7.

[39] Farina I, Fabbrocino F, Carpentieri G, Modano M, Amendola A, Goodall R, et al. On the reinforcement of cement mortars through 3D printed polymeric and metallic fibers. Compos Part B: Eng 2016;90:76−85.

[40] Farina I, Fabbrocino F, Colangelo F, Feo L, Fraternali F. Surface roughness effects on the reinforcement of cement mortars through 3D printed metallic fibers. Compos Part B: Eng 2016;.

[41] Singh R, Singh N, Fabbrocino F, Fraternali F, Ahuja I. Waste management by recycling of polymers with reinforcement of metal powder. Compos Part B: Eng 2016;105:23−9.

[42] Mori T, Tanaka K. Average stress in matrix and average elastic energy of materials with misfitting inclusions. Acta Metallurg 1973;21(5):571−4.

Numerical modeling of degraded microstructures

6

Abstract

In this chapter modeling of the degradation process of biomaterials and biological materials is discussed. As a sample a numerical approach implemented to model corrosion behavior of biodegradable composites and the bone loss of osteoporosis is provided. In the first section of this chapter, we are focused on the corrosion behavior of polylactic acid (PLA)/Mg composites as biomaterials. These composites are made of PLA and magnesium (Mg) with different volume fractions. The scanning electron microscopy (SEM) images of these composites were taken, and statistical reconstruction of the composite based on SEM images was done by the phase recovery algorithm. The three-dimensional (3D) structure of this composite was extracted with this reconstruction; then a 3D cellular automata model was developed to predict the corrosion of this composite. At the second section, we discuss about numerical modeling of osteoporosis to predict the bone loss at the mesoscopic scale (bone trabecula) in microgravity. The model can correlate the calculated bone degradation mechanism with data available in the literature showing the effective bone density loss measured experimentally. An optimization algorithm is used for an average bone microstructure distribution and long-term prediction. Extrapolation is made to link the local bone loss at the structural scale with the corresponding effective bone strength. The first part of the section details the extraction of the bone microstructure using microcomputed tomography images and numerical model development. Next the degradation and optimization schemes are detailed. Finally, some results are presented for long-term degradation.

6.1 Introduction

6.1.1 Degradation of PLLA/magnesium composite

Magnesium is present in 300 biological reactions that occur in the body and can adjust the enzyme in cells [1,2,3]. In 2018 Amerinatanzi et al. proposed a novel method to predict the biodegradation of magnesium alloy implants. In this method a continuum damage mechanism (CDM) finite element (FE) model was proposed to phenomenologically estimate the corrosion rate of a biocompatible Mg−Zn−Ca alloy. This model predicts the corrosion behavior of Mg-based alloys more precisely with a reduction in computational cost [4]. Magnesium corrodes in a chloride environment quickly; therefore a magnesium stent deteriorates rapidly in the body [5], and that could be a problem. There have been numerous investigations to use

---

Applied Micromechanics of Complex Microstructures. DOI: https://doi.org/10.1016/B978-0-443-18991-3.00004-0
© 2023 Elsevier Inc. All rights reserved.

polymers as a biodegradable artery stent. The proper capabilities of polylactic acid (PLA) with a low corrosion rate are proved in different studies [1,6].

A PLA/magnesium composite for orthopedic applications was developed by Cifuentes et al. (2012), and the mechanical and corrosion properties of this model was studied. Cifuentes et al. (2012) reported the in vitro degradation of biodegradable polylactic acid/Mg composites [7]. In 2017 Li et al. studied a PLA reinforced with two-dimensional (2D)-braided magnesium alloy wires with the braiding angle of 45 degrees. The mechanical properties and degradation were examined in the simulated conditions of the body [8]. Li et al. proposed a mathematical model to describe the mechanical and degradation behaviors of the Mg alloy wires/PLA composite in the consistent and staged dynamic environments [1,7].

There are two principal methods to examine the effects of implants in the human body. One is that the implant is planted in the environment of the body (in vivo); the other one is to place the implant in the simulated body fluid (in vitro). Since in vivo experiments are too expensive, the in vitro experiments are more favored in many cases. Different types of solutions are used to simulate body fluids from 0.9% NaCl solution, Hanks' balanced salt solution (HBSS), Earle's balanced salt solution (EBSS), and so on [9,10]. Some studies examine degradation of Mg in physiological saline (0.9% NaCl) solution, but the outcomes achieved with physiological saline solutions are incomparable or even conflicting from that obtained under physiological circumstances [11]. To reach closer physiological condition results, simulated body fluids (SBFs) and Hanks solution are generally utilized to determine the degradation rate of Mg. These solutions have a similar inorganic ion balance compared to plasma [1].

Corrosion inspections take a significant amount of time, so it would be helpful to have a numerical model to predict the behavior of corrosion in varying geometries in different time periods. There are many models already existing in the literature. The corrosion of Mg model based on continuum damage was developed by Grogan et al. [12]. In this model the finite element (FE) method was used to model the corrosion process under different mechanisms. Another study presented the sample degeneration with a moving boundary approach, which concentrates on the first hour of corrosion after immersion and neglects the formation of the protective film, Grogan et al. [13]. Bajger et al. [14] introduced a semiquantitative mathematical model based on transport-driven corrosion, which provided a computational tool to predict the degradation rate of the implants prior to the conduction of any in vitro or in vivo experiments [1,14].

There are also other numerical simulations of the corrosion implemented utilizing both Monte Carlo (MC) and cellular automata (CA) techniques [15]. Cellular automata is one of the most potent methods to simulate corrosion. The main feature of the method is its probabilistic nature that can represent the stochastic properties of electrochemical reactions in corrosion [1].

The first model was developed back in 1993, where a simple 2D passivation and depassivation model for pitting corrosion was described [16,17]. Models rapidly grew to include further aspects that affect corrosion, followed by 3D cellular automata models that described pitting corrosion [18]. 3D modeling is necessary to describe the corrosion behavior because in many cases a 2D model cannot explain localized corrosion. The principal mechanism of corrosion is anodic and cathodic reactions. An electric potential difference drives these reactions. Potentially, there can be two forms of corrosion, uniform and localized corrosion. Some of the previous models describe these reactions as a single location event, but the recent studies had defined the mechanism as two simultaneous electrochemical semireactions seeing the influence of two events at separate places but intrinsically dependent. These are randomly distributed over the metal surface to simulate the stochastic character of corrosion [1,19−21].

Stafiej et al. considered the impact of the spatially separated reactions on the incubation time of localized corrosion [22]. Vautrin et al. investigated the case when the anodic and cathodic reactions occur at the same place [23]. Pérez-Brokate et al. modeled a 3D discrete stochastic of occluded corrosion [24]. Guo et al. developed a cellular automata simulation for degradation of PLA with an accelerable reaction-diffusion model [1,25].

To use cellular automata to model corrosion, a 3D structure is required. Therefore reconstruction is a necessity to convert a 2D scanning electron microscopy (SEM) image of the samples to a 3D model. It is shown that some of the effective characteristics of a random heterogeneous material are well corresponded with a particular formalism called N-point statistics [1,26,27].

There are two approaches to reconstruct a 3D model. One is using experiments like X-ray computed tomography (CT) or focused ion beam (FIB)/SEM tomography. Also, numerical methods and statistical reconstruction using SEM and transmission electron microscopy (TEM) images can be used [28]. Garmestani et al. introduced a semi-inverse MC reconstruction of two-phase heterogeneous materials using two-point functions [29]. Baniassadi et al. developed a 3D reconstruction technique with MC methodology based on n-point statistical functions [30]. Hasanabadi et al. developed a novel approach for reconstruction of heterogeneous materials from 2D cross-sections via phase-recovery algorithm [1,31].

6.1.2 Bone degradation

Space exploration has always been a challenge, particularly for long-duration space missions. These long-term missions create new physical and mental challenges. In space or in microgravity, our body biology reacts differently than when under gravity forces on the ground. These gravity forces play a major role on the physiology of our body's organs as living beings have adapted to their existing environment on earth for millennia. For example due to a decrease or disappearance of gravity forces (body weight), bone loss is observed and can be a serious threat for astronauts in long-duration space missions. Gravity has a substantial role in keeping bone strength to an optimized condition that will support our body weight. Empirical research studies have showed that the bone mineral density (BMD) decreases every month by an average of 1% to 2% [32−34] in space. Hence without gravity the bone will degrade and adapt to its new "mechanically unloaded" environment [2].

During the past 50 years, spaceflights have shown that both long-term and short-term missions induce bone loss. However, the amount of bone loss varies as a function of the bone type and mechanical load environment. The bodyweight is not the same throughout the different bones of our body. For example lower limb bones, the pelvis, and the vertebral column sustain more weight. Consequently, their structures are stronger than other bones. Clinical and experimental observations [34,35] show considerable change in the bone loss of femur, vertebral column, and pelvis. However, these studies do not show any significant changes in the density of other bones such as the humerus, radius, and ulna as they do not sustain high-gravity loads compared to the other bones. However, this is probably due to the time length of the proposed study as it is expected that with lower mechanical loads the bone density would decrease over time. A decrease in gravity and consequently a decrease in the applied pressure on the skeleton cause significant changes in bone density [35,36]. Bed rest studies [37,38] show that a 5- to 12-week period of bed rest causes 3% to 4% loss in the femoral neck and spine. Other works [39] indicate that the amount of bone loss during a 17 weeks bed rest reaches 5% in the femur, 4% in the lumbar spine, and 2% in the tibia. In addition the variation of cortical and trabecular bone loss is different. Comparing these two parts of the bone the amount of volumetric change, volumetric bone mineral density (vBMD), mass, and density are much higher in the trabecular bone [40]. A specific study on the femoral neck by Lang [41] reports a 16.5% bone loss in the trabecular bone for only about 9.4% loss in the cortical bone in identical mechanical load conditions [2].

Bone is a living material and renews itself continuously along the life. Each year 25% of the trabecular bone and 3% of the cortical bone are renewed [42]. As such, bone cells are being continuously activated to either construct new bones or resorb old bones under the various mechanical loads of life. In this process three types of cells are principally in charge of this bone remodeling, namely, osteoclasts, osteoblasts, and osteocytes [43]. Osteoclasts are the cells in charge of the resorption process. They are multinucleated cells and originate from stem cells of the bone marrow. Osteoblasts are the cells in charge of the bone-forming process. Both osteoclasts and osteoblasts are located at the bone surface for their respective degradation and formation of the bone structure [44,45]. On the contrary to the two other types, octeocytes are almost immovable. They have a stellate shape with various biological processes [46]. They manufacture collagen and other substances that make up the bone extracellular matrix [47]. They are also, and more importantly, believed to send mechanobiological signals to osteoblasts and osteoclasts and drive the bone remodeling process through the mechanosensation of the various externally applied mechanical stimuli [48]. In the bone formation process, new bones are formed in response to applied mechanical load, the density of the bone increases, and consequently the strength of the bone is enhanced. On the contrary, in the mechanically "unloaded" state the resorption process occurs with a bone density decrease and loss of strength. Overall the bone remodeling process during average normal life conditions and hence the corresponding bone effective density are the results of a competition, driven by the osteocytes, between the osteoblasts creating the bone and osteoclasts resorbing it under certain levels of applied mechanical loads. This bone density variation depends specifically on the rates of effective bone formation and resorption processes as quantified by Wang [49,50] and Thudium [2,49,50].

In the past decades, many theoretical numerical models have been tried over the years to represent the bone density evolution as a function of time. It mainly started with the very well known Wolff's law [51], stating a structural optimization of the bone microstructure as a function of the applied mechanical loads. Carter [52,53] followed by Huiskes [54] stated the existence of a mechanical equilibrium (for daily life mechanical loads) around which the bone density would be constant (only everyday living remodeling occurs without density variation) through the existence of a so called "lazy zone" for bone strain localized between about 100 and 300 and about 1500−3000 microstrain [2,55].

Over the past 50 years, various definitions of macroscopic bone stimulus have been proposed to be at the origin of the bone remodeling process. They range from adaptative response to physiological approaches, optimal response, and cellular automata [56−65]. Some focus on more local definitions of the mechanisms being at the origin of the bone remodeling process such as cellular sensitivity to fluid transports [66−71] or accounting for more sophisticated constitutive material models for the bone [72−77]. It has come reasonably clear lately that the integration, at some stage, of biological influence in the bone remodeling process was necessary in order to integrate phenomena that cannot be modeled otherwise just by a strain mechanical energy. In essence if the bone is biologically empty, one can apply all possible mechanical loads; no bone remodeling will occur. This is particularly true in microgravity where the bone remodeling is only based on biology and not on mechanical loads. Hence in the past 10 years, new multiphysics models have emerged trying to integrate biology in the bone remodeling problem [78−84]. However, accounting for the local bone microstructure distribution remains a challenge as the material is highly heterogeneous and requires specific homogenization tools [31,85−87] able to account for the specific bone microstructure distribution at the meso and local scales [88−98], in the objective of being able to merge all these together [2,99].

In this chapter modeling of degradation of biomaterials and biological materials is explained in Section 6.3 and Section 6.4. As a sample a numerical approach is implemented to model corrosion behavior of biodegradable PLA/Mg composites and the bone loss of osteoporosis.

6.2 Numerical modeling of the degradation process of PLA/Mg composites

In this section a new approach to the study of corrosion behavior of PLA/Mg composites is presented. This model uses 3D Cellular Automata to predict the corrosion of these composites with different volume fractions. For modeling the 3D structure to use in our method the statistical correlation functions were exploited. These statistical reconstructed samples were used for calibration of the numerical model using the experimental data. It was shown that the samples with a higher volume fraction of magnesium have a higher corrosion rate, and the variations in this rate are not linearly correlated with the change of volume fraction. In the end the importance of statistical reconstruction was demonstrated by comparing the results of a reconstructed model and a randomly generated model, and it was shown that a random generated model could not be used to model the corrosion behavior of these composites [1].

6.2.1 Sample preparation

Hot extrusion is one of the most prevalent means to produce objects with a fixed cross-sectional profile. This process is conducted at high temperature, which keeps the materials from hardening along with making the procedure of driving the material through the die more straightforward. However, hot extrusion has a couple of disadvantages, and the biggest one is the cost of machinery and its maintenance. Thus an internal mixer was used to mix and disperse the particles of magnesium in the polymer. In this method the raw material needed is much lesser compared to the hot extrusion. All the raw materials were dried before mixing for 8 hours at the temperature of 70°C. In Fig. 6.1 the produced composites with different volume fractions of Mg are shown [1].

Several methods and instruments can be used to probe a microstructure; one is using SEM to analyze the main surfaces and fractures. There are some groundworks to be done for investigating these surface fractures. For a sample that is not destructed the examined surface should be polished; for this matter, alumina powder or diamond paste is used. For a destructed sample, polishing is not admissible. Because of the transparency of PLA the bottom layers are observed in the microscopy images, so to analyze the surface of the samples, SEM is needed [1].

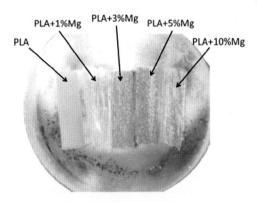

Figure 6.1 Produced composite with different volume fractions of Mg right before taking the SEM images [1].

6.2.2 Statistical reconstruction of samples

The statistical reconstruction technique is the realization of a microstructure from its statistical correlation function which is extracted from SEM images. In this research, reconstruction has been performed using approximation of full-spectrum two-point correlation functions (TPCFs) of a 3D microstructure. In this research, phase-recovery algorithms are used to reconstruct 3D microstructures from 2D images of PLA/Mg composites. Statistical TPCFs were calculated from SEM images as input statistical functions for reconstruction of 3D Eigen microstructures. The phase recovery method is a fast and efficient method for reconstruction of multiphase random heterogenous materials. More detailed information of the reconstruction procedure can be found in our previous publications [1,31,100].

Discretizing grid points of the Eigen 2D images and specifying to every point only one of the N-phases is the starting point in the phase recovery algorithm. This distribution is designated by $\chi^n(s)$ and the characteristic function defined by Eq. (6.1), where the superscript n identifies the phase number and s identifies the number of each grid that defines the 2D images [1].

$$\chi^{(i)}(x) = \begin{cases} 1 \ x \ in \ phase \ i, \\ 0 \ Otherwise, \end{cases} \tag{6.1}$$

For each position, one-point or volume fraction of every phase n can be defined by Eq. (6.2) and Eq. (6.3).

$$C_1^n = \frac{1}{S} \sum_{s=0}^{S-1} \chi^n(s) \tag{6.2}$$

$$\sum_{n=1}^{N} \chi^n(s) = 1, \quad \chi^n(s) \in \{0, 1\} \tag{6.3}$$

Likewise, TPCFs can be represented by Eq. (6.4), where the superscripts n and n' indicate the phases of interest and r is the correlation vector which represents the statistics of the cut section [1].

$$C_2^{nn'}(\mathrm{r}) = \frac{1}{S} \sum_{s=0}^{S-1} \chi^n(s) \chi^{n'}(s + r) \tag{6.4}$$

Fast Fourier transform (FFT) of the cut section function is defined by Eq. (6.5), where $|F_k^n|$ is the amplitude and θ_k^n is the phase of the Fourier transform [1].

$$F_k^n(\mathrm{r}) = \mathscr{F}(\chi^n(s)) = \frac{1}{S} \sum_{s=0}^{S-1} \chi^n(s) e^{2\pi i s k/S} = \frac{1}{S} F_k^n e^{i\theta_k^n} \tag{6.5}$$

By employing FFT and applying the convolution theorem to Eq. (6.4) and assuming periodicity of the structure, Eq. (6.6) is achieved [1].

$$F_k^{nn'} = \mathscr{F}\left(C_2^{nn'}(r)\right) = \frac{1}{S}\left|F_k^n\right|e^{-i\theta_k^n}\left|F_k^{n'}\right|e^{i\theta_k^{n'}}$$

(6.6)

For $n = n'$ the terms associated with phase angles are left out and the fundamental relationship between FFT of TPCF and FFT of the one-point correlation function of a particular vector, k, is defined by Eq. (6.7), where $\tilde{F}_k^n$ is the complex conjugate of F_k^n [1].

$$F_k^{nn'} = \frac{1}{S}\tilde{F}_k^n F_k^n = \frac{1}{S}\left|F_k^n\right|^2$$

(6.7)

This equation is the base of the phase recovery algorithm in cut section reconstruction. F_k^{nn} can be calculated by utilizing TPCFs of all points of the representative elementary volume (RVE), and with this equation, $\left|F_k^n\right|$ is obtained. For the first iteration an initial cut section is considered; the algorithm starts; and in each iteration, $\left|F_k^n\right|$ is updated, but the phase remains consistent. This loop is iterated till a prespecified criterion is satisfied. $\chi^n(s)$ is determined for every point s and designated phase n with recovery of the phase θ_k^n. For a two-phase cut section, $\chi^n(s)$ is calculated using Eq. (6.2). Phase recovery requires the entire spectrum of TPCFs for 3D RVE; therefore 3D TPCF should be approximated based on two original 2D cut sections using Eq. (6.8) [1,31].

$$C_2^{11}(x_1^1, x_3^1) = \frac{C_2^{11}(x_1^1, x_2^1)C_2^{11}(x_2^1, x_3^1)}{v_1} + \frac{C_2^{12}(x_1^1, x_2^2)C_2^{21}(x_2^2, x_3^1)}{v_2}$$

(6.8)

6.2.3 Modeling corrosion

Three principal reactions happen when a metal is corroded [24]. They are called spatially separated electrochemical (SSE) reactions, spatially joint (SJ) reactions (SJ), and diffusion. In this model the simplified electrochemical reactions (anodic and cathodic) are studied, where only H^+ and OH^- ions are considered [1].

6.2.3.1 Spatially separated electrochemical reactions

In an acidic or neutral environment ($pH \leq 7$) the metal oxidation occurs as an anodic semireaction. This reaction is followed by cation hydrolysis [1].

$$Mg \rightarrow Mg^{2+} + 2e^-$$

(6.9)

$$Mg^{2+} + H_2O \rightarrow Mg(OH_{aq})_2 + 2H^+$$

(6.10)

The two-step semireaction can be abbreviated in the following equivalent chemical formulation:

$$Mg^{2+} + H_2O \rightarrow Mg(OH_{aq})_2 + 2H^+ + 2e^- \tag{6.11}$$

In a basic environment $(pH > 7)$, metal oxidation occurs like the following chemical expression:

$$Mg + 2OH^- \rightarrow Mg(OH_{solid})_2 + 2e^- \tag{6.12}$$

The cathodic semireactions are expressed below. In acidic or neutral environments (pH ≤ 7) the cathodic reaction is mostly in the form of a decrease of hydrogen ions [1]:

$$2H^+ + 2e^- \rightarrow H_2 \tag{6.13}$$

Meanwhile, for a basic medium (pH > 7), we have

$$2H_2O + 2e^- \rightarrow H_2 + 2OH^- \tag{6.14}$$

6.2.3.2 Spatially joint reactions

We assume that the anodic and cathodic semireactions occur at the same site; we use the SJ reactions. By this reaction the local pH remains constant. In an acidic or neutral medium (pH ≤ 7) the aggregates of cathodic and anodic reactions are [1]

$$Mg + 2H_2O \rightarrow Mg(OH_{aq})_2 + H_2 \tag{6.15}$$

In the basic medium, we have

$$Mg + 2H_2O \rightarrow Mg(OH_{solid})_2 + H_2 \tag{6.16}$$

Also, there is the possibility that MOH_{solid} dissolutes in a basic medium:

$$Mg(OH_{solid})_2 \rightarrow Mg(OH_{aq})_2 \tag{6.17}$$

6.2.3.3 Diffusion

SSE reactions generate H^+ and OH^- ions. When these two interact with each other a neutralization happens [1].

$$H^+ + OH^- \rightarrow H_2O \tag{6.18}$$

6.2.3.4 Cellular automata modeling

A cellular automata is a model studied in computer science, mathematics, physics, complexity science, biology, and microstructure modeling. A cellular automaton consists of a regular lattice of cells, each in one state, such as on and off. The lattice can be in any number of dimensions. Each cell has a neighborhood that can alter the state of that cell. We assign an initial condition (time $t = 0$) by selecting a state for each cell. A new generation is formed (advancing one by one) according to some fixed rule that defines the updated state of each cell in terms of the current state of the cell and the states of the cells in its neighborhood. Typically the rule for switching the state of cells is identical for each cell and does not change over time and is applied to the entire grid concurrently, though there could be exceptions, such as the stochastic cellular automata and asynchronous cellular automata [1].

The neighborhood of a cell is the adjacent cells. The two most common types of neighborhoods are the von Neumann neighborhood and the Moore neighborhood [1].

The von Neumann neighborhood (or 4-neighborhood) is classically represented on a 2D square lattice composed of a central cell and its four adjacent cells. The Moore neighborhood on a 2D lattice is composed of nine cells, a central cell and the eight cells around it. These neighborhoods are depicted in Fig. 6.2 [1].

6.2.3.5 Cellular automata model of corrosion

Here, a 3D lattice represents the system. In this system, a solid can have several states. If the metal (M) is in contact with the solution, it gets reactive. We show Mg and PLA with red cubes and blue cubes, respectively. The light blue cubes and light red cubes represent PLA and Mg when they are reactive and have contact with the solution. The neutral solution is shown with gray cubes in Fig. 6.3 [1].

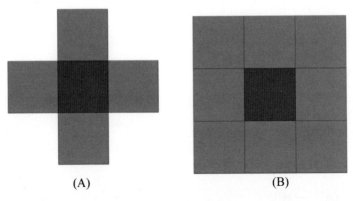

Figure 6.2 (A) von Neumann neighborhood in a 2D lattice where each cell has four neighbors; (B) Moore neighborhood in a 2D lattice where each cell has eight neighbors [1].

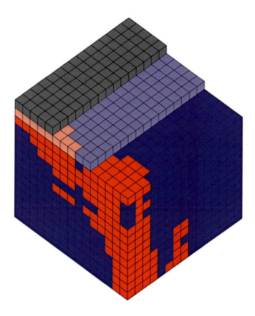

Figure 6.3 3D lattice where red cubes represent Mg and blue cubes represent PLA and better connectivity is considered in our model, in which each cube has 26 neighbors [1].

6.2.3.6 Spatially separated electrochemical reactions

In the previous section, we showed that the acidification process could occur in two different reactions based on the pH of the local environment. We model those reactions with cellular automata as below [1]:

In an acidic or neutral environment (pH ≤ 7),

$$\text{Reactive}_1 \rightarrow \text{Acid}_1 \left(\text{Mg} + 2\text{H}_2\text{O} \rightarrow \text{Mg(OH}_{\text{aq}})_2 + 2\text{H}^+ + 2e^-\right) \quad (6.19)$$

In a basic environment (pH > 7), metal oxidation occurs. In the form of the following chemical expression,

$$\text{Reactive}_1 + \text{Basic}_2 \rightarrow \text{Passive}_1 + \text{Neutral}_2 \, (\text{Mg} + 2\text{OH}^- \rightarrow \text{Mg(OH}_{\text{solid}})_2 + 2e^-) \quad (6.20)$$

The subscript 2 refers to a randomly picked neighbor. The cathodic reactions are represented as follows:

In an acidic or neutral environment (pH ≤ 7),

$$\text{Reactive}_1 \,(\text{Passive}) + \text{Acid}_2 \rightarrow \text{Reactive}_1 \,(\text{Passive}) + \text{Neutral}_2 \left(2\text{H}^+ + 2e^- \rightarrow \text{H}_2\right) \quad (6.21)$$

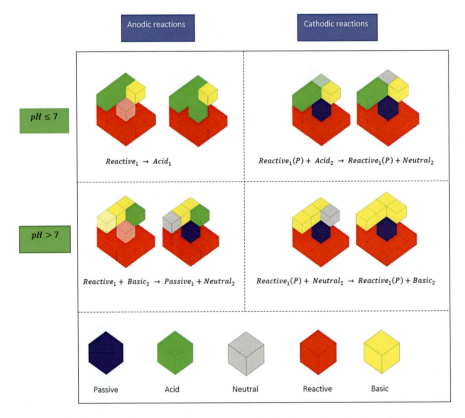

Figure 6.4 Schematic of spatially separated electrochemical reactions in the cellular automata model [1].

Meanwhile, for a basic medium (pH > 7), we have

$$\text{Reactive}_1(\text{Passive}) + \text{Neutral}_2 \rightarrow \text{Reactive}_1(\text{Passive}) \\ + \text{Basic}_2 (2H_2O + 2e^- \rightarrow H_2 + 2OH^-) \quad (6.22)$$

In these reactions, Reactive includes Mg, Acid designates H^+, Basic designates OH^-, $Mg(OH_{solid})_2$ is considered passive, and $Mg(OH_{aq})_2$ is considered neutral. These reactions are presented in Fig. 6.4. In this figure the central cube is the main cube that is expressed with index 1 in the equations. The light-red (pink) cube displays a reactive cube when it is the central cube [1].

6.2.3.7 Spatially joint reactions

The SJ reactions are modeled in cellular automata with these methods [1]:

$$\text{Reactive} \rightarrow \text{Neutral} \ (Mg + 2H_2O \rightarrow Mg(OH_{aq})_2 + H_2) \quad (6.23)$$

In the basic medium, we have

$$\text{Reactive} \rightarrow \text{Passive} \left(\text{Mg} + 2\text{H}_2\text{O} \rightarrow \text{Mg(OH}_{\text{solid}})_2 + \text{H}_2\right) \tag{6.24}$$

$$\text{Passive} \rightarrow \text{Neutral} \left(\text{Mg(OH}_{\text{solid}})_2 \rightarrow \text{Mg(OH}_{\text{aq}})_2\right) \tag{6.25}$$

6.2.3.8 Diffusion

In this section, we assume that if an acidic or basic site interacts with a neutral site, these two sites will switch place. In cellular automata terms, it implies [1]

$$\text{Acid}_1 + \text{Neutral}_2 \rightarrow \text{Neutral}_1 + \text{Acid}_2 \tag{6.26}$$

$$\text{Basic}_1 + \text{Neutral}_2 \rightarrow \text{Neutral}_1 + \text{Basic}_2 \tag{6.27}$$

Meanwhile, if H^+ and OH^- ions interact with each other, a neutralization occurs:

$$\text{Acid}_1 + \text{Basic}_2 \rightarrow \text{Neutral}_1 + \text{Neutral}_2 \left(\text{H}^+ + \text{OH}^- \rightarrow \text{H}_2\text{O}\right) \tag{6.28}$$

The diffusion reactions are presented in Fig. 6.5. In this figure the central cube is the main cube that is expressed with index 1 in the equations. The light-yellow or green cube in the center represents the cube with index 1 in equations. The light-yellow or green cube in the neighborhood represents the cube with index 2 in equations [1].

In a cellular automata simulation, when a time step is complete, some metal sites are corroded. There could be a case that a metal site is not corroded, but all the other 26 neighbors are corroded, or the site is not percolated with the main solid. In this case, after each time step, we consider those sites to degenerate, and the algorithm should eliminate these sites at the end of each time step. The example of this process is illustrated in Fig. 6.6 [1].

To have a stochastic model, we assign a probability to each reaction. The SSE reactions have a probability independent of pH called P_{sse}; however, for the SJ reactions, not only they are related to local PH but also they are a function of the parameter called N_{exc}, which is the difference between the acidic and basic sites in the 26 Moore neighbors. This parameter describes the intensity of the acidic environment. The probabilities of the SJ reactions are shown in Table 6.1. Diffusion reactions do not have a probability; instead, we have a parameter called N_{diff}, which is the number of diffusions take place in each time step. In our model, we assumed this number of diffusion to be 26, that is, the number of Moore neighborhoods so that in each time step, all of the 26 neighbors have a possibility to be selected to diffuse once [1].

Our lattice size is $100 \times 100 \times 100$. The P_{sse} is different for magnesium and PLLA because Mg corrodes 5 to 10 times faster than PLLA. In this work, we have the amount of corroded Mg after 30 days, so we focus on the Mg corrosion, and therefore modeling the PLLA corrosion is not necessary here. The probability of the SJ reactions is shown in the table below. The experiment is done in 30 days, so

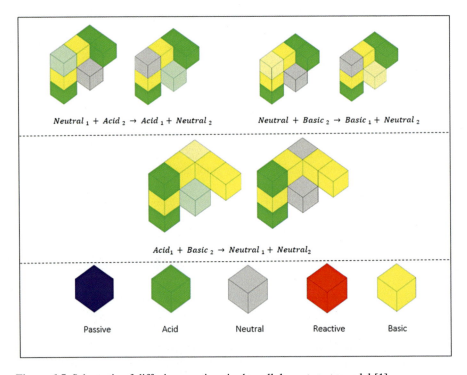

Figure 6.5 Schematic of diffusion reactions in the cellular automata model [1].

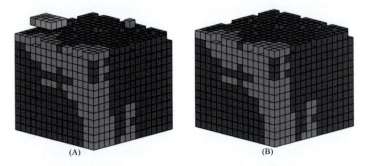

Figure 6.6 Cellular automata model after a time step before removing islands (A). Cellular automata model after a time step right after removing islands (B) [1].

Table 6.1 The probabilities of the SJ reactions [1].

	pH < 7	pH = 7	pH > 7
Reactive → Neutral	0	0	0
Reactive → Passive	0	0.001	1
Passive → Neutral	$\frac{1}{26} \times Nexc$	0	0

Numerical modeling of degraded microstructures

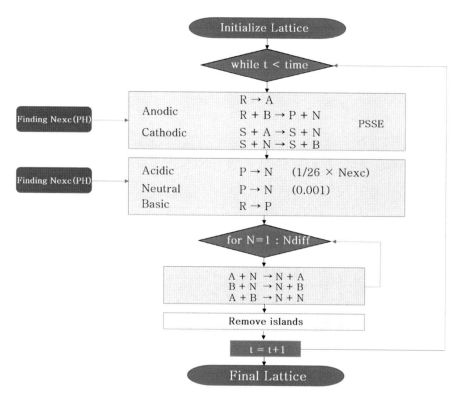

Figure 6.7 Flowchart of the cellular automata algorithm to model corrosion [1].

we have 720 hours, and we consider each time step to be an hour. P_{sse} for both Mg and PLLA should be calibrated with the experimental results in this duration. We calibrate these parameters with the result of the stent with 10% volume fraction and should validate the model predictions with those of experiments for the sample with the volume fraction of 5% [1].

The flowchart of the algorithm is presented in Fig. 6.7.

In the first time step, we have no acidic or basic site, so the environment is neutral; in this step, the anodic reaction (Reactive$_1$ → Acid$_1$) can take place, but the cathodic reaction (Reactive$_1$(Passive) + Acid$_2$ → Reactive$_1$(Passive) + Neutral$_2$) cannot occur because there is no acidic site yet. In this case, we consider that the reaction (Reactive$_1$(Passive) + Neutral$_2$ → Reactive$_1$(Passive) + Basic$_2$) can take place instead, so the problem could be solved [1,17].

6.2.4 Results

6.2.4.1 Corrosion test experimental setup

In this experiment the implant is placed in Hanks' buffer solution in a specific period. This solution contains the elements that can allow cells to stay alive. In Table 6.2 the elements and their mass concentration are listed [1,101].

Table 6.2 Elements and their mass concentration in Hanks' buffer solution [1].

Element	Mass concentration (gr/L)
NaCl	8
KCl	0.4
$CaCl_2$	0.14
$NaHCO_3$	0.25
$C_6H_6O_6$	1
$MgCl_2 0.6H_2O$	0.1
$MgSO_4 0.7H_2O$	0.06
$KH_2PO_4.H_2O$	0.06
$Na_2HPO_4 0.7H_2O$	0.06

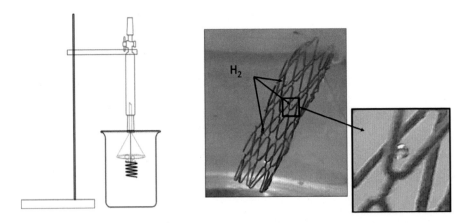

Figure 6.8 Hydrogen measurement system [1].

When the implant is placed in the liquid, due to presence of magnesium in the composite and the chloride solution, hydrogen gas will be released [102]. The amount of released hydrogen is directly related to decayed to magnesium by the equation below [1]:

$$Mg + H_2O \rightarrow MgO + H_2 \quad (6.29)$$

This volume of hydrogen can be measured accurately using a funnel and buret, and the weight loss of magnesium can be calculated. With this method, one may measure the volume and mass of the magnesium that is eroded. The released hydrogen by the surface of the sample is collected and measured, as shown in Fig. 6.8 [1].

The area exposed to the solution, volume, density, and mass of magnesium in the composite samples are presented in Table 6.3 [1].

Hydrogen is released for every sample that includes Mg (except the sample with 100% PLA). The trend for hydrogen release is raising for each sample; this increase shows that after 30 days, magnesium is still present in the composite and has not been fully decayed. The rate of this release improves when the volume fraction of

Table 6.3 Properties of the composite samples [1].

Volume fraction of magnesium (vf%)	0	1	3	5	10
Area exposed to the solution (cm^2)	12.04				
Total volume (mL)	0.98927				
Initial total mass (gr)	1.2366	1.241	1.2499	1.2589	1.2786
Density (gr/mL)	1.25	1.2545	1.2635	1.2725	1.2925
Total mass of magnesium (mg)	0	16.8176	50.4528	84.088	168.1759

Table 6.4 The rate of hydrogen releases from PLLA/Mg composites after 30 days [1].

Volume fraction of the composite (%)	0	1	3	5	10
Final volume of hydrogen (mL)	0	0.35	1.45	5.4	18.8
Rate of hydrogen release ($\frac{mg}{ml.day}$)	0	0.0007	0.0028	0.0141	0.0556

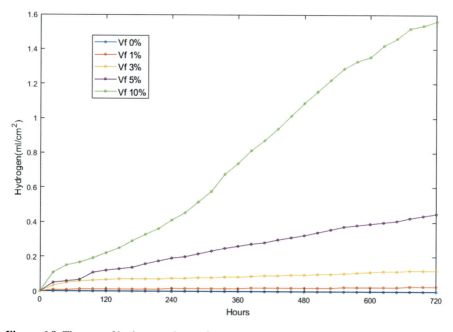

Figure 6.9 The rate of hydrogen releases from PLLA/Mg composites after 30 days [1].

Mg increases. We can consider the rate of hydrogen release to be linear, so the slope of the diagram below is calculated and given in Table 6.4 [1].

As shown in Fig. 6.9, it is revealed that with the increase of the volume fraction of Mg from 1% to 3% the amount of released hydrogen is up to 4 times. When the

Mg volume fraction is 5% and 10%, this amount will rise to 20 and 80 times, respectively. This difference is caused by a phenomenon called percolation threshold. This is when the Mg particles get connected and form a lattice [1].

6.2.4.2 Reconstruction

The SEM images of the stent have been taken before placing the stent in the solution. With image processing, SEM images have been converted to binary ones, which assign each pixel to either Mg or PLLA. These images are shown in Fig. 6.10 [1].

Reconstructed models of two images are shown in the form of a 3D lattice to use for the corrosion modeling and are represented in Fig. 6.11 [1].

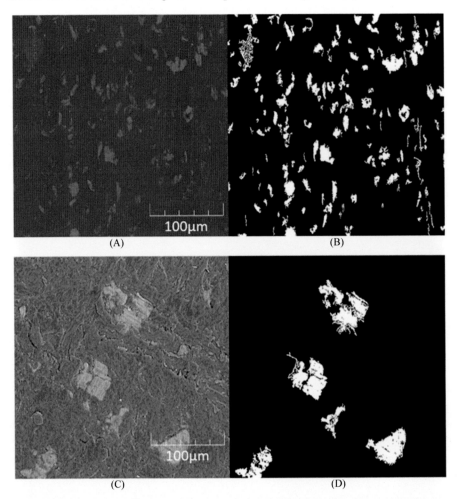

Figure 6.10 (A) SEM image of the stent with 10% Mg, (B) binary image of the stent with 10% Mg, (C) SEM image of the stent with 5% Mg, and (D) binary image of the stent with 5% Mg [1].

Numerical modeling of degraded microstructures

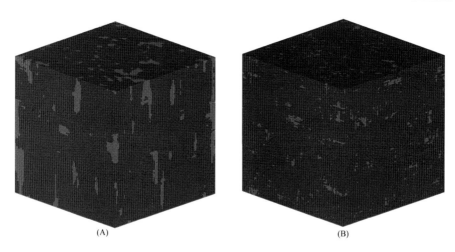

Figure 6.11 (A) Reconstructed 3D lattice of the stent with 10% Mg and (B) reconstructed 3D lattice of the stent with 5% Mg [1].

6.2.4.3 Cellular automata

As presented in the experiments section, the hydrogen release rate (rate of corrosion or in our cellular automata model, the possibility of the Psse reactions) changes relative to the volume fraction of Mg in the composite [1].

In the next stage, we try to find this parameter in our model with 10% volume fraction and see if we can validate the result with this constant for the composite with 5% volume fraction. Based on the experiments, for the composite with 10% Mg, after 30 days, 31.3584 mg of Mg is corroded. In the initial conditions the total mass of Mg in this sample was 168.1759 mg, so after 1 month, 18.65% of Mg is corroded. We should find the Psse for Mg in our cellular automata model so that after 720 time steps, 18.65% of Mg is corroded. With trial and error, *Psse* is calculated to be 0.1259 (the corrosion of magnesium in the sample with 10% volume fraction is shown in Fig. 6.11), so Psse for Mg is [1]

$$Psse_{Mg} = 0.008624 \qquad (6.30)$$

The result of the model for the sample with 5% of Mg is shown in Fig. 6.12. In both samples with 5% and 10% of Mg the amount of Mg that is corroded in the end is less than 20%. In the cellular automata algorithm the corrosion occurs from the first layer on the top of the microstructure. This means that about 20%–25% of the layers from the top are corroded, so the structure that was modeled is the one in Fig. 6.11. In Fig. 6.13 only the top 40 layers are shown, and in Fig. 6.12 only the top 20 layers are shown so that the top layers that are corroded are clearer in these figures [1].

In this case, 2.7350% of Mg is corroded. In the experiment, 2.716% of Mg was corroded; therefore with the calibration of parameters in a sample with 10% of Mg, we were successful in verifying the results for a sample with 5% volume fraction [1].

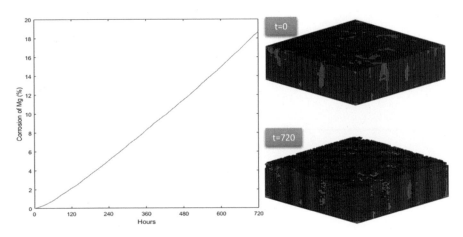

Figure 6.12 The corrosion of magnesium in the sample with 10% volume fraction [1].

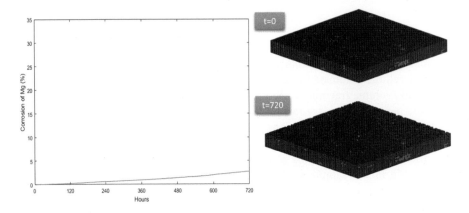

Figure 6.13 The corrosion of magnesium in the sample with 5% of volume fraction [1].

Now we will show the importance of reconstruction for modeling the structure to use it in cellular automata. We consider a random lattice with 5% and 10% volume fractions. In this case, since the Mg particles are randomly distributed and are not connected the rate of corrosion will decrease. In Fig. 6.14 only Mg particles are displayed to show the percolation threshold effect. In the reconstructed model the Mg particles get connected and form several clusters; thus the corrosion rate is higher than the random lattice. Each cluster is shown with a unique color [1].

In these figures, each cluster is distinguished by a color. It is obvious that in the reconstructed model the corrosion rate is higher than the arbitrary case. If we run the proposed model with previous parameters, we have the below results (Fig. 6.15 and Table 6.5) [1].

The error margin in the sample with 5% volume fraction is less than that in the sample with 10% volume fraction. The main reason for this occurrence is that the

Numerical modeling of degraded microstructures

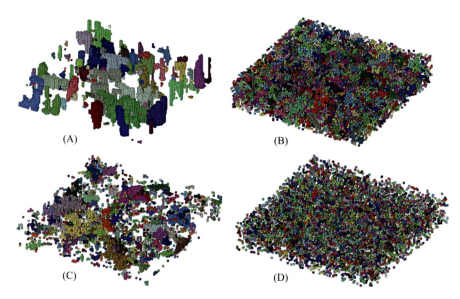

Figure 6.14 (A) Reconstructed lattice with 10% Mg, (B) random lattice with 10% of Mg, (C) reconstructed lattice with 5% Mg, and (D) random lattice with 5% of Mg [1].

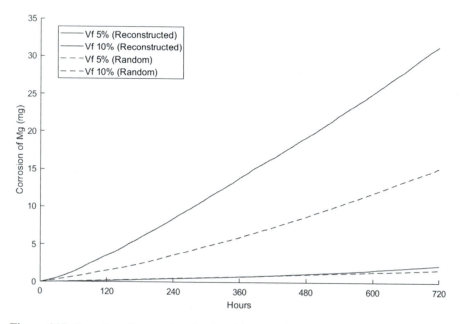

Figure 6.15 Corrosion of magnesium for the arbitrary and reconstructed models [1].

Table 6.5 Corrosion of magnesium for the arbitrary and reconstructed models [1].

Mg volume fraction	5%	10%
Corrosion in the reconstructed model (mg)	2.3	31.45
Corrosion in the arbitrary model (mg)	1.7266	15.27
Corrosion in the experiment (mg)	2.28	31.36

Figure 6.16 Randomly generated lattice with a volume fraction of 25% [1].

amount of corrosion is too small. In cellular automata, we consider that the corrosion starts from the top layer, so the first layer is corroded with the same rate in both cases. In this case, it means that one percent of the sample is eroded with the same pace, and the connectivity does not play a role in the first layer, but when the algorithm advances to the next layers, connectivity becomes vital and has an impact on the rate of corrosion [1].

For the volume fractions around 25% (and higher) the lattice becomes almost percolated (as observed from Fig. 6.16), so it is expected to see the same results for corrosion rate for both reconstructed and random structures, but if we want to find the mechanical characterization for the proposed model, then reconstruction is a necessity.

6.3 A numerical model of bone density loss in osteoporosis

In this section a new approach is proposed to predict bone density loss in microgravity. It is based on an optimization model identified on experimental data and able to predict bone degradation at the trabecular scale. This model aims at the prediction of long-term bone density loss in microgravity. The investigation is implemented on real human bone microstructures. The bone loss is determined as a function of time. We show that bone loss does not happen at a constant rate and depends on the bone microstructure distribution, and the size of the bone surface available at the mesoscale [2,103,104].

6.3.1 Model objectives

There currently exist no experimental data to precisely quantify the bone density variation kinetics at the trabecular scale. At best, bone density variation is measured at the macroscopic patient scale over time. However, since it is agreed that bone density variation is patient-dependent, there exist an influence of the bone microstructure distribution and bone biology on this evolution for each person. Hence we propose to develop a model that should account for the bone microstructure distribution to propose an effective bone density evolution as a function of time and based on macroscopic experimental data available in the literature. The development methodology is divided in four different steps as follows [2]:

1. Macroscale: identify the effective bone degradation at the macroscopic scale from experiments for given environments.
2. Macroscale to mesoscale: identify at the local trabecular scale (for a given patient) the bone degradation mechanisms as a function of the macroscopic identification.
3. Mesoscale to macroscale: predict bone degradation at the local scale for the given bone microstructure and homogenize at the macroscopic scale.
4. Macroscale: predict long-term macroscopic bone degradation.

The principal objective here is to validate a theoretical model that will be identified on macroscopic experimental data and be able to account for the patient's bone microstructure distribution to predict long-term bone degradation [2].

6.3.2 3D reconstruction of the bone microstructure

Trabecular bone samples were extracted from iliac crest bone biopsies and then defatted and imaged with a cone-beam system Skyscan 1072 (Skyscan, Brucker, Kontich, Belgium). The images were acquired at 80 kV and 100 μA, with an angular rotation of 0.45 degree. 3D images were reconstructed with a voxel size of 10.77 μm using the manufacturer's reconstruction algorithm, which is based on a customized Feldkamp algorithm [2].

The experimental CT scans (see an example on Fig. 6.17A) were used to create the FE numerical models. A 3D stitching of the micro-CT data was done to construct the 3D virtual model of trabecular bone microstructures. For this a MATLAB® code has been developed using the CT images as input data. The reconstructed model (Fig. 6.17B) is meshed based on the resolution of the CT images. The FE mesh is constructed on the CT-scan voxels, leading to a cube containing about 1 million voxels per sample [2].

The initial samples are constituted only by the dried bone microstructure and empty space around them. To numerically represent the bone density evolution (at the bone border surface), we assume that the sample is inside a real bone and hence empty space is filled with the bone marrow. For this the reconstructed numerical FE model is composed of two phases, which are the bone and bone marrow in contact with each other (Fig. 6.17B) [2].

The mechanical properties of each voxel are assigned based on the Hounsfield unit (HU) of each pixel in the CT-scan images. Each phase (bone and bone marrow)

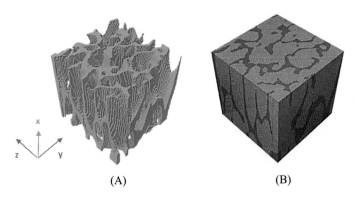

Figure 6.17 Example of a trabecular bone 3D model constructed with a MATLAB code. All samples of this work are obtained from experimental CT scans. (A) An experimental sample at the voxel scale. (B) The constructed numerical model from MATLAB with both the dark-gray voxels being the bone trabeculae and the light-gray voxels corresponding to the bone marrow [2].

and more specifically the definition of the microstructure distribution of the bone are defined on the values allocated to the Hounsfield unit. A normalized zero value corresponds to the bone marrow, while a normalized value of 1 corresponds to the bone. Hounsfield values that are between 0 and 1 will be located around the interface between the bone and bone marrow. We hypothesize that the materials are homogeneous and isotropic and that they have a linear elastic behavior with a Young's modulus of the bone and bone marrow of 15 GPa and 1 kPa, respectively. The Poisson's ratio is set to 0.3 for both materials [2].

6.3.3 Determination of the bone quality in microgravity

Bone density is related to its remodeling occurring throughout our entire life as being a competition between bone formation (through osteoblasts cells) and bone degradation (through osteoclasts cells). For constant mechanical load, this density varies according to age. It increases during childhood, is more or less constant during adulthood, and decreases after about 50 years old (due mainly to osteoporosis). Comparatively, bone loss occurs in microgravity due to a decrease in the applied mechanical load (change in gravity conditions) and loses its mineral contents rapidly to adapt to the new living conditions [2].

Currently the bone strength is usually measured using bone mineral density (BMD) based on dual-energy X-ray absorptiometry (DXA). This type of measurement is usually performed to follow the bone status of astronauts. However, BMD is an expression of bone strength that expresses the amount of bone mineral content divided by a surface area for which the measured BMD is representative of about 70% of the bone strength [105]. This does not correspond to an accurate parameter for true bone density in terms of volume fraction, which is the bone mass divided by the volume of the bone. Some works have proposed a simple conversion of the

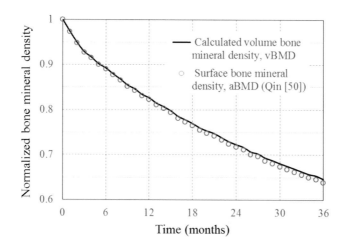

Figure 6.18 Percentage of the bone mineral density loss aBMD based on surface measurements and volumetric bone mineral density loss vBMD. The circles are the fit through the experimental data obtained from surface measurements by Qin [107]. The black line represents the volumetric bone density loss obtained by our calculation using the interpolation by Liu et al. [2,108].

volumetric BMD to bone volume fraction [106], but until now, there does not seem to be any method to precisely relate the BMD to the bone volume fraction. Hence, as the volume fraction is the parameter used here in the numerical simulations and FE modeling a more direct link is required between BMD and bone quality [2].

In Qin [107] a comparative study is proposed between microgravity bone loss and ageing bone loss. Based on this comparison, it is estimated that for a 12 months space mission the bone loss would be about 18% of its initial amount of mineral content. Liu et al. [108] proposed a linear relation between the volumetric BMD (vBMD, determined by a two-compartment model method) and area BMD (aBMD, obtained from DXA). Following this approach, we applied this linear relation to evaluate the amount of vBMD compared to aBMD for the bone loss estimated by Qin et al. [107]. From these works, we show that the amount of bone loss measured by aBMD as a function of time is remarkably close to the calculated vBMD variation with a difference lower than 1% (see Fig. 6.18). Hence the results indicate that it is possible to use the estimated bone mineral density loss aBMD based on surface measurements as a direct comparison with the volumetric mineral bone density loss vBMD [2].

6.3.4 Theoretical model

In the following a new strategy is proposed to represent the bone density variation in microgravity. Each voxel of our model (Fig. 6.17) is defined based on the Hue value (Hounsfield unit) corresponding to a given bone density within the element (voxel). At the trabecular scale the voxels located inside the trabecula will have the highest density (normalized to 1) and those in the bone marrow will have the lowest

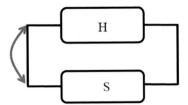

Figure 6.19 Simplified model to transform hard (H, bone) and soft (S, bone marrow) phases to represent bone formation or bone degradation at the bone surface [2].

density (normalized to 0). For voxels located around the surface of the bone trabecula, their density will take different values between 0 and 1. For a bone density increase, these voxels will have their density increase, while for bone loss, it will decrease. Hence within a given voxel, this density is assumed to be a percentage of the voxel volume represented as a hard phase, while the rest is the soft phase. A bone density of 0.3 in a given voxel will be constituted of 30% volume hard phase and 70% volume soft phase. The bone density change is then theoretically represented as a phase exchange between the hard phase (bone) and soft phase (bone marrow) within the voxel, and this is for all voxels inside the model [2].

For each element located on the border between the bone and bone marrow a phase exchange is defined. With this approach, both reconstruction and resorption phenomena are possible as a function of the intensity of the applied mechanical load (see Fig. 6.19). No empty space is assumed between the hard and soft phases, and each phase is able to transform into one another [2].

As a reversible exchange is defined between the hard (H) and soft (S) phases, the evolution of each phase is governed by the following kinetic equations [2]:

$$\dot{C}_H = -K_{HS}C_H + K_{SH}C_S \quad (6.31)$$

and

$$\dot{C}_S = K_{HS}C_H - K_{SH}C_S \quad (6.32)$$

where C_H and C_S are the volume fractions of the hard and soft phases, respectively. The coefficients K_{HS} and K_{SH} express the rate of exchange between phases. In the general case, these parameters are dependent on the applied mechanical stimulus and other biological issues as they describe the activity of the bone cells. K_{SH} denotes the coefficient affected by the osteoblast activity, and K_{HS} is indicative of the osteoclast activity. Since the soft and hard phases can be both present within the same voxel, the summation of the volume fraction of these two phases is equal to 1 [2].

$$C_H + C_S = 1 \quad (6.33)$$

Moreover, derived from Eq. (6.3) the summation of the rate of these coefficients in healthy conditions is equal to 0 [2].

$$\dot{C}_H + \dot{C}_S = 0 \quad (6.34)$$

Numerical modeling of degraded microstructures 279

Solving Eq. (6.34) as a function of time the volume fraction of a bone voxel changes at each time increment as a function of the coefficients K_{HS} and K_{SH} as follows [2]:

$$(C_H)_{n+1} = \frac{(C_H)_n + K_{SH}.\Delta t}{1 + (K_{SH} + K_{HS}).\Delta t} \tag{6.35}$$

Δt is the time increment between t_{n+1} and t_n. As it was presented that the osteoblasts and osteoclasts are located at the surface of the bone structure, the remodeling happens at the surface of the bone [45]. Thus the remodeling equation should be solved only for the voxels placed at the surface of the bone [2].

From the many models existing in the literature presented above, it is clear that the bone density evolution depends on the locally developed mechanical stimulus and will be different for each voxel of the bone surface on a mechanically loaded structure. In addition, each voxel of the bone surface may have one or more of its faces in contact with the soft (bone marrow) phase which will influence its evolution kinetics. Hence the size area in contact with the bone marrow for each voxel face makes the evolution rate vary throughout the voxels. When the surface increases the density rate increases. To account for the variability of the interface area for each voxel the variables K_{HS} and K_{SH} are defined as a function of the surface area such that [2]

$$K_{HS} = \text{Contact area} \times A_{HS} \tag{6.36}$$

and

$$K_{SH} = \text{Contact area} \times A_{SH} \tag{6.37}$$

where A_{HS} and A_{SH} are variables that need to be determined through an optimization procedure due to the fact that the evolution rates are dependent on the size of the interface surface per voxel and also on the locally developed mechanical stimulus [2].

For microgravity conditions, no applied mechanical load exists. Hence the bone degradation kinetics does not depend on the locally developed mechanical stimulus. In the following, we assumed, without specific knowledge available, that biology distribution was homogeneously distributed within the bone microstructure, corresponding to an averaged healthy patient. This means that the bone degradation kinetics will be homogeneous everywhere on the microstructure and only dependent on the bone microstructure and surface distribution. The degradation kinetics is then directly identified with the model parameters from the effective experimental degradation rate [2].

6.3.5 Optimization algorithm

The determination of the model parameters, in the remodeling governing Eqs. (6.36) and (6.37) for each voxel, was done using a particle swarm optimization (PSO) algorithm. PSO is a population-based algorithm developed by

Eberhart et al. [109] in 2001. It was initially inspired by the movement of plural biological organisms like birds. The algorithm is randomly initialized with a population of dual variables. In the current study, each voxel contain two variables, A_{HS} and A_{SH}. To find the optimized solution of the problem the remodeling process is performed for all the voxels with an iterative process. For each dual variable (particle) and iteration the best experience (minimum value) is extracted and defined as the personal best (PBest) among all particles, and the dual variable (particle) which has the smallest value (smallest personal best — PBest) is introduced as the global best (GBest). A schematic flowchart of the PSO algorithm is presented in Fig. 6.20 [2].

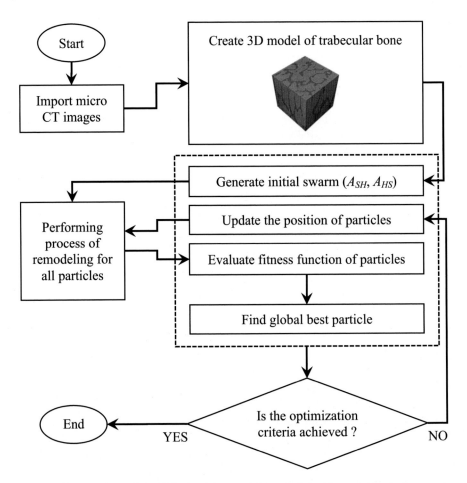

Figure 6.20 Flowchart of the PSO algorithm used for optimizing the variables in the remodeling equations. Swarms contain two variables of the equations. The best swarm determined through the optimization algorithm shows that somehow the trend of bone loss from numerical simulation has minimum discrepancy with the predicted trend of bone loss by Qin [2,107].

Through the iterative process the optimal solutions for the parameters A_{HS} and A_{SH} are found. Here the experimental data presented by Qin [107] were used for optimization of the degradation kinetics of the bone. They are representative of an average bone degradation rate over a 36 months period. After parameter identification from the optimization process and comparison between the calculated degradation rate and experimental data, we show that the difference between this optimized process and the real macroscopic bone degradation (from Qin's data) is very small (see Results Section 6.4.3 below) and will not lead to any misinterpretation in the overall degradation mechanism. Hence it is assumed that the optimization process is valid for the entire sample [2].

Once the optimization process was finished and the best parameters A_{HS} and A_{SH} were determined, we used the proposed model presented in paragraph 3.1 to calculate the bone degradation mechanism and long-term prediction.

This prediction could only be validated at a later stage as there is currently no experimental information on the local degradation kinetics at the bone trabecular scale nor over long periods of time. However, in our previous works [87], we evaluated the influence of bone microstructure distribution on the developed mechanical energy for bone remodeling using a statistical reconstruction method and showed that for a given effective bone density the influence of the bone microstructure distribution remains small for a given static load. Here, although no external mechanical load is present, we are particularly interested in this small patient-dependent variability as it will impact long-term predictions. Of course, this effect would be increased by the presence of a mechanically loaded microstructure where the bone degradation kinetics would depend on the intensity of the locally developed stimulus. This will need to be addressed separately [2].

6.3.6 Results and discussion

6.3.6.1 Bone microstructure data

Using the provided experimental micro-CT images, 10 models of the trabecular human bone were generated. For these created models the bone volume fractions (bone volume/total volume of sample) varied from 13.64% to 31.16%. The bone surface depends on the microstructure distribution of the bone and its volume fraction. For the studied samples, we observed that by increasing the volume fraction the amount of bone surface area increased in general (see Table 1, later down). Of course, this is completely patient-dependent, but it provides a general trend that affects both the degradation rate and bone stiffness as a function of time [2].

6.3.6.2 Effect of the surface contact area on the rate of degradation

The bone density variation occurs at the surface of the bone and is calculated using the experimental degradation kinetics proposed by Qin [107]. A larger bone surface (voxel area) in contact with the bone marrow increases the degradation rate. As all

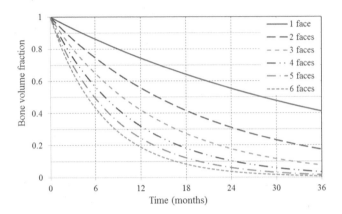

Figure 6.21 Bone loss per voxel as a function of time using our degradation model based on the number of faces in contact with the bone marrow [2].

voxels have the same size, we investigated this effect on one voxel of the model as a function of the number of faces in contact with the bone marrow being considered as a descriptor of the contact area. The results are presented in Fig. 6.21. As expected, with an increase in the number faces in contact with the bone marrow a faster degradation is observed, while for only one face the degradation is the slowest [2].

Since it was assumed that the effective bone density change was the same as the one at the local voxel size the only influencing parameters in the bone microstructure degradation at this scale will be the density of bone within the voxel and the size of the contact area with the bone marrow, whereas at the global scale, it will be the number of bone voxels in the sample and their contact surfaces constituting the bone surface for a given sample [2].

6.3.6.3 PSO optimization results

The long-term prediction of bone loss in microgravity should be done on an "average" sample in order to be representative of an "average" person since the bone microstructure varies from one person to another. Hildenbrand and Anderson [110,111] showed that the standard volume fraction of the femur trabecular bone is around 27%. Hence to use the general trend of bone loss presented by Qin [107], we choose among the available samples the one with a volume fraction equal to 26.12% (corresponding to an elderly woman aged 70). The results presented in the following are based on this sample [2].

The initial population used for the PSO optimization process is composed of 100 particles initialized randomly. The procedure is implemented for 50 optimization iterations. The time steps are divided into a 1 month period for performing the remodeling process over a 36 months (3 years) period corresponding to the duration of Qin's [107] experimental data. The results of the optimization process are presented in bold in Table 6.6 [2].

Table 6.6 Input and output parameters of the PSO optimization process for the experimental degradation study applied to the "average" bone sample [2].

Nb. particles	Nb. iterations	Time step	A_{HS} (1/mm^2)	A_{SH} (1/mm^2)	Avg discrepancy error
100	50	1 month	*1*	*280*	0.45%

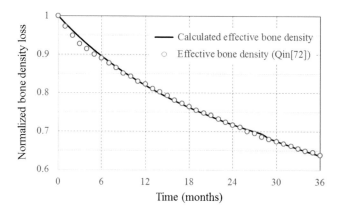

Figure 6.22 Comparison of the bone density loss obtained from the PSO optimization work at the scale of the sample and the fit through the experimental data by Qin [107]. The black line is obtained from the degradation model developed at the voxel scale for all voxels at the bone surface and homogenized over the whole sample [2].

From these identified parameters the bone degradation was calculated and the prediction of the effective (macroscopic) bone density degradation was plotted as a function of time and compared with Qin's results [107]. Fig. 6.22 shows both the predicted trend of our model and the bone density loss by Qin [107]. We can observe a very good superposition of both curves [2].

The optimization process was performed to validate the proposed predictive theoretical numerical model with the data available in the literature. With this validation, it is then possible to extend the degradation process and predict the evolution of the bone microstructure for a long-term living in microgravity conditions. The model was used to extrapolate the bone degradation over a 12 years period and observe the bone microstructure evolution. The results of this prediction are presented in Fig. 6.23 for the studied sample. After a period of 144 months (12 years) the trabecular bone has lost about 82% of its initial volume fraction. This corresponds, at the trabecular scale and for an average starting volume fraction equal to 26%, to a final density of only 4% of the initial bone volume fraction [2].

A better understanding of the degradation mechanism regarding the bone microstructure evolution is shown in Fig. 6.24. The results show three different sections within the same "average" sample. The voxels are displayed in various colors for

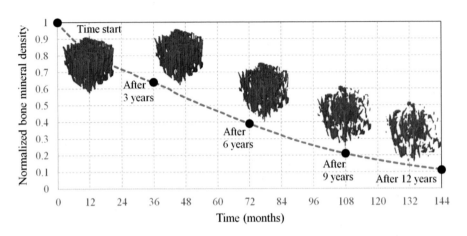

Figure 6.23 Prediction of the bone microstructure degradation and changes in the bone volume fraction over a period of 12 years (144 months) [2].

three ranges of normalized bone densities: (1) white between 0.9 and 1, (2) gray between 0.4 and 0.9, and (3) black between 0.1 and 0.4 [2].

Of course, several simplifying hypotheses were made in the current study, also simplifying the interpretation toward an exact evaluation in real life. These indicate mainly, but not exhaustively, the fact that the degradation mechanism remains phenomenological based on experimental data. However, this brings strength to our approach as the local bone mechanobiology is still not well understood and would have provided some approximation errors in our analyses. Nevertheless, although a phenomenological degradation over 3 years gives an almost exact prediction, there is no guaranty that this trend continues for very long durations as most probably, when the bone density reaches a lower threshold, new biological events will develop in order for the bone density not to degrade completely. Another aspect is the assumption that the macroscopic degradation kinetics is the same as the local degradation kinetics. More work is required to better correlate these two effects. However, it feels confident that under the hypothesis of isotropy and homogeneity of the materials and without externally applied mechanical load (no strain energy within the structure) the bone degradation kinetics should remain constant everywhere. The last point is related to the patient's variability. As different persons have different bone densities, microstructure distributions, and related bone biology, there is no reason why the bone loss should be equal between each person. It is reminded that the prediction provided in the current work is based on an average degradation curve (Qin [107]) and with more experimental data available on different groups of people the model could be adjusted for better long-term predictions [2].

6.3.6.4 Influence of the available bone surface on the rate of degradation at the sample scale

After Fig. 6.21 we can observe that there is a large variation of bone degradation kinetics as a function of the bone surface area. This is also dependent on the bone

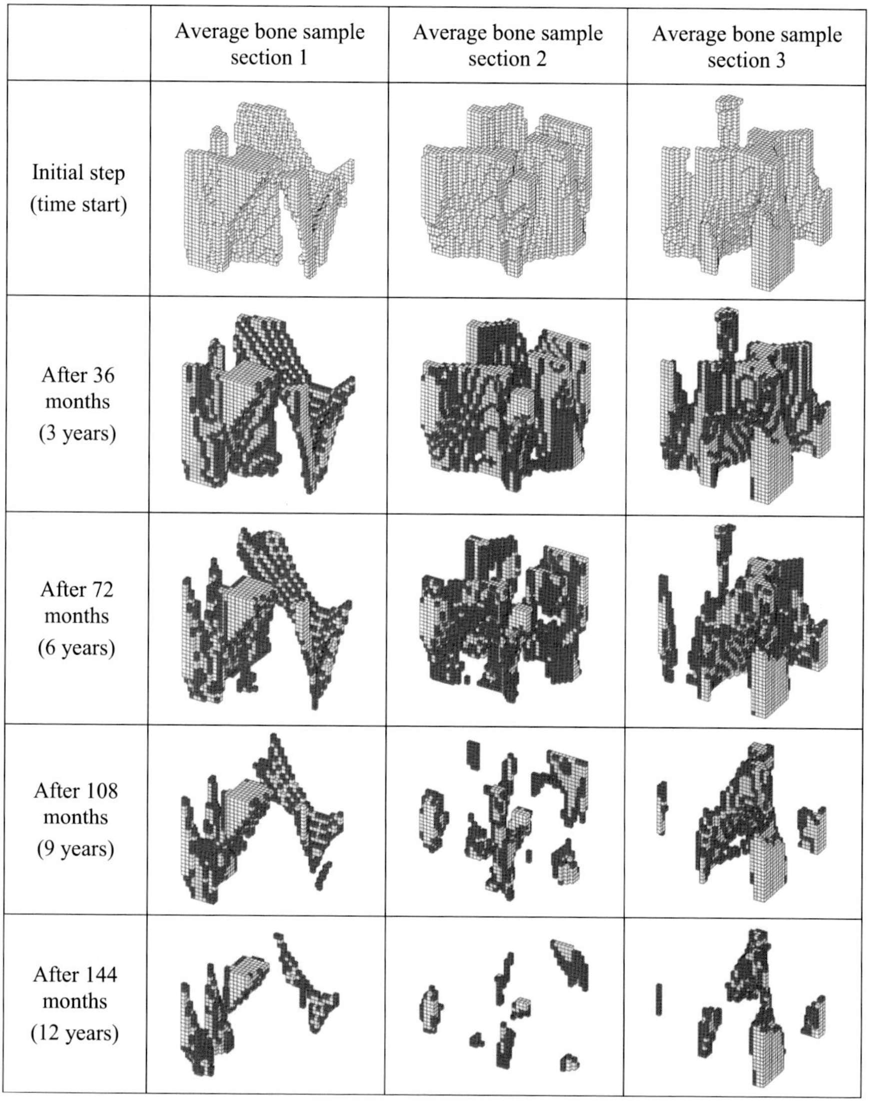

	Average bone sample section 1	Average bone sample section 2	Average bone sample section 3
Initial step (time start)			
After 36 months (3 years)			
After 72 months (6 years)			
After 108 months (9 years)			
After 144 months (12 years)			

Figure 6.24 Simulation of the degradation of bone voxels in different sections of the microstructure. Each column corresponds to a different section of the same "average" sample. Evolution is shown for every 36 months. The white voxels are densities between 0.9 and 1, gray voxels are densities between 0.4 and 0.9, and black voxels are densities between 0.1 and 0.4 [2].

density in the sample and the microstructure distribution (which is patient-dependent and highly heterogeneous). Since the average bone density is measured macroscopically, it is interesting to highlight the degradation rate for different samples with similar initial bone volume fractions but different bone surface areas as it will correspond to different patient cases. The simulations of bone loss were

Table 6.7 Generated samples to compare bone density loss as a function of time depending on (1) similar bone volume fractions (bold) and different bone surface areas and (2) similar bone surface areas (bold) and different bone volume fractions [2].

Sample N°	Initial bone surface (mm^2)	Initial bone volume fraction	Final bone volume fraction
1	12.590	**0.1364**	0.1059
2	14.311	**0.1367**	0.1021
3	13.193	**0.1524**	0.1202
4	14.839	**0.1526**	0.1171
5	**17.572**	**0.2415**	0.1990
6	20.438	**0.2416**	0.1919
7	**12.448**	0.1365	0.1059
8	**12.594**	0.1209	0.0910
9	**17.668**	0.3116	0.2686
10	**17.776**	0.2280	0.1845

performed for six samples with three different couples of similar initial bone volume fractions but with different bone surface areas. On the contrary, we also compared five samples (two different groups) of similar bone surface areas but different initial bone volume fractions. The different cases are presented in Table 6.7. The bold values are the comparative initial values for either initial bone volume fraction or initial surface area [2].

The comparative similar initial bone volume fractions were respectively 13.6%, 15.2%, and 24.1%. For these the difference in the available bone surfaces between similar samples varied from 1.64 to 2.87 mm^2. For similar initial bone surfaces the comparative cases were around 12.5 and 17.6 mm^2. The prediction of bone loss was made using the above proposed model with optimized coefficients. The results are presented in Fig. 6.25 [2].

We can observe, as it was presented previously on a single voxel, that the bone degradation rate at the sample scale is directly dependent on the amount of bone surface available. A larger bone surface has generated a faster degradation. Hence not only the initial bone density and the bone microstructure distribution play a critical role in the degradation kinetics but also the amount of bone surface plays a role. Comparatively, we can also observe that this tendency seems fairly constant for each similar initial bone density (Fig. 6.25A) [2].

For other cases where we have different initial bone densities with similar bone surface areas (Fig. 6.25B), we observe an inverse effect. The degradation kinetics seems to be similar for each case, which plays in favor of the surface area. The bone degradation kinetics is therefore very much patient-dependent, and the influencing parameters such as initial bone density and surface area, together with microstructure distribution, are important parameters that should be accounted for in order to obtain more predictive long-term evolutions [2].

These interpretations are of course based on the hypotheses of similar degradation kinetics for the whole structure of the samples being assumed homogeneous in the

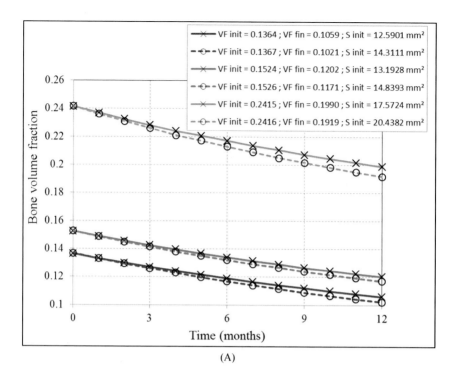

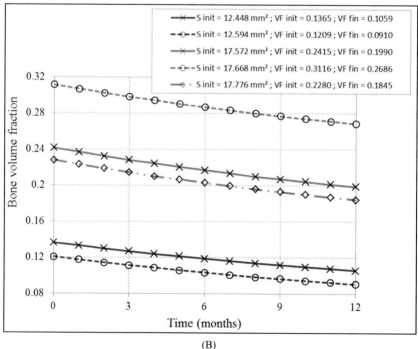

Figure 6.25 Comparative bone loss as a function of time for different samples with (A) similar bone volume fractions and different bone surface areas and (B) similar bone surface areas and different bone volume fractions [2].

288 Applied Micromechanics of Complex Microstructures

material properties and therefore having similar degradation kinetics. We used averaged macroscopic experimental data (Qin [107]) with our predictive model. For as long as the studied bone remains within this "averaged" trend group the prediction will remain valid. However, for bones (patients) that will be out of this "averaged" trend (due to different bone mechanobiology) a new identification (or bone degradation kinetics) should be made in order to have adequate long-term prediction [2].

6.3.6.5 Effective elastic modulus at the sample scale

It was presented in Fig. 6.23 and Fig. 6.24 a long-term prediction of the bone mineral density loss on one "average" sample as a function of time. It is of course especially important to know the bone microstructure degradation as it will impact directly the potential recovery after coming back to normal loading conditions. However, it does also directly impact the strength (stiffness) of the bone. As we have supposed here that the bone behavior was linear elastic, we look at the decrease of the Young's modulus as a function of time related to the bone volume fraction. It is a useful quantitative evaluation of patients' bone strength or astronauts' bone strength for long-time space flight. For this, we link our model's prediction of the microstructure degradation to the macroscopic bone stiffness. We used the predictive model to calculate the corresponding effective elastic modulus with a decrease in the volume fraction of the hard phase. The results of the simulations are presented in Fig. 6.26A and B [2].

Fig. 6.26A and B show the general trend decrease in the bone stiffness as a function of time, which is related to the number of remaining voxels degrading in the sample and also to the loss of connectivity between different regions of the sample. Two main conclusions can be extracted from the table in Fig. 6.26A. The decreasing rate of Young's modulus, bone volume fraction, and number of voxels as a function of time is highly nonlinear with a slow initial evolution and a fast increase toward the end. Also the degradation mechanism over the microstructure as a function of the creation of voxel clusters (per range of volume fraction) seems to follow the trend of Young's modulus to volume fraction degradation. A decrease in the volume fraction leads to a descending trend in the elastic modulus. As the bone degradation is occurring everywhere, it will ultimately create clusters of bones by a reduction of the connectivity between the different zones of the structure, hence reducing the effective bone strength. Instead of having a giant cluster of the hard phase, there are several smaller clusters loosening the structure. Since our degradation model is homogeneous over the whole structure, it will ultimately lead to a complete loss of bone density as a function of time. Nevertheless, this is an assumption as it seems also clear that when reaching a minimum threshold the bone biology will change and react in a different way in order not to let the degradation continue until complete vanishing of the structure. This should be included at a later stage of this work [2].

To highlight the Young's modulus to bone volume fraction degradation, different degradation points are presented in Fig. 6.26B (circles and blue curve). We

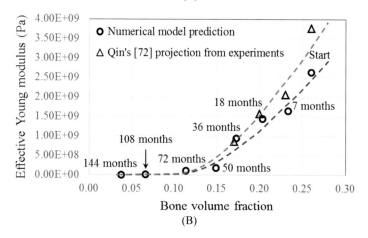

Figure 6.26 (A) Results obtained from the degradation analysis on the "average" sample as a function of time. As a function of degradation time the total numbers of voxels remaining in the sample are presented together with the number of voxels inside the largest voxel cluster. The number of voxels is also presented for different bone volume fractions and as a function of time together with the corresponding effective Young's modulus [2]. (B) Change in the macroscopic Young's modulus in the X-direction in time for the "average" sample as a function of the bone volume fraction degradation [2].

can observe a certain discrepancy in the decreasing trend believed to be related to the anisotropic distribution of the bone sample. The sample is probably not "big enough" to show a "smooth" degradation trend as compared to effective homogeneous degradation. Of course, this is sample related. The bone degradation over an entire bone should not show this variation. Nevertheless, the overall trend of degradation seems respected as a function of time. We also note that creating bone clusters within the sample not only creates important drops in the overall bone stiffness but also argues on the fact that real physical bone degradation may not be constant everywhere (as assumed here) in the structure by favoring some local places and not others in order to lose the complete bone stiffness in one step. Hence smoother bone stiffness degradation could be expected by localizing the biological degradation with a more physical mechanobiological model.

However, the inconvenience of this approach is that it requires precise knowledge of the mechanobiological couplings at the local scale, which is not the case yet. These are strengths and weaknesses of these two approaches and should be investigated further at a later stage [2].

Finally, also presented in Fig. 6.26B are projected data from Qin's [107] experiments (triangle and red curves). These are extracted from the experimental BMD loss as a function of time related to the corresponding bone volume fraction and knowing that the BMD corresponds to about 70% of the bone stiffness [105]. When normalizing these data with our corresponding bone sample density, we projected the "experimental" Young's modulus degradation and compared with our "predicted" Young's modulus degradation. It appears that despite the sample microstructure variability and simplifying hypotheses that were made in order to extract the microstructure bone degradation, there is a very good correlation between the two curves. The main causes at the origin of this result are (1) the model parameter identification on the experimental data, (2) the degradation phenomenon assumed homogeneous over the entire bone microstructure, and (3) the bone sample density and microstructures that are reasonably close to each other. One could argue that it feels obvious that the model predicts what it was identified for. However, the same model with identified parameters was used for the 10 different samples and was able to predict the degradation mechanisms for each sample separately and observe the variability between each sample. Hence we could say that the proposed model is able to extract and predict patient-dependent bone degradation. This in turn could be used for the prediction of various patient cases knowing their specific bone microstructure distribution [2].

6.4 Conclusion remarks

In this chapter a numerical approach has been implemented to model the degradation process of two types of heterogeneous microstructures. The in vitro degradation process of PLA/Mg composites experimentally has been measured, and a numerical approach has been used to model corrosion behavior of PLA/Mg composites. The numerical model uses 3D cellular automata to predict the corrosion of 3D reconstructed microstructures. The statistical correlation function has been used to reconstruct 3D microstructures from the SEM images. The numerical approach has been calibrated using the experimental data [1].

The bone microstructure is one of the most critical biological materials that have recently attracted increasing attention. In this chapter a new simplified approach was proposed to predict bone density loss in microgravity based on the transformation of hard (H, bone) and soft (S, bone marrow) phases to represent the bone formation or bone degradation bone surface. A prediction is made for long-term bone degradation based on simplifying hypotheses. The degradation mechanisms are presented at the trabecular scale. The bone stiffness loss was predicted with good accuracy as compared with normalized experimental projections [2].

References

[1] Shahmohmmadi A., Baghani M., Shariat Panahi M., Wang K., Hasanpur E., Baniassadi M. Computational modeling of degradation process on the mechanical performance of Polylactic acid/Magnesium composite. In: Proceedings of the institution of mechanical engineers, Part L: journal of materials: design and applications 2020;235(1). Available from: https://doi.org/10.1177/1464420720948253.

[2] Bagherian A, Baghani M, George D, Rémond Y, Chappard C, Patlazhan S, et al. A novel numerical model for the prediction of patient-dependent bone density loss in microgravity based on micro-CT images. Contin Mech Thermodyn 2019; 32(3):927−43. Available from: https://doi.org/10.1007/s00161-019-00798-8.

[3] Swaminathan R. Magnesium metabolism and its disorders. Clin Biochem Rev 2003; 24(2):47−66.

[4] Amerinatanzi A, Mehrabi R, Ibrahim H, Dehghan A, Shayesteh Moghaddam N, Elahinia M. Predicting the biodegradation of magnesium alloy implants: modeling, parameter identification, and validation. Bioeng (Basel) 2018;5(4):105. Available from: https://doi.org/10.3390/bioengineering5040105.

[5] Zartner P, Cesnjevar R, Singer H, Weyand M. First successful implantation of a biodegradable metal stent into the left pulmonary artery of a preterm baby. Catheterization Cardiovasc Intervent 2005;66(4):590−4.

[6] Sharkawi T, Cornhill F, Lafont A, Sabaria P, Vert M. Intravascular bioresorbable polymeric stents: a potential alternative to current drug eluting metal stents. J Pharm Sci 2007;96(11):2829−37. Available from: https://doi.org/10.1002/jps.20957.

[7] Cifuentes SC, Lieblich M, Saldaña L, González-Carrasco JL, Benavente R. In vitro degradation of biodegradable polylactic acid/Mg composites: influence of nature and crystalline degree of the polymeric matrix. Materialia 2019;6:100270. Available from: https://doi.org/10.1016/j.mtla.2019.100270.

[8] Li X, Chu C, Zhou L, Bai J, Guo C, Xue F, et al. Fully degradable PLA-based composite reinforced with 2D-braided Mg wires for orthopedic implants. Compos Sci Technol 2017;142:180−8. Available from: https://doi.org/10.1016/j.compscitech.2017.02.013.

[9] Yang L, Hort N, Willumeit R, Feyerabend F. Effects of corrosion environment and proteins on magnesium corrosion. Corros Eng Sci Technol 2013;47(5):335−9. Available from: https://doi.org/10.1179/1743278212y.0000000024.

[10] Zhao D, Wang T, Guo X, Kuhlmann J, Doepke A, Dong Z, et al. Monitoring biodegradation of magnesium implants with sensors. JOM 2016;68(4):1204−8. Available from: https://doi.org/10.1007/s11837-015-1775-z.

[11] Hadzima B, Mhaede M, Pastorek F. Electrochemical characteristics of calcium-phosphatized AZ31 magnesium alloy in 0.9% NaCl solution. J Mater Sci Mater Med 2014;25(5):1227−37. Available from: https://doi.org/10.1007/s10856-014-5161-0.

[12] Grogan JA, O'Brien BJ, Leen SB, McHugh PE. A corrosion model for bioabsorbable metallic stents. Acta Biomater 2011;7(9):3523−33. Available from: https://doi.org/10.1016/j.actbio.2011.05.032.

[13] Grogan JA, Leen SB, McHugh PE. A physical corrosion model for bioabsorbable metal stents. Acta Biomater 2014;10(5):2313−22. Available from: https://doi.org/10.1016/j.actbio.2013.12.059.

[14] Bajger P, Ashbourn JMA, Manhas V, Guyot Y, Lietaert K, Geris L. Mathematical modelling of the degradation behaviour of biodegradable metals. Biomech Model Mechanobiol 2017;16(1):227−38. Available from: https://doi.org/10.1007/s10237-016-0812-3.

[15] Malki B, Baroux B. Computer simulation of the corrosion pit growth. Corros Sci 2005;47(1):171−82. Available from: https://doi.org/10.1016/j.corsci.2004.05.004.

[16] Balázs L, Gouyet JF. Two-dimensional pitting corrosion of aluminium thin layers. Phys A: Stat Mech its Appl 1995;217(3−4):319−38. Available from: https://doi.org/10.1016/0378-4371(95)00048-c.

[17] Meakin P, Jossang T, Feder J. Simple passivation and depassivation model for pitting corrosion. Phys Rev E Stat Phys Plasmas Fluids Relat Interdiscip Top 1993;48 (4):2906−16. Available from: https://doi.org/10.1103/physreve.48.2906.

[18] Van der Weeën P, Zimer AM, Pereira EC, Mascaro LH, Bruno OM, De Baets B. Modeling pitting corrosion by means of a 3D discrete stochastic model. Corros Sci 2014;82:133−44. Available from: https://doi.org/10.1016/j.corsci.2014.01.010.

[19] Lan K-C, Chen Y, Hung T-C, Tung H-M, Yu G-P. Simulation of the growth of oxide layer of stainless steels with chromium using cellular automaton model: verification and parameter study. Computat Mater Sci 2013;77:139−44. Available from: https://doi.org/10.1016/j.commatsci.2013.04.037.

[20] Li L, Li X, Dong C, Huang Y. Computational simulation of metastable pitting of stainless steel. Electrochim Acta 2009;54(26):6389−95. Available from: https://doi.org/10.1016/j.electacta.2009.05.093.

[21] Wang H, Han E-H. Computational simulation of corrosion pit interactions under mechanochemical effects using a cellular automaton/finite element model. Corros Sci 2016;103:305−11. Available from: https://doi.org/10.1016/j.corsci.2015.11.034.

[22] Stafiej J., Taleb A., Vautrin-Ul C., Chaussé A., Badiali J.P. Simulation of corrosion processes with anodic and cathodic reactions separated in space. In: Marcus P, Maurice V, editors. Passivation of metals and semiconductors, and properties of thin oxide layers. Amsterdam: Elsevier Science; 2006, p. 667−672.

[23] Vautrin-Ul C, Taleb A, Stafiej J, Chaussé A, Badiali JP. Mesoscopic modelling of corrosion phenomena: coupling between electrochemical and mechanical processes, analysis of the deviation from the Faraday law. Electrochim Acta 2007;52(17):5368−76. Available from: https://doi.org/10.1016/j.electacta.2007.02.051.

[24] Pérez-Brokate CF, di Caprio D, Féron D, de Lamare J, Chaussé A. Three dimensional discrete stochastic model of occluded corrosion cell. Corros Sci 2016;111:230−41. Available from: https://doi.org/10.1016/j.corsci.2016.04.009.

[25] Guo C, Niu Y. Cellular automaton simulation for degradation of poly lactic acid with acceleratable reaction-diffusion model. ACS Biomater Sci Eng 2019;5(4):1771−83. Available from: https://doi.org/10.1021/acsbiomaterials.9b00015.

[26] Safdari M, Baniassadi M, Garmestani H, Al-Haik MS. A modified strong-contrast expansion for estimating the effective thermal conductivity of multiphase heterogeneous materials. J Appl Phys 2012;112(11):114318.

[27] Sanfeld A. Introduction to the thermodynamics of charged and polarized layers. Wiley; 1968.

[28] Sheidaei A, Baniassadi M, Banu M, Askeland P, Pahlavanpour M, Kuuttila N, et al. 3-D microstructure reconstruction of polymer nano-composite using FIB−SEM and statistical correlation function. Compos Sci Technol 2013;80:47−54.

[29] Garmestani H, Baniassadi M, Li D, Fathi M, Ahzi S. Semi-inverse Monte Carlo reconstruction of two-phase heterogeneous material using two-point functions. Int J Theor Appl Multiscale Mech 2009;1(2):134−49.

[30] Baniassadi M, Ahzi S, Garmestani H, Ruch D, Remond Y. New approximate solution for N-point correlation functions for heterogeneous materials. J Mech Phys Solids 2012;60(1):104−19. Available from: https://doi.org/10.1016/j.jmps.2011.09.009.

[31] Hasanabadi A, Baniassadi M, Abrinia K, Safdari M, Garmestani H. 3D microstructural reconstruction of heterogeneous materials from 2D cross sections: a modified phase-recovery algorithm. Computat Mater Sci 2016;111:107−15. Available from: https://doi.org/10.1016/j.commatsci.2015.09.015.

[32] Keyak J, Koyama A, LeBlanc A, Lu Y, Lang T. Reduction in proximal femoral strength due to long-duration spaceflight. Bone 2009;44(3):449−53.

[33] Vico L, Collet P, Guignandon A, Lafage-Proust MH, Thomas T, Rehaillia M, et al. Effects of long-term microgravity exposure on cancellous and cortical weight-bearing bones of cosmonauts. Lancet. 2000;355(9215):1607−11. Available from: https://doi.org/10.1016/s0140-6736(00)02217-0.

[34] LeBlanc A, Schneider V, Shackelford L, West S, Oganov V, Bakulin A, et al. Bone mineral and lean tissue loss after long duration space flight. J Musculoskelet Neuronal Interact 2000;1(2):157−60.

[35] Ruggiu A, Cancedda R. Bone mechanobiology, gravity and tissue engineering: effects and insights. J Tissue Eng Regen Med 2015;9(12):1339−51. Available from: https://doi.org/10.1002/term.1942.

[36] Amin S. Mechanical factors and bone health: effects of weightlessness and neurologic injury. Curr Rheumatol Rep 2010;12(3):170−6. Available from: https://doi.org/10.1007/s11926-010-0096-z.

[37] LeBlanc A, Schneider V, Krebs J, Evans H, Jhingran S, Johnson P. Spinal bone mineral after 5 weeks of bed rest. Calcif Tissue Int 1987;41(5):259−61. Available from: https://doi.org/10.1007/BF02555226.

[38] Zerwekh JE, Ruml LA, Gottschalk F, Pak CY. The effects of twelve weeks of bed rest on bone histology, biochemical markers of bone turnover, and calcium homeostasis in eleven normal subjects. J Bone Min Res 1998;13(10):1594−601. Available from: https://doi.org/10.1359/jbmr.1998.13.10.1594.

[39] Leblanc AD, Schneider VS, Evans HJ, Engelbretson DA, Krebs JM. Bone mineral loss and recovery after 17 weeks of bed rest. J Bone Min Res 1990;5(8):843−50. Available from: https://doi.org/10.1002/jbmr.5650050807.

[40] Ram R.R. Hierarchy of bone structure report. 2006.

[41] Lang TF, Leblanc AD, Evans HJ, Lu Y. Adaptation of the proximal femur to skeletal reloading after long-duration spaceflight. J Bone Min Res 2006;21(8):1224−30. Available from: https://doi.org/10.1359/jbmr.060509.

[42] Hardy R, Cooper MS. Bone loss in inflammatory disorders. J Endocrinol 2009;201 (3):309−20. Available from: https://doi.org/10.1677/JOE-08-0568.

[43] Kumar G. Orban's oral histology & embryology. Elsevier Health Sci; 2014.

[44] Veni MAC, Rajathi P. Interaction between bone cells in bone remodelling. J Acad Dental Educ 2017;2:1−6.

[45] Florencio-Silva R, Sasso GRdS, Sasso-Cerri E, Simões MJ, Cerri PS. Biology of bone tissue: structure, function, and factors that influence bone cells. BioMed Res Int 2015;2015.

[46] Sugawara Y, Kamioka H, Honjo T. Tezuka K-i, Takano-Yamamoto T. Three-dimensional reconstruction of chick calvarial osteocytes and their cell processes using confocal microscopy. Bone 2005;36(5):877−83.

[47] Hadjidakis DJ. Androulakis, II. Bone remodeling. Ann N Y Acad Sci 2006;1092 (1):385−96. Available from: https://doi.org/10.1196/annals.1365.035.

[48] Klein-Nulend J, Bakker AD, Bacabac RG, Vatsa A, Weinbaum S. Mechanosensation and transduction in osteocytes. Bone 2013;54(2):182−90. Available from: https://doi.org/10.1016/j.bone.2012.10.013.

[49] Wang Q, Seeman E. Skeletal growth and peak bone strength. Prim Metab Bone Dis Disord Miner Metab 2013;127−34.

[50] Thudium C. Development of novel models for studying osteoclasts. Div Mol Med Gene Ther 2014.

[51] Wolff J. Das Gesetz der Transformation der Knochen Hirschwald. Berlin; 1892.

[52] Carter DR. Mechanical loading histories and cortical bone remodeling. Calcif Tissue Int 1984;36(1):S19−24. Available from: https://doi.org/10.1007/BF02406129, Suppl 1.

[53] Carter D, Orr T, Fyhrie DP. Relationships between loading history and femoral cancellous bone architecture. J Biomech 1989;22(3):231−44.

[54] Huiskes R, Weinans H, Grootenboer HJ, Dalstra M, Fudala B, Slooff TJ. Adaptive bone-remodeling theory applied to prosthetic-design analysis. J Biomech 1987;20 (11−12):1135−50. Available from: https://doi.org/10.1016/0021-9290(87)90030-3.

[55] Frost HMJTar. Bone "mass" and the "mechanostat": a proposal. 1987; 219 (1):1−9.

[56] Hegedus D, Cowin S. Bone remodeling II: small strain adaptive elasticity. J Elast 1976;6(4):337−52.

[57] Burr DB, Martin RB, Schaffler MB, Radin EL. Bone remodeling in response to in vivo fatigue microdamage. J Biomech 1985;18(3):189−200. Available from: https://doi.org/10.1016/0021-9290(85)90204-0.

[58] Weinans H, Huiskes R, Grootenboer HJ. The behavior of adaptive bone-remodeling simulation models. J Biomech 1992;25(12):1425−41. Available from: https://doi.org/10.1016/0021-9290(92)90056-7.

[59] Mullender M, Huiskes R, Weinans H. A physiological approach to the simulation of bone remodeling as a self-organizational control process. J Biomech 1994;27(11):1389−94.

[60] Lekszycki T. Modelling of bone adaptation based on an optimal response hypothesis. Meccanica. 2002;37(4−5):343−54.

[61] Tovar A. Bone remodeling as a hybrid cellular automaton optimization process. 2004.

[62] Ruimerman R, Hilbers P, van Rietbergen B, Huiskes R. A theoretical framework for strain-related trabecular bone maintenance and adaptation. J Biomech 2005;38 (4):931−41. Available from: https://doi.org/10.1016/j.jbiomech.2004.03.037.

[63] Lekszycki T. Functional adaptation of bone as an optimal control problem. J Theor Appl Mech 2005;43(3):555−74.

[64] Pivonka P, Zimak J, Smith DW, Gardiner BS, Dunstan CR, Sims NA, et al. Model structure and control of bone remodeling: a theoretical study. Bone 2008;43(2):249−63. Available from: https://doi.org/10.1016/j.bone.2008.03.025.

[65] Hart R. Bone modeling and remodeling: theories and computation. Bone Mech Handb 2001;1(42):31.1.

[66] Lemaire T, Naili S, Sansalone V. Multiphysical modelling of fluid transport through osteo-articular media. An Acad Brasiliera de Ciências 2010;82(1):127−44.

[67] Lemaire T, Capiez-Lernout E, Kaiser J, Naili S, Sansalone V. What is the importance of multiphysical phenomena in bone remodelling signals expression? A multiscale perspective. J Mech Behav Biomed Mater 2011;4(6):909−20.

[68] Sansalone V, Kaiser J, Naili S, Lemaire T. Interstitial fluid flow within bone canaliculi and electro-chemo-mechanical features of the canalicular milieu. Biomech Modeling Mechanobiol 2013;12(3):533−53.

[69] Sansalone V, Gagliardi D, Desceliers C, Haiat G, Naili S. On the uncertainty propagation in multiscale modeling of cortical bone elasticity. Comput Methods Biomech Biomed Engin 2015;18(Suppl 1):2054−5. Available from: https://doi.org/10.1080/10255842.2015.1069619.

[70] Lemaire T, Kaiser J, Naili S, Sansalone V. Three-scale multiphysics modeling of transport phenomena within cortical bone. Math Probl Eng 2015;2015.

[71] Madeo A, Lekszycki T. A continuum model for the bio-mechanical interactions between living tissue and bio-resorbable graft after bone reconstructive surgery. Comptes Rendus Mécanique 2011;339(10):625−40.

[72] Lekszycki T, Dell'Isola F. A mixture model with evolving mass densities for describing synthesis and resorption phenomena in bones reconstructed with bio-resorbable materials. ZAMM-J Appl Mathematics 2012;92(6):426−44.

[73] Madeo A, George D, Lekszycki T, Nierenberger M, Rémond Y. A second gradient continuum model accounting for some effects of micro-structure on reconstructed bone remodelling. Comptes Rendus Mécanique 2012;340(8):575−89.

[74] Andreaus U, Giorgio I, Lekszycki T. A 2-D continuum model of a mixture of bone tissue and bio-resorbable material for simulating mass density redistribution under load slowly variable in time. ZAMM-J Appl Mathematics Mech/Zeitschrift für Angew Mathematik und Mechanik 2014;94(12):978−1000.

[75] Scala I, Spingarn C, Rémond Y, Madeo A, George D. Mechanically-driven bone remodeling simulation: application to LIPUS treated rat calvarial defects. Mathematics Mech Solids 2017;22(10):1976−88.

[76] Giorgio I, Andreaus U, Scerrato D, dell'Isola F. A visco-poroelastic model of functional adaptation in bones reconstructed with bio-resorbable materials. Biomech Model Mechanobiol 2016;15(5):1325−43. Available from: https://doi.org/10.1007/s10237-016-0765-6.

[77] Giorgio I, Andreaus U, Dell'Isola F, Lekszycki T. Viscous second gradient porous materials for bones reconstructed with bio-resorbable grafts. Extreme Mech Lett 2017;13:141−7.

[78] Andreaus U, Colloca M, Iacoviello D. Optimal bone density distributions: numerical analysis of the osteocyte spatial influence in bone remodeling. Comput Methods Prog Biomed 2014;113(1):80−91. Available from: https://doi.org/10.1016/j.cmpb.2013.09.002.

[79] Bednarczyk E, Lekszycki T. A novel mathematical model for growth of capillaries and nutrient supply with application to prediction of osteophyte onset. Z für Angew Mathematik und Phys 2016;67(4):94.

[80] Lu Y, Lekszycki T. A novel coupled system of non-local integro-differential equations modelling Young's modulus evolution, nutrients' supply and consumption during bone fracture healing. Z für Angew Mathematik und Phys 2016;67(5):111.

[81] Allena R, Maini P. Reaction−diffusion finite element model of lateral line primordium migration to explore cell leadership. Bull Math Biol 2014;76(12):3028−50.

[82] George D, Allena R, Remond Y. Mechanobiological stimuli for bone remodeling: mechanical energy, cell nutriments and mobility. Comput Methods Biomech Biomed Engin 2017;20:91−2. Available from: https://doi.org/10.1080/10255842.2017.1382876 supl.

[83] George D, Allena R, Rémond Y. A multiphysics stimulus for continuum mechanics bone remodeling. Mathematics Mech Complex Syst 2018;6(4):307−19.

[84] George D, Allena R, Rémond Y. Integrating molecular and cellular kinetics into a coupled continuum mechanobiological stimulus for bone reconstruction. Contin Mech Thermodyn 2018;1−16.

[85] Rémond Y, Ahzi S, Baniassadi M, Garmestani H. Applied RVE reconstruction and homogenization of heterogeneous materials. John Wiley & Sons; 2016.

[86] Hasanabadi A, Baniassadi M, Abrinia K, Safdari M, Garmestani H. Efficient three-phase reconstruction of heterogeneous material from 2D cross-sections via phase-recovery algorithm. J Microsc 2016;264(3):384−93. Available from: https://doi.org/10.1111/jmi.12454.

[87] Sheidaei A, Kazempour M, Hasanabadi A, Nosouhi F, Pithioux M, Baniassadi M, et al. Influence of bone microstructure distribution on developed mechanical energy for bone remodeling using a statistical reconstruction method. Mathematics Mech Solids 2019; 1081286519828418.

[88] Sigmund O. On the optimality of bone microstructure. IUTAM symposium on synthesis in bio solid mechanics: Springer; 1999, p. 221−34.

[89] Rodrigues H., Jacobs C., Guedes J., Bendsøe M. Global and local material optimization models applied to anisotropic bone adaptation. IUTAM symposium on synthesis in bio solid mechanics: Springer; 1999, p. 209−20.

[90] Huiskes R, Ruimerman R, van Lenthe GH, Janssen JD. Effects of mechanical forces on maintenance and adaptation of form in trabecular bone. Nature 2000;405(6787):704−6. Available from: https://doi.org/10.1038/35015116.

[91] Nowak M. Structural optimization system based on trabecular bone surface adaptation. Struct Multidiscip Optim 2006;32(3):241−9.

[92] Jang IG, Kim IY. Computational study of Wolff's law with trabecular architecture in the human proximal femur using topology optimization. J Biomech 2008;41(11):2353−61.

[93] Goda I, Assidi M, Ganghoffer JF. A 3D elastic micropolar model of vertebral trabecular bone from lattice homogenization of the bone microstructure. Biomech Model Mechanobiol 2014;13(1):53−83. Available from: https://doi.org/10.1007/s10237-013-0486-z.

[94] Goda I, Ganghoffer J-F, Czarnecki S, Wawruch P, Lewiński T. Optimal internal architectures of femoral bone based on relaxation by homogenization and isotropic material design. Mech Res Commun 2016;76:64−71.

[95] Lee YH, Kim Y, Kim JJ, Jang IG. Homeostasis-based aging model for trabecular changes and its correlation with age-matched bone mineral densities and radiographs. Eur J Radiol 2015;84(11):2261−8. Available from: https://doi.org/10.1016/j.ejrad.2015.07.027.

[96] Spingarn C, Wagner D, Remond Y, George D. Multiphysics of bone remodeling: a 2D mesoscale activation simulation. Biomed Mater Eng 2017;28(s1):S153−8. Available from: https://doi.org/10.3233/BME-171636.

[97] Allena R, Cluzel C. Heterogeneous directions of orthotropy in three-dimensional structures: finite element description based on diffusion equations. Mathematics Mech Complex Syst 2018;6(4):339−51.

[98] Cluzel C, Allena R. A general method for the determination of the local orthotropic directions of heterogeneous materials: application to bone structures using μCT images. Mathematics Mech Complex Syst 2018;6(4):353−67.

[99] Goda I, Ganghoffer J-F, Maurice G. Combined bone internal and external remodeling based on Eshelby stress. Int J Solids Struct 2016;94:138−57.

[100] Izadi H, Baniassadi M, Hasanabadi A, Mehrgini B, Memarian H, Soltanian-Zadeh H, et al. Application of full set of two point correlation functions from a pair of 2D cut sections for 3D porous media reconstruction. J Pet Sci Eng 2017;149:789−800. Available from: https://doi.org/10.1016/j.petrol.2016.10.065.

[101] Hanks JH. Hanks' balanced salt solution and pH control. Methods Cell Sci 1975; 1(1):3−4.

[102] Song G, Atrens A, StJohn D. An hydrogen evolution method for the estimation of the corrosion rate of magnesium alloys. Magnes Technol 2001;2001:255−62.

[103] Shahmohammadi A, Famouri S, Hosseini S, Farahani MM, Baghani M, George D, et al. Prediction of bone microstructures degradation during osteoporosis with fuzzy cellular automata algorithm. Mathematics Mech Solids 2022. Available from: https://doi.org/10.1177/10812865221088520.

[104] Famouri S, Baghani M, Sheidaei A, George D, Farahani MM, Panahi MS, et al. Statistical prediction of bone microstructure degradation to study patient dependency in osteoporosis. Mathematics Mech Solids 2022. Available from: https://doi.org/10.1177/10812865221098777.

[105] Klibanski A, Adams-Campbell L, Bassford T, Blair SN, Boden SD, Dickersin K, et al. Osteoporosis prevention, diagnosis, and therapy. J Am Med Assoc 2001;285(6):785–95.

[106] Pennline JA, Mulugeta L. Mapping Bone Mineral Density Obtained by Quantitative Computed Tomography to Bone Volume Fraction. NASA 2017.

[107] Qin Y-X. Challenges to the musculoskeleton during a journey to Mars: assessment and counter measures. J Cosmol 2010;12:3778–80.

[108] Liu Y-L, Hsu J-T, Shih T-Y, Luzhbin D, Tu C-Y, Wu J. Quantification of volumetric bone mineral density of proximal femurs using a two-compartment model and computed tomography images. BioMed Res Int 2018;2018.

[109] Eberhart RC, Shi Y, Kennedy J. Swarm intelligence. Elsevier; 2001.

[110] Hildebrand T., Rüegsegger P.J.J.O.M. A new method for the model-independent assessment of thickness in three-dimensional images. 1997;185(1):67–75.

[111] Anderson IA, Carman JBJJOB. How do changes to plate thickness, length, and face-connectivity affect femoral cancellous bone's density and surface area? Investig Regul Cell Model 2000;33(3):327–35.

Microstructure hull and design

7

Abstract

In this chapter, three different approaches are exploited to design microstructures of heterogeneous materials. In the first section, microstructure hull and property hull are explained and used to design the optimum microstructures. A realization technique and neural network have been combined to calculate the microstructure and property hull for solid oxide fuel cell anode microstructures. Understanding hull space and closures helps us to search a microstructure in the realistic domains.

In the second section, an efficient micromechanical methodology is developed for tailoring random heterogeneous microstructures to achieve desired elastic and thermal properties. In the third section, the optimal design of porous and periodic microstructures through topology optimization of the associated periodic unit cell constitutes the topic of the second methodology for materials design. Finally, a new approach for designing hybrid Triply Periodic Minimal Surface (TPMS) microstructures is discussed.

7.1 Introduction

Designing materials with desired properties has become an unavoidable task. Moreover the advances in manufacturing methods, especially 3D printing, accelerate developing material design methods. These methods necessitate developing theoretical progress in the field's output of these attempts and have shown fascinating features that are rare in nature. For example, manufacturable microstructures with extreme thermoelastic properties [1], simultaneously programmable coefficient of thermal expansion (CTE) and Poisson's ratio [2], enhanced stiffness and energy absorption [3], tunable stiffness and buckling strength [4], tailored Poisson's ratio and Young's modulus [5], and tailored energy dissipation [6] are novel engineered materials that have been successfully designed.

In addition to additive manufacturing processes, recent technological advances (e.g., high-performance computer hardware) have revolutionized the field of materials design through the use of machine learning (ML) algorithms. For instance, designing multifunctional composite materials with desired thermal conductivity [7], designing materials with desirable electronic band gaps [8], designing supercompressible metamaterials [9], and designing bioinspired hierarchical composites [10] are recent achievements of ML algorithms for designing new materials.

In this chapter, we present practical examples of designing engineering materials with optimal properties. Different methodologies are presented in this chapter to address materials design for different types of microstructures.

Applied Micromechanics of Complex Microstructures. DOI: https://doi.org/10.1016/B978-0-443-18991-3.00002-7
© 2023 Elsevier Inc. All rights reserved.

Designing optimal microstructures for solid oxide fuel cell (SOFC) electrodes is a subtle task owing primarily to the multitude of the electro-chemo-physical phenomena taking place simultaneously that directly affect working conditions of an SOFC electrode and its performance. In Section 7.3.1 a novel design paradigm is presented to obtain the desired triple phase boundary length (TPBL), ionic/electronic phase conductivity, and gas diffusion properties for an SOFC cell. The method builds on top of a previously developed methodology for digital realization of generic microstructures. The obtained realizations are then used to predict TPBL, ion conductivity, and gas diffusion of their representing SOFC electrodes. The study follows by building a database obtained from these realizations to train a neural network that relates input geometrical parameters to these three properties. It is shown that the presented methodology allows one to obtain properties and microstructure hull for SOFC electrodes and finally achieve optimal microstructures. The results indicate that the gas diffusion for SOFC electrodes is strongly dependent on the geometry of the microstructure, while TPBL and ionic conductivity, to less extent, are related to the geometry. The analysis also shows that there exists a trade-off between ionic conductivity, TPBL, and diffusion factor [11].

In Section 7.3.2 a novel design paradigm is presented to obtain some geometry-related electrochemical and physical properties of an infiltrated SOFC electrode. A range of digitally realized microstructures with different backbone porosities and electrocatalyst particle loadings under various deposition conditions are generated. Triple phase boundary (TPB), active surface density of particles, and gas transport factor are evaluated in realized models based on the selected infiltration strategy. Based on this database a neural network is trained to relate the desired range of input geometric parameters to a property hull. The effect of backbone porosity, loading, distribution, and aggregation behavior of particles is systematically investigated on the performance indicators. It is shown that from the microstructures with an extremely high amount of TPB and particle contact surface density a relatively low gas diffusion factor should be expected; meanwhile, increasing those parameters does not have sensible contradiction with each other. Excessive agglomerating of particles has a negative effect on TPB density, but the distribution of seeds always has a positive effect. Direct search and genetic algorithm (GA) optimization techniques are used finally to achieve optimal microstructures based on assumed target functions for effective geometric properties [12].

A novel method for combining two microstructures and consequently obtaining new geometrical arrangement with intermediate properties is presented in Section 7.4.1. A series of two-phase microstructures is digitized, and two-point correlation functions (TPCFs) are obtained using discrete Fourier transform. Then by combining the autocovariance functions, TPCFs of the new geometry are obtained. In the subsequent step the new microstructure is realized using the phase recovery algorithm. By changing the combination factor from 0 to 1, it is possible to obtain various intermediate microstructures, whereas the properties of the combined structures gradually and smoothly change from the first microstructure to the second one. This study provides a straightforward approach for combining microstructures regardless of their morphology, state of anisotropy, and volume fractions. The proposed procedure makes it possible to bridge between candidate microstructures with certain optimal properties and compose hybrid optimal properties [13]. Multifunctional materials with a combination of optimal properties such as thermal conductivity and structural strength are of high interest to modern multidisciplinary engineering practices. The realization of binary or ternary optimal properties in a single composite material system often requires precise tailoring of its underlying microstructure. In Section 7.4.1 a novel method is presented to combine the properties and geometries of individual microstructures using TPCFs and finally obtain an optimal combination of properties for the outcome microstructure. When each of the input microstructures has an optimal property the method is capable of obtaining a microstructure with intermediate properties. For two microstructures, by gradually increasing the combination factor from 0 to 1 the outcome microstructure gradually transforms from one to another and, more importantly, its corresponding properties also follow the same gradual transition [13].

Traditional approaches to the design of multidisciplinary microstructures very often employ multifunctional topology optimization techniques [14–17]. These methods start from a full representation of a coupled PDE system representing the physics of the problem to determine the optimal distribution of the material phases. Aside from their complexity and many other limitations imposed by the numerical solution techniques used to solve such coupled systems, these methods are computationally expensive [17] and best work for optimizing a single property. The presented approach in combination with a topology optimization method can be employed to optimize a microstructure for each needed property separately and then bridge between the microstructures to obtain hybrid optimal properties. For this purpose, based on the designer's discretion the combination factor that determines the contribution of each of the microstructures can be obtained using a simple single variable optimization technique [13].

The first step in microstructural sensitive design for performance optimization is to represent the microstructure by a proper mathematical descriptor function [18–20], denoted here also by microstructural measure function. Some of the famous statistical microstructural measures that can be used for this purpose are surface correlation functions [19], lineal measures [19,21], chord-length density function [22], pore-size functions [19], two-point cluster functions [23], and n-point correlation functions [13,24–27].

Other notable mathematical microstructure descriptors that are developed to capture higher-order information are based on random set theory and the integral geometry approach [22]. In this framework, Minkowski functionals (MFs) provide a complete set of statistical measures applicable to the morphological characterization of disordered systems such as random heterogeneous microstructures. By relying on theorems from integral geometry [22,28], it is shown that MFs are directly related to four additive natural geometrical characteristics of R^3 subspaces, namely, volume, surface area, integral of mean curvature, and Euler characteristics [29]. In this family of statistical measures, scalar MFs are proven to be suitable to represent the shape of random media [30], especially the Boolean model of overlapping particles [31,32]. Arns et al. [33] presented a reconstruction procedure for the Boolean model of random composite media at any arbitrary particle fraction using the integral geometric approach and Minkowski functionals. Based on a single three-dimensional image, with a specific volume fraction, they derived the global Minkowski functional and reconstructed media for all volume fractions. The scalar Minkowski functionals are also successfully extended to the equivalent tensorial forms that also capture the state of the orientation and anisotropy of a disordered system [13,34].

Meanwhile, several microstructure descriptors can be found in the literature; a mathematical material optimization scheme, such as the one developed in the current work, can best rely on a descriptor that (1) cascades relevant geometrical attributes of a microstructure, (2) is easily obtainable from readily available methods, and (3) provides direct link to the microstructure properties. Among the list presented earlier, n-point correlation functions meet all these requirements. N-point correlation functions are known to well coordinate with the spectral and material knowledge frameworks [18,20,35,36]. Furthermore, their ability to directly estimate effective properties of random heterogeneous materials is proved [19,37,38]. In this case, it is shown that at least physical properties such as effective mechanical, thermal, electrical, and permeability of a wide range of random heterogeneous materials are strongly correlated with their microstructural n-point statistics [18,19,39,40]. These functions describe geometrical attributes of a microstructure by a set of distribution functions that represent more information about the microstructure as their order increases [18,19]. The first member of the group is the one-point or the probability of finding a random point in a specified phase that asymptotes to the volume fraction of that phase. Similarly, TPCFs indicate the probability of finding the head and tail of a vector in specified phases, and therefore they capture a first-order representation for the geometrical attributes of a microstructure. Using TPCFs, it is possible to reconstruct microstructures from limited 2D cross-sections [26,41,42], create and optimize microstructures [43], and even propose a method for identifying the optimized set of representative volume elements (RVEs) from an ensemble of available microstructure data sets [44]. TPCFs are easily calculated, using Fourier transform, making them perfectly suitable for reconstruction purposes using the phase recovery algorithm [45]. Therefore in the current work, TPCFs are used as the microstructure descriptor [13].

This study suggests to take TPCFs of two microstructures and mathematically combine them by a proper combination factor. TPCFs of the resultant microstructures are determined by considering a fraction of each initial TPCF. Jiao et al. [25,46] have shown that the convex combination of two realizable autocovariance functions (that is a linear function of TPCFs) is also an autocovariance for a realizable microstructure, and therefore these combined correlation functions can be used to construct the resultant microstructure. This is a straight forward and yet practical approach in using TPCFs and finding optimal microstructures without resorting to expensive topology optimization techniques [13].

Here, by reconstruction, we mean the realization of the microstructure by its TPCFs (or other suitable used measures in general). Some of such reconstruction methods are simulated annealing [47−49], random-field models [50], and gradient-based methods [51]. In the current study an efficient method based on the phase recovery algorithm is used for the reconstruction. The method inherits Saxton and Gerchberg's work [52] in signal processing that was first implemented by Fullwood et al. [45] for the reconstruction of microstructures based on TPCFs. Some of the distinctive properties of the adopted reconstruction approach (compared to other methods, e.g., simulated annealing) are its fast convergence, its ability to use nonisotropic TPCFs, and its capability for the reconstruction of anisotropic and multiphase media [45].

The main purpose of Section 7.4.1 is to propose a new efficient procedure to mathematically combine two heterogeneous microstructures without any concern about unequal volume fractions and anisotropy. Using the proposed method, it is possible to obtain a range of microstructures that gradually sweep the correlation space between parent microstructures. Geometrical arrangement and properties of the resultant microstructures smoothly change with variation of the combination factor from 0 (being identical to the first parent) to 1 (equivalent to the second parent). The proposed procedure can be used to eliminate some of the problems existing in current microstructure design methods that rely on repetitive-cell techniques and convert a multiobjective optimization procedure into some single-objective ones [13].

In Section 7.4.2 another methodology is presented for optimal microstructure design of random heterogeneous materials. The methodology includes statistical reconstruction based on the TPCF, thermomechanical homogenization, and optimization steps. The method relies on the tailoring anisotropy in a random heterogeneous microstructure to achieve desired effective elastic modules or thermal conductivity in one direction. Moreover the capability of the method to design a random heterogeneous microstructure with desired elastic modules in one direction and desired thermal conductivity in another direction is investigated. The study aims to present the computational algorithms developed for computational reconstruction, homogenization, and optimization steps. Discussion of some case studies is provided in Section 7.4.2 to demonstrate the feasibility of the method in designing a microstructure with desired thermomechanical properties with a minimal set of design variables. Finally the applicability of the method in efficient computational design of multifunctional materials is demonstrated. Case studies also highlight the advantages and limitations of the method, along with its computational cost [53].

Relying on the latest development of fast Fourier transform (FFT) methods and stochastic MSD approaches the current study presents a novel integrated methodology for the design of composites and the broader class of complex random heterogeneous materials. The method combines the simplicity and flexibility of the stochastic TPCFs for reconstruction with the numerical efficiency of the FFT homogenization; this method also uses a gradient-free algorithm for optimization. Without loss in generality the present study demonstrates the applicability of the method to the optimal design of multifunctional two-phase composites. Sections 7.2 and 7.3 collectively present key underlying FFT homogenization and reconstruction algorithms proposed. Section 7.4 demonstrates the feasibility and applicability of the method in several case studies [53].

In Section 7.5, two examples of the topology optimization are presented for material design. Here the attention is confined to two-phase heterogeneous materials in which the topology identification of manufacturable 3D periodic unit cells (3D-PUCs) is conducted by means of a topology optimization technique. The associated objective function is coupled with a 3D numerical homogenization approach that connects the elastic properties of the 3D-PUC to the target product.

The topology optimization methodology that is adopted in Section 7.5.1 is the combination of the SIMP (solid isotropic material with penalization) method and OCA (optimality criteria algorithm), referred to as SIMP-OCA methodology. The simple SIMP-OCA is then generalized to handle the topology design of 3D manufacturable microstructures of cubic and orthotropic symmetry. The performance of the presented methodology is experimentally validated by fabricating real prototypes of extremal elastic constants using additive manufacturing. Experimental evaluation is performed on two designed microstructures: an orthotropic sample with Young's moduli ratios of $E_2/E_1 = 2.5$ and $E_3/E_1 = 2$ and a cubic sample with a negative Poisson's ratio of -0.19. In all practical examples studied, laboratory measurements are in reasonable agreement with the prescribed values, thus corroborating the applicability of the proposed methodology [54].

Design and manufacture of new materials that have overall mechanical properties not usually seen in conventional materials have become more of a reality with the advent and advancement of additive manufacturing (AM) and 3D printing technologies. The continuing advances in AM processes have provided greater degrees of freedom to fabricate a new generation of structures with intricate patterns in a variety of scientific and technological applications, such as architectured porous biomaterials [55], stretching-dominated cellular truss structures [56], and multiscale cellular structures [57], to mention a few. Taking the capacity of AM technologies for granted the primary focus of Section 7.5 is presenting an efficient and easy-to-implement algorithm for optimal design of porous and periodic microstructures that are manufacturable by 3D printing and will have the mechanical properties specified beforehand [54].

In microstructural design of heterogeneous materials, any periodic structure can be represented by its constituting PUC. Such a form of representation allows for scaling the problem of overall microstructural design down to the topology optimization of the corresponding PUC based on the prescribed effective properties. Topology optimization is therefore a useful tool for designing such periodic microstructures that can be constructed from the juxtaposition of identical unit cells. Properly speaking, in the context of periodic microstructures, topology optimization makes the generation of almost any manufacturable configuration within the design space possible. Working with the topology concept provides higher degrees of freedom (compared to sizing and shape optimization), thus enabling fabrication of lightweight materials of improved structural properties such as stiffness [54].

Following the introduction of homogenization theory for structural design and optimization [15,58] and also the development of the numerical homogenization technique using the finite element analysis by Guedes and Kikuchi [59] for the materials design, Sigmund [60] and Hassani and Hinton [61] pioneered the concurrent implementation of topology optimization and numerical homogenization for materials design purposes. As a matter of fact the investigation by Sigmund [62] marks the start of the research studies on materials design that have evolved around the idea of topology optimization of a constituting unit cell, which has gradually turned into a promising procedure in materials design [54].

Among the solution algorithms proposed for the topology optimization part of the materials design, Sigmund et al. [63]. employed the sequential linear programming (SLP) for the design of 3D piezocomposites with extremal properties. Guest and Prévost [17] deployed the method of SIMP [64−67] coupled with the method of moving asymptotes (MMA) [68] to obtain 3D porous microstructures with optimized stiffness and fluid permeability by defining a multiobjective optimization problem. In another study, Zhang et al. [69]. introduced an efficient approach using a strain energy-based method to calculate the effective elastic properties of topologically optimized 2D and 3D cellular solids. Challis et al. [70]. employed the levelset method to propose their own topology optimization approach for the design of two-phase, 3D isotropic composites with a maximized bulk modulus and conductivity. Radman et al. [71]. utilized the bidirectional evolutionary structural optimization (BESO) technique [72,73] to design 2D and 3D unit cells of cellular isotropic materials with maximum bulk and shear moduli. This technique was also used by Yang et al. [74] for designing 3D orthotropic materials with prescribed Young's moduli ratios. For the design of 3D elastic porous microstructures, Özdemir [75] made use of the topological derivatives based on the associated mathematical framework. However, the numerically obtained (optimal) topologies are not necessarily manufacturable unless this issue is appropriately taken into account [54].

To address the manufacturability of designed microstructures, Suresh [76] proposed an efficient microstructural topology optimization for 3D-printing applications. This approach is presented as a topology sensitivity technique based on the level-set method. Andreassen et al. [77] introduced their own approach based on MMA-type optimizers to design 3D manufacturable extremal elastic materials. In a recently proposed optimization algorithm by Gao et al. [78] for the manufacturable design of 3D microstructured materials the parametric level set method (PLSM) underlies the 3D energy-based homogenization method (EBHM). In a recent study published in Science, Chen et al. [79] also used material density interpolation formulation with an MMA-type optimizer for computing the space of mechanical properties by physically realizable extremal microstructures.

The very core of our topology optimization methodology is the SIMP method incorporated with OCA [80−82], hereafter denoted by SIMP-OCA, which is originally presented by Sigmund [83] together with the corresponding 99-line MATLAB® code. This popular code has been extended by Andreassen et al. [84] for the density filtering procedure based on the modified SIMP method [85]. To address the material design of the structural optimization code, Xia and Breitkopf [86] extended Sigmund's code [83] for the optimal topology design of 2D PUCs using an optimality criteria-type optimizer. They included periodic boundary conditions (PBCs) into their code to enhance its performance. Their generated code could design 2D materials with maximized bulk and shear moduli or a minimized

Poisson's ratio, without having to impose additional constraints. In view of its successful results, in this work the SIMP-OCA methodology is extended to 3D manufacturable porous and periodic microstructures with extremal elastic properties. More promisingly the complex 3D microstructures designed by the SIMP-OCA methodology have the adequate resolution to be ready for manufacturing. The performance of the adopted methodology is experimentally validated by fabricating real prototypes of extremal elastic microstructures using additive manufacturing. Of microstructural examples realized within the framework of this investigation, there are an orthotropic sample and a cubic one, which are elaborated further on in this chapter. They are both designed using the approach developed here and are fabricated by fused deposition modeling (FDM) and selective laser sintering (SLS) technologies, respectively (cf. for example, Kalpakjian and Schmid [87], for further details on these manufacturing processes) [54].

Finally, it should be stated that in this work, we take advantage of the standard (or basic) filtering method (i.e., sensitivity filtering) introduced by Sigmund [60,88] to design 3D-printable microstructures. In fact, using this filtering technique the proposed methodology allows for designing 3D-manufacturable microstructures of cubic and orthotropic symmetry while retaining a good level of performance [54].

In Section 7.5, topology optimization is used to design structures with tunable effective stiffness (Young's modulus). These changes are reversible, and the structure can become harder under strain as an external stimulus. These structures consist of a substructure or unit cell that has one or more gaps inside. Since the change of stiffness in these structures is a function of geometric parameters, the effect of some special geometric parameters is studied. After designing a building block or unit cell for the structure a periodic structure was made of these unit cells and numerical simulation and compression test are performed on it. Both results prove the effectiveness of the method to alter the effective Young's modulus of the structure. Also, GA is used to find the optimized structure with specific stress—strain curves [89]. Finally, new hybridization methods were proposed for Triply Periodic Minimal Surface (TPMS) structures, and a comprehensive study was conducted on the directional elastic modulus, shear modulus, and Poisson's ratio of seven TPMS structures [90].

7.2 Practical approach to estimate microstructure hull and closures

7.2.1 Sample 1 (hull space for SOFC microstructure)

Microstructural attribution of the conventional electrodes can significantly alter the performance of an SOFC device. Main factors that affect the electrochemical performance of SOFC electrodes are TBPL, ionic conductivity, and fluid diffusivity of the microstructure. These parameters are linked with the volume fractions of each phase and the 3D microstructural attributions of the phases [11].

In this chapter, several 3D microstructure realizations were carried out by varying the grain growth and porosities to cover a wide range of possible isotropic and anisotropic configurations. An in-house algorithm was developed to create a stack of cross-section images and calculate TPBL and average particle size from those microstructures. These 3D realizations were used to estimate the conductivity and tortuosity as desired properties for the microstructure. A combination of input and output parameters was suggested to train a neural network. A microstructure hull was developed, containing a large number (~ 10000) of hypothetical microstructures, to obtain the property closure of all possible microstructures for a range of input geometric parameters. It was observed that for the microstructures with a very high amount of TPBL a relatively low ionic conductivity and gas diffusion factor should be expected. In the other hand, only a limited number of microstructures can be found with high ionic conductivity and TPBL. Similarly, it was observed that for a high amount of diffusion factor, TPBL and ionic conductivity cannot be obtained simultaneously. Finally, to find the best microstructure a simple direct search scheme and an artificial neural network coupled with GA optimization scheme were used to find optimal microstructures from the members of the microstructure hull. Although the results strongly depended on the objective function, the proposed method deemed useful to determine the limitation of obtainable performance from the experimental works. The findings of this study can be used as guidelines for future experimental investigations [11].

7.2.1.1 *Microstructure generation and characterization*

The state of heterogeneity of an electrode including the size and distribution of phases is strongly influenced by the nucleation, growth, and initial distribution of its constructive ingredients [91,92]. In a virtual realization, heterogeneity is controlled by the nucleation and grain growth mechanisms as functions of time and morphology [93], and the state of heterogeneity can be quantified by statistical measures such as lineal-path and N-point correlation functions [91]. A Monte Carlo approach for the reconstruction of the electrode microstructure constitutes three steps: (1) generation, (2) distribution, and (3) growth of the cells. In the first step, several initial seed cells are randomly placed in a unit cell of the electrode. Upon initial seed placement a cellular automation algorithm can be used for the growth step. The growth procedure continues until the desired volume fraction for each phase is achieved. From a different perspective the growth step continues till all phases meet and fully occupy the initial grid. It is worth to mention that penetration between the phases should be avoided at all times throughout the initial distribution and growth of the cells [11,27].

7.2.1.1.1 TPB density

As many studies also report, for example, see Janardhanan [92], the density of active TPBs, that is, where the ionic conductor cluster and electric conductor are connected to each other, plays a major role in the overall microstructure of the SOFCs and significantly affects the efficiency of the SOFC. As interfaces between the phases, they play an essential role in the electrochemical performance and the power generation of the SOFC [92]. To evaluate this parameter in a microstructure realization, here a method introduced by Sebdani et al. [93] is adopted. At first the active cluster of each phase is recognized, and then the 26 neighbors of each active void voxel are investigated to find the location of TPBs following an algorithm based on the calculation of overall active TPBL [11].

7.2.1.1.2 Ionic and electric conductivity

Electrochemical performance of SOFC cathodes can be greatly affected by the electrical and ionic conductivity. In particular the relation between geometry of the electrode and polarization is discussed in [94]. Since the electrode geometry is one of the main factors in determining the conductivity of ions and electrons through the electrode, in this section, depending on the phase, electric and ionic conductivities are calculated for candidate 3D SOFC microstructures. For this purpose, each section of the microstructure is converted to a 2D image so that each phase is distinguished with a different color. This image stack can be recognized as a multiphase material using a thresholding process in a piece of reconstruction software. Using Avizo xlab, EFI Corporation, after converting each voxel into the volume elements a potential difference is applied between two opposite faces of the material sample, while other faces are treated as electrical insulators. By applying Ohm's law of conductivity the total electrical flux going through the input face can be computed by [11]

$$\frac{J}{S} = \sigma \frac{V_{\text{in}} - V_{\text{out}}}{L} \tag{7.1}$$

where J is the current density, S is the area of cross-section of a phase, V is the applied voltage, L is the length of the sample, and σ is the electrical conductivity. As the area may vary, the total electrical flux can be calculated by [11]

$$J = \int_s - \sigma \nabla v \, ds \tag{7.2}$$

Owing to the similarity between the ionic and electronic conduction, this equation is directly applicable to both cases using intrinsic ionic or electric conductivity [95,96]. Using a finite volume scheme the equation systems can be obtained and solved. In Rhazaoui et al. [97] the effective conductivity of the electron conductor phase (like Ni in the anode of SOFC or LSM in the cathode) has little or no impact on the current generation. In other words, electron transport is not rate-limiting and usually can be ignored. On the contrary the ionic conductivity can play an important role in the performance of the cell, especially at low temperature or for conductors with lower ionic conductivity. Here, to compute the effective conductivity the intrinsic conductivity of the ionic conductor phase is assumed to be 0.1 S/m [11,98].

7.2.1.1.3 Gas diffusion

Another effective parameter in the performance of SOFC electrodes is the gas diffusion, especially in high current densities. Important factors characterizing the gas diffusion in porous media are the porosity and tortuosity of the gas routes [99]. Due to its simplicity, Fick's law is commonly adopted to assess gas diffusion. For porous media, Fick's first law can be modified by introducing porous media factors as [11]

$$D_{ij}^{\text{eff}} = \frac{\varphi}{\tau} D_{ij}, \tag{7.3}$$

where D_{ij} is the binary diffusivity of the gas species, D_{ij}^{eff} is the effective binary diffusivity of the gas species, and φ and τ are the porosity and tortuosity, respectively. In this research, tortuosity is obtained from the effective thermal conductivity using Avizo xlab. For this purpose the thermal conductivity of the pore network, K_{eff}, is obtained by performing a thermal simulation and the tortuosity is calculated based on

$$\tau = \varepsilon \frac{K_{bulk}}{K_{eff}}, \tag{7.4}$$

where in this equation, we assume $K_{\text{bulk}} = 1$. As discussed by Zhao et al. [100], whenever the molecular distance of a gas is in the order of average pore size the Knudsen diffusion should be directly considered. In the current study, this effect is neglected to simplify the optimization scheme and the diffusion factor is defined regardless of the average pore size to estimate the gas transport capability in the porous electrodes by

$$\text{Diffusion factor} = \frac{\varepsilon}{\tau}. \tag{7.5}$$

7.2.1.2 Microstructure hull

In the traditional design process, material and geometry are usually varied iteratively to meet design objectives. A more efficient approach to design requires simultaneous material and geometry optimization. An analogous approach for the design of materials requires a microstructure hull that consists of a set of possible microstructures existing within a region characterized by certain distribution functions. In other words a closure in this sense includes all possible effective property values predicted by sweeping input parameters defining these distribution functions, and generally a solution is a subset of such a closure. In the current study, it is assumed that if two points are chosen in the property space, they correspond to exactly two points in the geometry space connected with a continuous path in both spaces. Although this may be judged as an over simplistic approach, it allows us to demonstrate the concepts. Let us select TPB density, ionic conductivity, and gas diffusivity as critical design objectives. Mathematically speaking a property closure for these parameters can be obtained using an arbitrary analytical or approximate method and their boundaries represent constraints for the optimization process [20,101].

7.2.1.2.1 Methodology

In this study, as described in Table 7.1 the realizations of 3D models are carried out by selecting four levels of porosity within the range of 28%–37% with equal volume fractions for the ionic and electronic conductors. Similarly, three levels are chosen for the rate of nucleation and growth in the transversal (V_Z) and axial (V_{XY}) directions for each phase [11].

Fig. 7.1 depicts one of the microstructures obtained. Using these input specifications and their combinations, among the feasible combination of these parameters, 72 different microstructures were obtained because some of the nucleation levels have not been used in the feasible combination. In the next step the active clusters of each phase and active TPB length were extracted [11].

For each microstructure, ionic conductivities of the solid phase and gas diffusion factor were evaluated. Using these input and output parameters, as shown in Fig. 7.2, a neural network was trained to predict the properties of the microstructures for a given set of input parameters. In the current study the extrapolation input and output parameters is limited carefully to account for the complex behavior of fuel cells. Because of the regressive nature of the problem a back propagation

Table 7.1 Input parameters for microstructure realization [11].

Porosity	nucleation	V_z	V_{xy}
28	0.005	0.05	0.05
31	0.01	0.075	0.075
34	0.1	0.1	0.1
37			

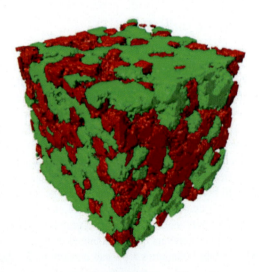

Figure 7.1 Digital reconstruction of the SOFC microstructure [11].

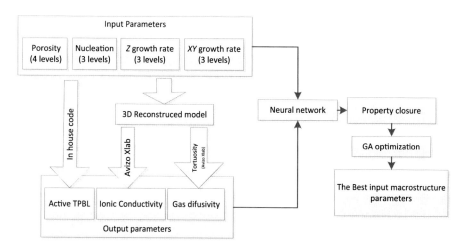

Figure 7.2 Schematic diagram showing different sections of the study and their relationship [11].

neural network (BPNN) based on four inputs (porosity, diffusion factor, axial and transversal growth rate of solid phases) and three normalized outputs (TPBL, ionic conductivity, and gas diffusion factor) was used from the neural network toolbox provided in MATLAB [11].

This model consists of an input layer, 2 hidden layers with 20 neurons, and an output layer. The value of the mean square error (MSE) and the regression for the validation data obtained were 0.00698 and 0.94894, respectively [11].

7.2.1.3 Optimization scheme

A well-trained neural network can be used to simulate the overall relationship between the microstructure and its properties within a limited range of input parameters. In this step, many hypothetical microstructures (~ 10000) are generated to obtain a microstructure hull using the pretrained neural network and the input parameter range listed in Table 7.2. Each parameter is divided into 10 levels in the range of prerealized microstructures. Based on these input parameters the properties of each microstructure are predicted by the neural network. The maximum and minimum values and their range of variation are shown.

Based on the obtained properties, it is clear that the diffusion factor and TPBL are respectively 11 and 4 times more sensitive to the geometry than the ionic conductivity with our assumptions [11] (Table 7.3).

A sample property closure obtained from the predefined input parameters is shown in Fig. 7.3. Each axis represents one of the properties and its feasible range within the selected design space [11].

Based on the preferred electrochemical and physical properties of the microstructure, which resulted in highest electrochemical sites or the highest reactant gas transport, only a limited number of these microstructures can be chosen. The region

Table 7.2 Microstructure input parameters used to obtain the microstructure hull [11].

	Input parameters			
	Porosity	**Nucleation**	**Z growth rate**	**XY growth rate**
Min	0.23	0.001	0.01	0.01
Max	0.50	0.1	0.1	0.1
Step	0.03	0.011	0.01	0.01
Variation range	0.27	0.099	0.099	0.09

Table 7.3 Microstructure properties predicted by the neural network for the microstructure hull [11].

	The range of microstructure properties		
	Diffusion factor	**TPBL**	**Ionic conductivity**
Min	0.052647	0.016694	0.002681
Max	0.128457	0.035060	0.009493
Variation range	0.075809	0.018366	0.006812

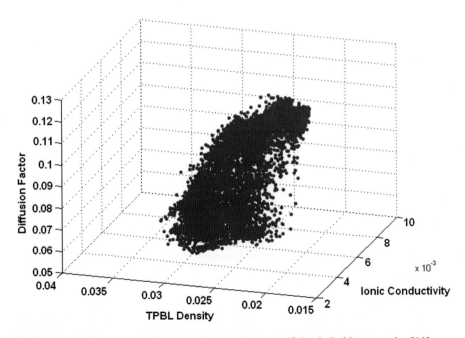

Figure 7.3 Microstructure hull; each axis represents one of the desirable properties [11].

of interest in the hull can be distinguished from the property closure by setting a target function. This target function is a linear (or nonlinear) combination of the microstructure properties that depends on the rate-limiting properties and their interactions in real microstructures. For example, high TPBL is useful when the ionic conductivity is enough to transport the ions of the electrochemical reactions [102]. Ionic conductivity, in addition to the geometry, depends on the intrinsic conductivity of the materials and temperature. Another example can be related to the diffusivity of the reactant gas into the electrode, which can be a rate-limiting phenomenon only in high current densities in the electrode, and a large number of TPBs and higher ionic conductivity cannot compensate the limitation of the gas diffusions in the performance of the cell [103]. As a result the target function should be estimated by some other in-process parameters like temperature and current density that are independent of the geometric investigations [11].

From a different viewpoint the target function can be illustrated by a free-form surface in 3D space of the microstructure hull that separates the desirable microstructures from the rejected ones. For example, if the target function is considered to be polynomial, the critical boundary is in the form of a flat plane that intercepts each 3D axis based on the coefficient of each variable in the target function. The intersection between that plane and the microstructure hull is a border for decision making that divides the accepted or rejected microstructures. The enclosed space between the microstructure hull boundaries and target function plane contains the appropriate microstructures. If there is no overlap between these restricted spaces, it means that the microstructure with preferred properties does not exist due to the geometric limitation. If there is an overlapping area an optimization method like GA can be used to explore the corresponding inputs parameters of these optimum microstructures. In this study, these processes are performed by the neural network toolbox and GA multiobjective optimization toolbox in MATLAB software [11].

7.2.1.4 Results and discussion

An ideal microstructure for the electrode is located in the region with high TPBL, ionic conductivity, and gas diffusion factor. To identify such a microstructure the property closure obtained from different microstructures within the microstructure hull is identified as shown in Fig. 7.3. In the following, in Figs. 7.4–7.6, microstructure properties are mutually compared; meanwhile, color specifies the third parameter [11].

As shown in Fig. 7.4, microstructures with the highest amount of ionic conductivity and TPBL have a lower gas diffusion capability (blue points) and therefore a restricted performance in high current densities [11].

A comparison between gas diffusion factor and ionic conductivity for hypothetical microstructures is shown in Fig. 7.5. Several microstructures with different amounts of gas diffusion factor exist for the selected range of ionic conductivity, but a high value of TPBL is only obtained when the diffusion factor is low. Additionally, as depicted in Fig. 7.5 a microstructure with a very high amount of ionic conductivity and diffusion factor does not exist [11].

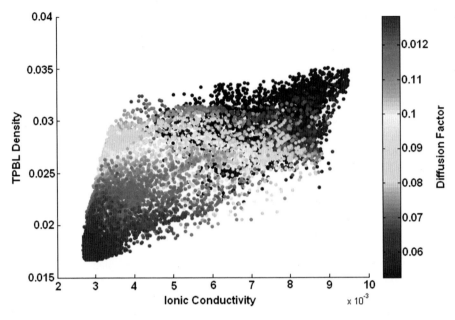

Figure 7.4 Microstructure property closure comparing ionic conductivity with TPBL. The diffusion factor is specified by gray scale colorbar [11].

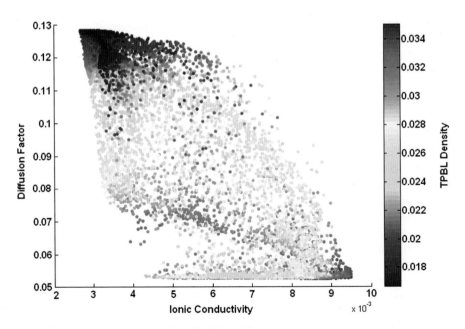

Figure 7.5 Microstructure property closure comparing ionic conductivity and diffusion factor. Here, TPBL is specified by color [11].

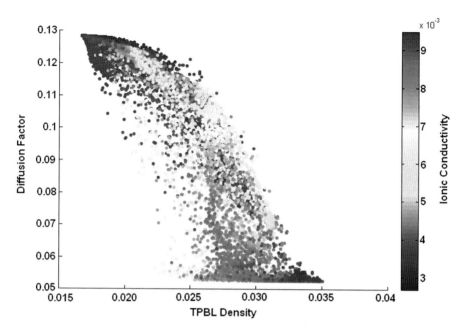

Figure 7.6 The position of different microstructures in property closure (TPBL vs diffusion factor variation, higher ionic conductivity specified grayscale colorbar) [11].

Finally the diffusion factor was compared with TPBL. As shown in Fig. 7.6, microstructures with a high diffusion factor have a high TPBL and ionic conductivity. This observation may be explained by comparing porosity of the microstructures. Normally the porosity has a direct effect on the gas diffusion factor, and a higher porosity limits the volume fraction of other two solid phases. A lower amount of ionic and electronic conductor materials initially decreases the TPBL and restricts the ion and electron transport in those phases [11].

Based on these figures a microstructure with the highest TBPL, ionic conductivity, and gas diffusion factor simultaneously cannot be achieved. However, microstructures with a high amount of TPBL and ionic conductivity, regardless of gas diffusion, are feasible as shown in Fig. 7.4. To solve this problem a desirable level of diffusion factor can be considered. For example, if the acceptable diffusion factor is set to >0.1 (like a microstructure with the porosity of 33% and a pore phase tortuosity less than 3) as observed in Fig. 7.7, microstructures with optimal properties are within the circle [11].

The goal of the current study is to discover the upper and lower bounds of feasible properties for a given set of input parameters (porosity, volume fractions, average particle size, and the average shape of particles) and generate corresponding microstructures. Additionally the relation between the input and output parameters can be explored [11].

For example, Fig. 7.8 illustrates the properties of microstructures in the range of microstructure input parameters (porosity, axial, and transversal growth rate of

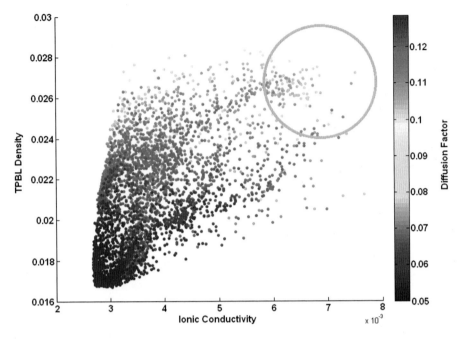

Figure 7.7 Microstructure properties with a high diffusion factor [11].

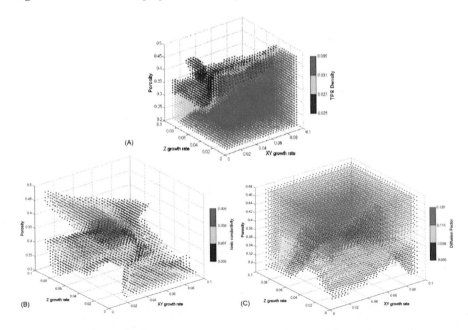

Figure 7.8 TPB density (A), ionic conductivity (B), and diffusion factor (C) range in grayscale colorbar an assumed microstructure hull with different porosities and axial or transversal growth rate of constructive grains [11].

Microstructure hull and design 317

constructive grains) as described in Table 7.2. Microstructures with the top third level of TPB density are color-coded to red, green, and blue in Fig. 7.8A, respectively. The ionic conductivity and diffusion factor of those microstructures are shown Fig. 7.8B and C. In these figures the region of high TPB density and diffusion factor can be simply recognized, but the ionic conductivity, as shown, indicates a complicated behavior in the range of input parameters [11].

The next goal for the current study is to find optimal microstructures and their corresponding input parameter values. A simple approach is adopted here by defining a normal polynomial objective function based on the given parameters such as [11]

$$\text{Objective function} = W_1 \times (\text{ionic conductivity}) + W_2 \times (\text{TPBL}) + W_3 \times (\text{diffusion factor}).$$

For this objective function, each parameter has a weighting factor. Adjusting these factors for fuel cells or other electrochemical devices can be complicated as the effect of each parameter depends on some external issues like temperature, current density, fuel and oxidant gas compounds, and pressure [104,105]. For example, by varying temperature the rate-limiting term can be changed and no longer plays a deterministic role in the performance. Determining these weighting factors requires comprehensive knowledge and is beyond the scope of the current study. Here, to simplify the problem a simple optimization problem is defined based on taking an arbitrary weighting factor ($W = 1$) for all output parameters. Proper adjustment of the weighting factor is left to a future study [11].

Using a neural network model to predict properties a direct search in microstructure hull is performed to find the extremums of the objective function. As listed in Table 7.4 a microstructure with 23% porosity, a diffusion coefficient of about 0.001, an axial growth rate of 0.02, and a transversal growth rate of 0.1 is an optimal microstructure with the maximum possible ionic conductivity and TPBL diffusion factor [11].

A more efficient approach to finding optimal microstructures is to resort to some optimization algorithm like GA. For the multiobjective GA search method the target function should be converted into a fitness cost function simply by performing an

Table 7.4 Characteristics of an optimal microstructure obtained based on the direct search approach [11].

Input parameters for microstructure generation	Porosity	0.23
	nucleation	0.001
	z growth rate	0.02
	xy growth rate	0.1
Microstructure properties	ionic conductivity	0.0095
	TPBL	0.0348
	Pore tortuosity	0.0533
	Target function (normalized output)	1.9739
	Target function	1.0174

Table 7.5 The optimum microstructure based on multiobjective GA search [11].

Input parameters for microstructure generation	Porosity	0.2092
	Nucleation	0.0012
	Z growth rate	0.0585
	XY growth rate	0.0783
Microstructure properties	Ionic conductivity	0.0091
	TPBL	0.0349
	Pore tortuosity	0.0526
	Target function (normalized output)	1.9301
	Target function	0. 966

inversion on the value (or the sign) of the target function. The search domain is limited to the input variable ranges proposed earlier to avoid impracticable extrapolation [11].

Using multiobjective GA, in each generation of the process an optimal microstructure can be obtained. Table 7.5 reports one of these microstructures that meet the criteria of variation limits in the optimization algorithm. The final property of the proposed microstructure is close to that of the microstructure identified by the simple optimization scheme, however, with different values for the input parameters [11].

7.2.2 Sample 2 (hull space for infiltrated SOFC microstructures)

Microstructural attribution of conventional and infiltrated electrodes can significantly alter the performance of SOFC electrodes. Density of active electrochemical sites [TPBs and double phase boundaries (DPBs)] and the fluid diffusivity of reactant gases into the microstructure are the main factors to determine the electrochemical performance of electrodes. Although in some cases the backbone ionic conductivity or electronic conductivity of particles can be rate-limiting in the electrochemical process, in this study, to simplify the optimization process, those are considered in an acceptable range. These parameters are linked with the backbone microstructure properties and also the infiltration parameters like loading, distribution of particles on the backbone, and their agglomeration behavior. In this study, several 3D microstructures (more than 130) were realized by varying the backbone porosity, particle loading, and deposition parameters to cover a wide range of possible configurations of infiltrated microstructures. An in-house algorithm is developed to calculate TPB and DPB density of particles, and the stack of virtual cross-section images is used to evaluate the mass transport factor in the microstructures. A combination of input and output parameters is suggested to train a neural network. A microstructure hull is developed, containing a large number (~ 9000) of hypothetical microstructures, to obtain the property closure of all possible microstructures for a range of input geometric parameters and further optimization goals [12].

It is observed that for the microstructures with an extremely high amount of TPB and surface density of infiltrated particles a relatively low gas diffusion factor should be expected; meanwhile, increasing those parameters does not have sensible contradiction with each other. Similarly, it is observed that both loading and porosity always have a positive effect on the contact surface density of particles. In addition to that, dispersion of particles on the surface of the backbone by adjusting the seeding ratio can significantly enhance the TPB density. In addition to that an optimum point is observed in controlling the ratio of particle agglomeration to get the maximum TPB density. Other microstructure constructive parameters like the effect of anisotropy in backbones and conductivity of different phases and the evaluation of these results with some experimental results are discussed in other studies [12].

Finally, to find the best microstructure a simple direct search scheme is used to find the optimal microstructure parameters among the members of the microstructure hull with different scenarios. Then an artificial neural network coupled with GA optimization scheme is developed to search the design space to find the best feasible geometric properties for an infiltrated electrode based on an optimization scenario. Finally, it should be noted that although the results strongly depended on the definition of objective function in different scenarios, the proposed method deemed useful to determine the limitation of available performance in experimental works and complicated interactions among microstructural parameters [12].

7.2.2.1 Microstructure generation and characterization

To achieve realistic microstructures, in this study, we attempt to computationally replicate the real process of infiltrated electrode fabrication. In the first stage a porous backbone microstructure is generated based on the Monte Carlo approach that is composed of three steps: (1) generation, (2) distribution, and (3) growth of the cells. In the first step, several initial seed cells are randomly placed in a unit cell of the electrode. Upon initial seed placement the growth step starts following a cellular automation algorithm. This procedure continues till the desired volume fraction for each phase is achieved. From a different perspective the growth step continues till all phases meet and fully occupy the grid. Penetration between the phases is avoided at all times throughout the initial distribution and growth of the cells [12,27].

One of the most important aspects of the model is the size of the RVE to obtain the microstructural behavior. To establish an applicable link between the RVE properties and full electrode model, statistical methods like correlation function diagrams have been used in [106−108]. They concluded that the RVE should cover at least 7×7 particles in each side and a particle diameter should be divided at least into 20 voxels to guarantee adequate precision in the evaluation of effective properties. In our study the line intercept method is used to extract the average particle size of the solid phase in the backbone. This parameter is evaluated in the range of $30−10$ voxel edge length in the backbone porosity of 25% up to 64%, if the RVE size considered is $150 \times 150 \times 150$ voxels. This assumption fortunately meets the criterion that a minimum number of particles (~ 7) should be fit into the edge of RVE with a minimum number of voxels (~ 20) to divide those particles. As a result the average particle diameter of the backbone microstructure would be in the range of $500−750$ nm, which is in the normal range of powder grading in SOFC materials; also, it corresponded to the range of $20−30$ nm for the length of voxel edge that would be the minimum size of deposited particles [12].

After the generation and characterization of the backbone, as discussed in another study, particles would be added into the realized microstructures. Unlike the previously mentioned modeling approaches which assumed the particles in the form of spheres or single cubic voxels, this proposed method can create a wide variety of geometric shapes on the surface of the backbone electrode in the form of aggregated particles. At first, several infiltrate particles controlled by seeding ratio are deposited on the active surface of the backbone as preliminary seeds. This process can be accomplished completely randomly or as a function of some parameters like backbone material or the surface geometry. In the next step the rest of particles are deposited around those seeds depending on the modeling adjustments. Agglomeration and dispersion of particles are controlled by controlling the probability of particle deposition only on the surface of predeposited seeds or on the backbone surface near the predeposited seeds. Those modeling parameters like seeding and agglomeration ratio can be evaluated from digitized 2D cross-section images with enough resolution to detect the average shape and amount of deposited particle phase on the surface of the backbone. Image processing software can be used to correlate the mathematical rules with the height, diameter, and contact angle of assumed skullcaps of particles on the surface of electrodes. When the realization of the models is completed a wide range of geometric characteristics are evaluated from those models [12].

To narrow down the scope of this study the most effective geometric parameters are determined based on the material characteristics of the electrode backbone and particles. Generally, there are three strategies in infiltration; in the first one the backbone is made of electronic conductor materials and the particles are made of ion conductor materials. The second strategy is vice versa; the scaffold is made of ion conductor materials and the particles are electronic conductors. In those strategies the most important reaction sites are TPBs as the most important performance indicator, which provide enough ionic conductivity to transport the ions of the electrochemical reactions [102]. In the third strategy the backbone is made of mixed ionic electric conductor materials (MIECs) and a small part of electrochemical reactions can occur on DPBs on the contact surface of scaffold and gas routes. In this kind of electrodes, which depend on the backbone material, the infiltrate particles can be made of electronic conductor materials to elevate electrical conductivity or ionic conductor materials to enhance the ionic conductivity of the microstructure. In addition to that, they can also be made of the MIEC materials to raise the reaction sites on the contact surface of the electrode with reactant gases. However, recently a large number of researchers added highly electrochemically active materials (like Pd, Ru,...) to these electrodes to enhance the oxygen reduction on the surface of backbones via spillover mechanism. They could alleviate the polarization resistance by increasing the electrochemically active surface density of the electrode [109−112]. Based on these findings, depending on the preferred strategy the effective geometric properties for estimating the active density of electrochemical sites would be changed. In this study, based on the third strategy, the active surface density of particles and their TPB with the scaffold and the gas routes is selected as the electrochemical performance indicators. In addition to the abundance and

Microstructure hull and design

effectiveness of reaction sites the diffusivity of the reactant gas is another important parameter, especially in high current densities or high-temperature working conditions. It can be a rate-limiting phenomenon in the electrode, and a large number of TPBs and higher ionic conductivity cannot compensate the limitation of the gas diffusions in the performance of the cell [103]. Altogether, we concentrated on input/output geometric properties of particles on the surface of backbones and assumed that the backbone conductivity is not in a range which can affect the electrochemical performance significantly specially in the last infiltration strategy. The ionic conductivity of the backbone and the electric conductivity of deposited particles are considered in other research works. For instance, in [11], we realized a range of traditional SOFC electrodes with different porosities and geometric properties in ionic or electric phases. In that study the ionic conductivity of electrode microstructures is considered as one of the outputs of the models along with TPB density and gas transport factor in the optimization process. Also in another study, we realized a range of virtually infiltrated SOFC electrodes with directional backbones to correlate the backbone conductivity with its porosity and geometric anisotropy in the models [12].

In conclusion, in each infiltration strategy the target function should be determined based on the role of materials in the microstructure. In addition to that, in-process parameters like temperature and current density are nongeometric parameters which can have a decisive role to adjust the importance of each variable in that function [12].

7.2.2.1.1 Contact surface density of particles

In addition to TPB calculation a number of researchers suggested that DPBs between deposited particles and active reactant gas routes can play an important role in electrochemical reactions, especially when highly active electrochemical particles are deposited on the surface of MIEC backbones [103,113]. The implemented algorithm in this study first identifies all active contact situations of each voxel between different phases. Then the active surface density of deposited particles as well as other contact conditions is extracted by sorting special columns in the identification matrix of realized microstructures [12].

7.2.2.1.2 Gas transport factor

Another effective parameter in the performance of SOFC electrodes is the gas diffusion, especially in high current densities or high temperatures. Important factors characterizing the gas diffusion in porous media are the porosity and tortuosity of the gas routes [99]. Due to its simplicity, Fick's law is commonly adopted to assess gas diffusion. For porous media, Fick's first law can be modified by introducing porous media factors as Eq. (7.6) [12].

$$D_{ij}^{\text{eff}} = \frac{\varphi}{\tau} D_{ij} \tag{7.6}$$

where D_{ij} is the binary diffusivity of the gas species, D_{ij}^{eff} is the effective binary diffusivity of the gas species, and φ and τ are the porosity and tortuosity, respectively.

In this study, tortuosity is obtained from the effective thermal conductivity calculation using Avizo xlab, EFI Corporation. For this purpose the thermal conductivity of the pore network (K_{eff}) is obtained by performing a thermal simulation and the tortuosity is calculated based on Eq. (7.7) [12].

$$\tau = \varepsilon \frac{K_{bulk}}{K_{eff}} \tag{7.7}$$

where we assume $K_{bulk} = 1$.

The Knudsen diffusion is another mechanism in porous microstructures. As discussed in [99,100] the Knudsen number (Kn) is a key factor to determine the influence of Knudsen diffusion mechanism in porous microstructures. That is defined as the ratio of molecular distance of the passing gas to the average pore size of porous materials. If Kn is much greater than 10, collisions between gas molecules and the porous electrode are more dominant than the collisions between gas molecules, resulting in negligible molecular diffusion and viscous diffusion. If Kn is much smaller than 0.1, collisions and interactions between gas molecules become dominant, and Knudsen diffusion becomes negligible compared with molecular diffusion and viscous diffusion. In the current study the mean free path of the reactant gas (H2) at normal operational temperature and working pressure is calculated to be about 0.4 μm. Whenever the average pore size in realized backbones is evaluated to be about 0.6 μm, the Kn would be about 1.7, which is much smaller than 10. Hence although the Knudsen diffusion can slightly affect the gas diffusivity in this range, to simplify the optimization scheme the gas transport factor can be defined regardless of the average pore size and is estimated by Eq. (7.8) [12].

$$\text{Diffusion factor} = \frac{\varepsilon}{\tau} \tag{7.8}$$

7.2.2.2 Microstructure hull

In traditional design, material and geometry are usually varied iteratively to meet design requirements. A more efficient approach is to design the required material and geometry simultaneously. Following this approach for the design of material a microstructure hull is needed, which consists of a set of possible microstructures existing within a region characterized by certain distribution functions obtained from the microstructure. In other words a closure in this sense includes all possible effective property values predicted by sweeping input parameters defining these distribution functions, and generally a solution is a subset of this closure. In this study, it is assumed that if two points are chosen in the property space, they correspond to two points in the geometry space. Mathematically speaking a property closure for these parameters can be obtained using an arbitrary analytical or approximate method and their boundaries represent constraints for the optimization process [20,101]. The TPB density, contact surface density, and gas diffusivity are selected as critical design objectives in this study to form the property hull [12].

In this study the realization of 3D models is carried out by choosing six levels of porosity within the range of 26%−64% and the same nucleation and directional growth rate of backbone microstructures. The volumetric loading of particles is changed in the range of 1%−25% in 11 levels. The seeding ratio that controls the distribution of particles on the surface of the backbone is changed in the range of 3%−25% in 10 levels. The seeding behavior which controls the deposition probability of seeds on the surface of the backbone is set in two levels, free seeding or being a function of contact surface density on the backbone. At last the agglomeration ratio is changed from 0.2 to 0.8 in seven levels to simulate the aggregation of particles on the surface of the scaffold. As mentioned before, by converting those realized models to 3D microstructures as shown in Fig. 7.9, new opportunities will be available [12].

Following these input specifications and their combinations, among the feasible combinations of these parameters, 131 different microstructures are realized and then characterized in regard of their properties. TPB density and contact surface densities are evaluated directly from the mathematical model, but the gas diffusion factor is evaluated from the 3D realized models. Using these input and output parameters, as shown in Fig. 7.10, a neural network is trained to predict the properties of the microstructures for a given set of input parameters. In the current study the extrapolation in input and output parameters is limited carefully to account for the complex behavior of fuel cells. Because of the regressive nature of the problem a back propagation neural network (BPNN) based on five inputs (backbone porosity, particle loading, seeding factor, seeding type, and surface coverage behavior) and three normalized outputs (TPB density, particle contact surface density, and gas diffusion factor) is used from the neural network toolbox provided in MATLAB [12].

This model consists of an input layer, a hidden layer with 20 neurons with the tansig transfer function, and an output layer with 3 neurons with the linear transfer function.

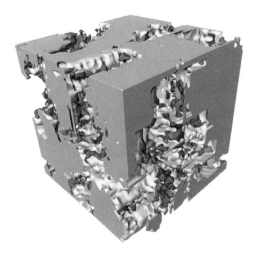

Figure 7.9 A small part of realized microstructures (backbone porosity: 47% and volumetric loading: 5%) [12].

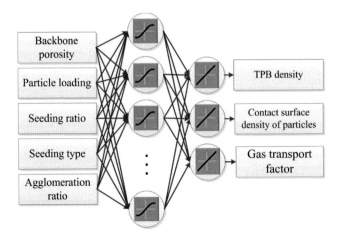

Figure 7.10 The neural network structure to predict microstructure properties [12].

70% of data is dedicated for neural network training, and the remaining data are used for testing and validation of the model. The mean square error (MSE) and the regression for the validation data are obtained as 1.4E-4 and 0.99, respectively [12].

7.2.2.3 Optimization scheme

When a neural network model is trained properly, it can be used as a practical tool to predict the output parameters in terms of input variables. In the other words, it can be possible to correlate the microstructure hull to property closure by sweeping feasible input variables in an acceptable range. Due to complex behavior of the fuel cell, this range should not exceed the previously reported input parameters to avoid improper extrapolation of data. This process enables researchers to predict the variation of output parameters in some specific domains where there are no data to analyze and provide a search engine for using further optimization techniques as shown in Fig. 7.11 [12].

This well-trained neural network is used to simulate the overall relationship between the microstructure and its properties within a limited range of input parameters. A large number of hypothetical microstructures ($\sim$12,600) are created to obtain a microstructure hull using the pretrained neural network, and the input parameter range listed in Table 7.6 [12].

Porosity and loading are divided into 10 levels; seeding and surface agglomeration are divided into nine and seven levels, respectively, in the range of prerealized microstructures, and the seeding type is considered in the free or confined method. Based on these input parameters the properties of each microstructure are predicted by the neural network. Among these models, 9070 microstructures are in the range of acceptable properties with allowable positive values. The maximum and minimum property values and their range of variation are shown in Table 7.7 [12].

Microstructure hull and design

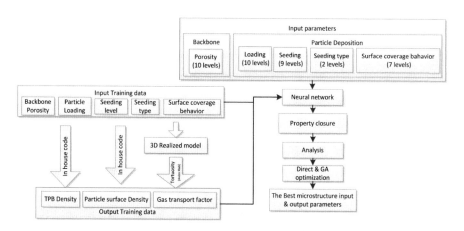

Figure 7.11 Schematic diagram showing different sections of the study and their relationship [12].

Table 7.6 Input parameters for microstructure realization [12].

	Max	Min	Variation
Porosity volume fraction	0.642	0.300	0.342
Loading	0.250	0.010	0.240
Seeding	0.250	0.025	0.225
Seeding type	1.000	0.000	1.000
Surface agglomeration ratio	0.800	0.200	0.600

Table 7.7 Microstructure properties predicted by the neural network for the microstructure hull [12].

	Max	Min	Variation range
TPB density (L/μm^2)	0.331	0.000	0.331
Contact surface density of particles (L/μm)	0.722	0.001	0.721
Gas diffusion factor	0.495	0.000	0.495

Based on the obtained properties, it is clear that the gas diffusion factor and particle contact surface density are 1.5 and 2.2 times more sensitive to the geometry than the TPB density with our assumptions, respectively [12].

A sample property closure obtained from the predefined input parameters is shown in Fig. 7.12. Each axis represents one of the properties and its feasible range within the selected design space [12].

Based on the preferred electrochemical and physical properties of the microstructure, which result in the highest electrochemical sites or the highest reactant gas transport, only a limited number of these microstructures can be chosen. The region of

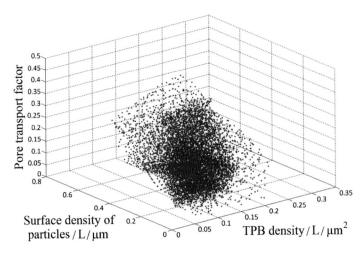

Figure 7.12 Microstructure property hull; each axis represents one of the desirable properties [12].

interest in the hull can be distinguished from the property closure by setting a target function. This target function can be a linear (or nonlinear) combination of the microstructure properties that can be extracted from the role of materials and the rate-limiting reactions in the real microstructure in the operational condition [12].

From a different viewpoint the objective function can be illustrated by a free-form surface in 3D space of the microstructure hull that separates the desirable microstructures from the rejected ones. For example, if the objective function considered is a linear combination of parameters, the critical boundary is in the form of a flat plane that intercepts each 3D axis based on the coefficient of each variable in the objective function. The intersection between that plane and the microstructure hull is a border for decision making that divides the accepted or rejected microstructures. The enclosed space between the microstructure hull boundaries and objective function plane contains the appropriate microstructures. If there is no overlap between these restricted spaces, it means that the microstructure with preferred properties does not exist due to the geometric limitation. If a property characteristic associated with a microstructure is located in overlapping area, geometric properties of that model can be explored using the direct search method. However, those properties necessarily would not be the maximum achievable ones. If there is not any microstructure in that area, an optimization method like GA can be used to explore the input geometric parameter domain to find feasible microstructures or other corresponding input parameters of microstructures with better properties. In GA an initial population of potential solutions will be randomly generated, then the fitness function values will be determined, and those solutions are proportionally ranked. The new parent population will be selected among the highly ranked solutions. Some genetic operators like crossover and mutation also will be used to combine parents and random change in new generations up to meeting stopping criteria. In this study, these processes are performed by the neural network toolbox and GA multiobjective optimization toolbox in MATLAB software [12].

7.2.2.4 Results and discussion

Based on our assumptions in effective geometric properties an ideal microstructure should have the highest active electrochemical sites (TPB and contact surface density of catalyst particles) and the maximum gas transport factor. Also, depending on the working condition of the electrode the priority of these parameters might be varied, and a combination of those parameters can be more useful to identify the best microstructures. Thus a property closure would be detected based on different microstructures within the microstructure hull. To clarify this concept, in Fig. 7.13, microstructure properties are mutually compared when the third parameter is specified with color [12].

As shown in Fig. 7.13A and B, microstructures with the highest amount of reaction sites (TPB or particle surface density) have a lower gas diffusion capability and may restrict the performance in some cases. A comparison between TPB density and

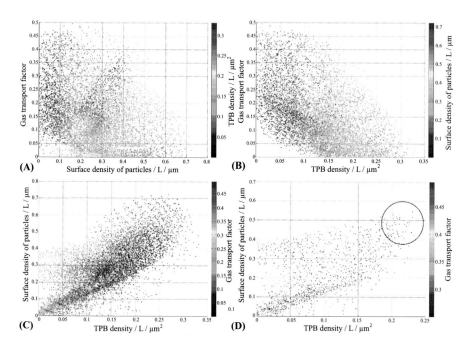

Figure 7.13 Microstructure property closure comparing the gas transport factor with the surface density of particles (A) and TPB density (B). The surface density of microstructures with the full range (C) and high range of gas transport factor (D) [12].

particle surface density for hypothetical microstructures is shown in Fig. 7.13C. Several microstructures with a high amount of TPB density and particle surface density exist, and there is no inconsistency between these properties in the microstructure, but as shown by blue points, these microstructures have a very low amount of diffusion factor. Based on these figures a microstructure with an extremely high amount of both reaction sites and gas diffusion factor is not available. However, high amounts of TPB density and contact surface density of catalyst particles, regardless of gas diffusion, are feasible as shown in Fig. 7.13C. To find a way to choose the best microstructure a desirable level of gas transport factor can be considered. For example, if the acceptable amount of gas transport factor is set to more than 0.25 (like a microstructure with the porosity of 50% and a pore phase tortuosity less than 2), feasible TPB and particle surface density ranges of microstructures can be observed, as shown in Fig. 7.13D. The microstructures located in the circle have the most obtainable reaction sites and acceptable levels of gas diffusion [12].

In addition, to discover the upper and lower bounds of feasible properties for a specific range of microstructures, it is possible to establish a relationship between the microstructure geometric properties and constructive variables of realized models (backbone porosity, loading, seeding, and agglomeration ratio of particles). For instance, Fig. 7.14A illustrates the variation of TPB density against the porosity and particle volumetric loading; meanwhile the other parameters are set fixed. It can be observed that TPB density has a maximum point by changing the particle loading and the porosity can have a positive effect to some extent. In Fig. 7.14B, it is shown

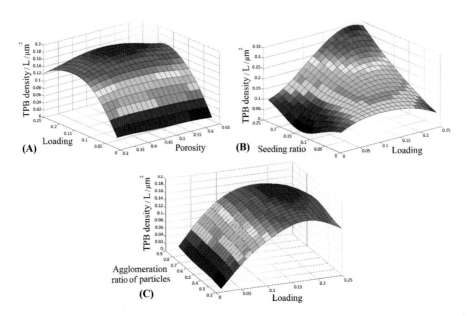

Figure 7.14 TPB density variation related to the porosity and loading (A) related to agglomeration ratio of particles and loading (B), and related to Seeding ratio and loading (C) [12].

Microstructure hull and design

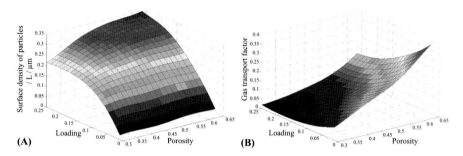

Figure 7.15 Surface density of particles (A) and pore transport factor (B) related to porosity and loading [12].

that in highly loaded microstructures the seeding ratio can enhance TPB density considerably. Also, as shown in Fig. 7.14C, both high agglomeration and dispersion of particles on the surface of the backbone have a negative effect on maximum TPB density [12].

The effect of particle loading and porosity on the surface density of deposited particles as an electrochemical factor and the gas transport factor as a physical factor is shown in Fig. 7.15A and B, respectively. It can be observed that the backbone porosity can extend the overall surface density on particles, especially in a high amount of volume fraction of deposited particles. On the other hand, the gas transport factor is continuously decreased by adding particles, although its reduction rate is more influenced by declining backbone porosity [12].

As discussed before the next goal of the current study is finding the optimal microstructure properties and their corresponding input constructive parameters. A simple approach is adopted here by defining a simple linear combination of performance indicators as an objective such as Eq. (7.9) [12].

$$\text{Objective function} = W_1 \times (\text{TPB density}) + W_2 \times (\text{Surface density of particles}) + W_3 \times (\text{Gas transport factor}) \quad (7.9)$$

In this equation the selected geometric properties have a weighting factor (W_{1-3}) based on their role in the performance of the electrode. Adjusting the role of each characterized property is a complicated task, especially in electrochemical devices like fuel cells, whereas they depend on some external issues [12].

As discussed before the intrinsic property of the constructive material, working conditions (temperature, pressure, and current density), and fuel and oxidant gas composition are the effective items in determining the impact factor of geometric properties [104,105]. For example, in conventional electrodes which consist of separate electrons and ion conductor phases the rate of electrochemical reactions and power generation are causally related to TPB density. In infiltrated microstructures, as described before, it depends on the infiltration strategy. For example, if the backbone is made of ion conductor materials, the particles should be made of an electron conductor phase to create reaction sites at their interfaces with backbone and gas routes

[92]. In this case the TPB density can play the major role via increasing W_1 in that function. On the other hand, if the backbone is made of mixed ionic electric conductor materials, in addition to TBP, the contact surface density of the electrode and infiltrated particles can have different effects on the rate of electrochemical reactions based on the catalytic activity of those materials [114]. In this case, W_2 can be considered as a critical parameter in the overall efficiency of the electrode. In addition to the density of reaction sites, varying temperature and current density can change the rate-limiting item [99]. To consider this state, W_3 would determine the impact of reactant gas diffusion capability in electrode performance [12].

Generally, determining these weighting factors requires comprehensive knowledge and is beyond the scope of the current study because of the intense discussion in that context. To simplify the optimization problem, five different scenarios over the normalized values of properties are designed. As shown in Table 7.8 the weighting factor is changed in those scenarios to determine the most effectice parameters based on different assummptions in material roles or working conditions [12].

The objective function of these scenarios in combination with the proposed neural network model can directly search the microstructure hull to find the best obtainable properties for infiltrated electrodes. Table 7.9 reports the optimum constructive parameters of realized microstructures and their corresponding geometric properties based on different scenarios using the direct search method in the property hull [12].

Table 7.8 Different optimization scenarios [12].

scenario	Effective factor(s)	Objective function weights		
		W_1	W_2	W_3
1	TPB density	1	0	0
2	Particle surface density	0	1	0
3	Gas diffusion factor	0	0	1
4	TPB and particle surface density	1	1	0
5	TPB, particle surface density, and gas diffusion factor	1	1	1

Table 7.9 The comparison of evaluated optimum microstructures in direct search and the multiobjective GA method based on different scenarios [12].

Optimization scheme		GA	Direct
Microstructural parameters	Backbone porosity	0.632	0.490
	Loading	0.180	0.225
	Seeding	0.240	0.25
	Seeding behavior	0.663	0
	Agglomeration ratio	0.345	0.200
Geometric property	TPB density (L/μm^2)	0.308	0.286
	Particle surface density (L/μm)	0.716	0.694
	Gas transport factor	0.034	0.001

Microstructure hull and design

The results indicate that if the active reaction site is considered the TPB, the best microstructural parameters should have 56% backbone porosity, 20% particle loading, the highest level of random seeding, and the agglomeration ratio of 70%. Based on the second scenario, if the contact surface density of particles plays the most important role, the loading should be in the highest level and the particles should be agglomerated as much as possible. In the third scenario the highest backbone porosity and agglomeration ratio are obtained in the lowest level of loading to facilitate the gas diffusion. The fourth scenario is a combination of the first and second ones to consider the maximum overall active reaction sites in the optimum model. Fig. 7.16 illustrates one of the section views and a fully realized optimum microstructure based on this scenario. Finally the last scenario is designed to assign a similar weight for all output properties to simultaneously maximize them [12].

Whereas the direct search method can only search among the realized microstructures, there are more efficient approaches to find optimal microstructures using metaheuristic optimization algorithms like GA. To use the multiobjective GA search method an objective function is designed based on one of the above-mentioned scenarios. At first the above-mentioned objective function is converted to a fitness cost function by performing an inversion on the value (or the sign). Then to avoid impracticable extrapolation a search domain is defined to limit the input variable into the proposed ranges. Using multiobjective GA, in each generation of the process an optimal microstructure is obtained. In GA an initial population (50) of potential solutions were randomly generated, then the fitness function values were determined, and those solutions were proportionally ranked. The new parent population was selected among the highly ranked solutions. As mentioned before, some genetic operators like crossover and mutation also are used to combine parents and random change in new generations up to meeting stopping criteria (reaching 100 generation or function tolerances of 1e-4).

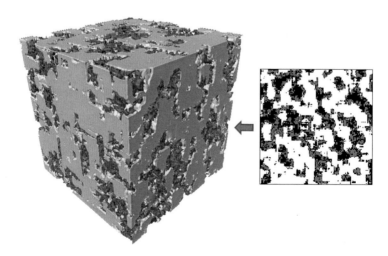

Figure 7.16 Realized optimum microstructure (left) and a sample cross-section view based on different scenarios [12].

332

Table 7.10 The optimum microstructure based on the direct search method in the microstructural and property hull [12].

Scenario		1	2	3	4	5
Microstructural parameters	Backbone porosity	0.566	0.528	0.642	0.490	0.642
	Loading	0.200	0.250	0.100	0.225	0.250
	Seeding	0.25	0.25	0.20	0.25	0.25
	Seeding behavior	0	0	1	0	0
	Agglomeration ratio	0.700	0.200	0.800	0.200	0.600
Geometric property	TPB density ($L/\mu m^2$)	0.331	0.268	0.011	0.286	0.198
	Particle surface density ($L/\mu m$)	0.429	0.722	0.160	0.694	0.531
	Gas transport factor	0.036	0.026	0.495	0.001	0.400

All these processes are performed by the neural network toolbox combined with the GA multiobjective optimization toolbox in MATLAB software [12].

Table 7.10 reports one of these microstructures that meet the function tolerance in the optimization algorithm under different scenarios (maximizing the TPB and particle surface density). A higher amount of TPB and particle surface density are achieved in this proposed microstructure compared to the last one that is found by direct search. It can be an interesting point that in this method the optimum microstructure has a lower loading and more gas transport factor than the one found via the direct search method in the same scenario [12].

7.3 Practical approach for materials design of random heterogeneous materials

7.3.1 Interpolation of random heterogeneous microstructures

This study proposes a new efficient procedure for combining two heterogeneous microstructures without any concern about unequal volume fractions, anisotropy, and morphology. Using statistical measures such as TPCF and autocovariance function, two microstructures were described mathematically. By a convex combination of the autocovariance functions, some new sets of TPCFs were obtained. A powerful phase recovery algorithm was introduced to realize the combined TPCFs and obtain the intermediate microstructure. The evaluation of the obtained microstructures corresponding to the combination factor revealed that the properties and appearance of the resultant microstructures also resemble those of a combination of the two initial microstructures. It was shown that by incrementing the combination factor the properties and also the morphology of the combined microstructure gradually change from the first microstructure to the second one. The procedure was examined for various microstructures such as isotropic and anisotropic and with variable volume fractions. An error analysis revealed that the proposed procedure may contain a maximum error equal to 5% based on TPCF criteria [13].

Microstructure hull and design

7.3.1.1 Basic concept

N-point correlation functions can be used as mathematical descriptors of a material's microstructure. For a multiphase microstructure, let us specify the microstructure function as [13]

$$x_s^n = \begin{cases} 1 & s \text{ in phase } n \\ 0 & \text{otherwise} \end{cases}, \tag{7.10}$$

where n and s are the intended phase and position, respectively. The medium considered here is a digitized 3D cubic microstructure. The grid points are numbered from 0 to $S - 1$, and for a two-phase medium the phases are specified with n and n'. Using Eq. (7.10) the one-point correlation function for phase n is defined as [13]

$$c_1^n = \langle x_s^n \rangle = \frac{1}{S} \sum_{s=0}^{S-1} x_s^n, \tag{7.11}$$

where $\langle \ldots \rangle$ symbol represents an ensemble average. The asymptotic value for one-point correlation for each phase is equivalent to the volume fraction. One-point correlation functions can be used for establishing useful upper and lower bound properties of heterogeneous materials using the variational technique [13,115].

TPCFs provide the first order of geometrical specifications of a microstructure. For a macroscopically homogeneous and limited microstructure, TPCF is defined as

$$c_2^{nn'}(r) = \frac{1}{S} \sum_{s=0}^{S-1} x_s^n \, x_{s+r}^{n'}, \tag{7.12}$$

where r is the vector for which the correlation function is calculated. It should be noted that for macroscopically homogeneous media the absolute positions of the grid points are irrelevant, so the correlation functions are defined in terms of relative differences between the position vectors of the points [13].

For a two-phase microstructure the possible TPCFs for the vector r, depending on the phases intended for the head and tail of the vector (see Fig. 7.17), are $c_2^{nn'}$, $c_2^{n'n}$, $c_2^{n'n'}$, and c_2^{nn}, which are linearly dependent due to the normality conditions. The relationship between the two-point correlations satisfies [13,116]

$$c_2^{nn'} + c_2^{n'n} + c_2^{n'n'} + c_2^{nn} = 1. \tag{7.13}$$

$$c_2^{nn'} = c_2^{n'n}. \tag{7.14}$$

$$c_2^{nn'} + c_2^{nn} = v_n. \tag{7.15}$$

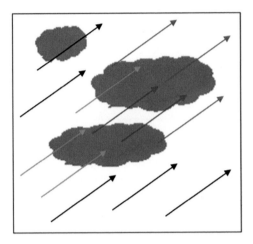

Figure 7.17 Schematic representation of possible TPCFs for a two-phase microstructure [13].

where v_n is the volume fraction of the phase n. Therefore there is only one independent set of TPCFs for every microstructure and all other sets can be determined in terms of an independent set.

In addition to Eqs. (7.13)–(7.15), there are other limit constraints for any arbitrary phases p and q as follows [13,116]:

$$\lim_{r \to 0} c_2^{pq}(r) = \begin{cases} v_p & \text{if } p = q \\ 0 & \text{if } p \neq q \end{cases}, \quad p.q = n.n' \tag{7.16}$$

$$\lim_{r \to \infty} c_2^{pq}(r) = \begin{cases} v_p^2 & \text{if } p = q \\ v_p v_q & \text{if } p \neq q \end{cases}, \quad p.q = n.n'. \tag{7.17}$$

The calculation of TPCFs using the microstructure function can be carried out by FFT [18,45]. The FFT of the microstructure function, Eq. (7.10), can be expressed as [13]

$$X_k^n = \mathcal{F}(x_s^n) = \frac{1}{S} \sum_{s=0}^{S-1} x_s^n \; e^{(2\pi i s k/S)} = \frac{1}{S} |X_k^n| e^{i\theta_k^n}. \tag{7.18}$$

where $|X_k^n|$ is the amplitude and θ_k^n is the phase of the Fourier transform.

Employing the periodicity assumption for the microstructure and using the convolution theorem [45] the FFT of the TPCF can be defined as [13]

$$C_k^{nn'} = \mathcal{F}(c_2^{nn'}(r)) = \frac{1}{S} |X_k^n| e^{i\theta_k^n} \; |X_k^{n'}| e^{-i\theta_k^{n'}}. \tag{7.19}$$

Microstructure hull and design

Therefore if n and n' are identical, Eq. (7.19) is converted to an important equation that interrelates the microstructure function to the amplitude of FFT of TPCF by [13]

$$C_k^{nn} = \frac{1}{S}|X_k^n|^2 = \frac{1}{S}\widetilde{X}_k^n X_k^n. \qquad (7.20)$$

where $\widetilde{X}_k^n$ is the complex conjugate of X_k^n. Therefore in order to calculate TPCFs for every microstructure, given Eq. (7.10) for all points, it is sufficient to simply multiply $\widetilde{X}_k^n$ and X_k^n, divided by S. Then $c_2^{nn}(r)$ can be determined by the inverse FFT of C_k^{nn} [13].

Representation of microstructures by TPCFs provides the possibility of merging two or more microstructures using an autocovariance function that is defined for the vector r as below [13]

$$f_2^n(r) = \frac{c_2^{nn}(r) - \nu_n^2}{\nu_n(1 - \nu_n)}. \qquad (7.21)$$

Autocovariance $f_2^n(r)$ is a linear and scaled function of the $c_2^{nn}(r)$ that varies between 1 and 0 for $r = 0$ and $r \to \infty$, respectively. It should be noted that every hypothetical distribution function cannot be considered as a TPCF or autocovariance function of a realizable microstructure. A valid correlation function for a realizable microstructure must satisfy limit conditions given by Eqs. (7.16) and (7.8). In addition to these general constraints a realizable set of desired TPCFs must meet other necessary conditions such as non-negativity, inequality, and slope at origin, some of which are discussed by Torquato et al. [19,117,118]. It is demonstrated in [25,46] that the a convex combination of two realizable autocovariance functions for statistically homogeneous media also satisfies all known necessary conditions and therefore can be realized as a combined microstructure via its respective TPCFs [13].

If $f_2'^n(r)$ and $f_2''^n(r)$ are the autocovariance functions of two realizable microstructures and α and $f_2^n(r)$ are the combination factor and autocovariance of the merged microstructure, respectively, $f_2^n(r)$ can be expressed as [13]

$$f_2^n(r) = \alpha f_2'^n(r) + (1 - \alpha)f_2''^n(r). \qquad (7.22)$$

Also the volume fraction of the combined microstructure is given by

$$\nu_n = \alpha \; \acute{\nu}_n + (1 - \alpha)\grave{\nu}_n. \qquad (7.23)$$

where $\acute{\nu}_n$ and $\grave{\nu}_n$ are the volume fractions of the initial microstructures. It is clear that $f_2^n(r)$ smoothly decays from 1 to 0 for $r = 0$ and $r \to \infty$, respectively [13].

7.3.1.2 Microstructure construction

The reconstruction process is the realization of a microstructure based on its assigned statistical correlation functions. TPCFs used as inputs for the reconstruction procedure are provided by Eqs. (7.21)–(7.23). The phase recovery algorithm first proposed for signal processing [52] is used here for the reconstruction. This method is based upon Eq. (7.20). Given a set of TPCFs for a microstructure and using this equation the amplitudes of the microstructure functions are determined. Further, it is sufficient to recover the phases [in Eq. (7.18)] using the phase recovery algorithm to determine the microstructure functions completely. Therefore the problem is to find the x_s^n for all points [considering the constraint that x_s^n must be only 0 or 1 according to Eq. (7.10)] given the full set of TPCFs [13].

The phase recovery algorithm used here follows the algorithm proposed by Fienup [119] for the reconstruction of an object from the modulus of its Fourier transform first utilized by Fullwood et al. [45] for microstructure reconstruction. This method has four steps, as illustrated in Fig. 7.18: (1) An initial random microstructure, x_s^nite, is guessed for the first iteration and the FFT of the microstructure function, X_k^nite, is taken using Eq. (7.18); (2) the modulus of the FFT, $|X_k^n$ite$|$, is replaced with the square root of the C_k^{nn}ref multiplied by S or $|X_k^n$ref$|$ using Eq. (7.20), and the phases are preserved; (3) using the inverse of Eq. (7.18) the inverse Fourier transform of $|X_k^n$ref$|e^{i\theta_k^n}$ite is taken and considered as new x_s^nite; (4) some constraints in real space [Eq. (7.10)] are imposed to the obtained x_s^nite by rounding its value. The realized microstructure in step (4) is used as input for step (1). After step (4) the error between the TPCFs of the realized microstructure and $c_2^{nn}(r)_{ref}$ is calculated, and if the convergence criteria are satisfied the algorithm stops and the x_s^nite of that iteration is considered as the final constructed microstructure. In the suggested reconstruction procedure the rational error is defined as the average of the differences between the referenced and constructed TPCFs for all vectors and is expressed as [13]

$$\text{Error} = \frac{1}{S} \sum_{r=0}^{r=S-1} \frac{|c_2^{nn}(r)_{ite} - c_2^{nn}(r)_{ref}|}{c_2^{nn}(r)_{ref}}. \tag{7.24}$$

In contrast to other methods such as simulated annealing that use radial correlation functions in which orientation information is averaged out [49], phase recovery uses the orientational TPCFs. Therefore using this method a microstructure can be accurately reconstructed from its correlation function alone up to a translation or inversion [45] provided that each point is occupied only by one phase [Eq. (7.10)] [13].

7.3.1.3 Results and discussion

In this section a series of representative examples are presented to demonstrate the capabilities of the proposed procedure in combing various microstructures, even with anisotropic media and unequal volume fractions. For anisotropic microstructures, it is demonstrated that by gradually increasing the combination factor the related properties evolve; meanwhile the anisotropies interchange from one microstructure to another. All examples in this section are two-phase void/solid microstructures (except for mechanical property examination in Section 7.4.2), and each phase is considered impermeable to anything that can be conducted by the other phase [13].

Microstructure hull and design

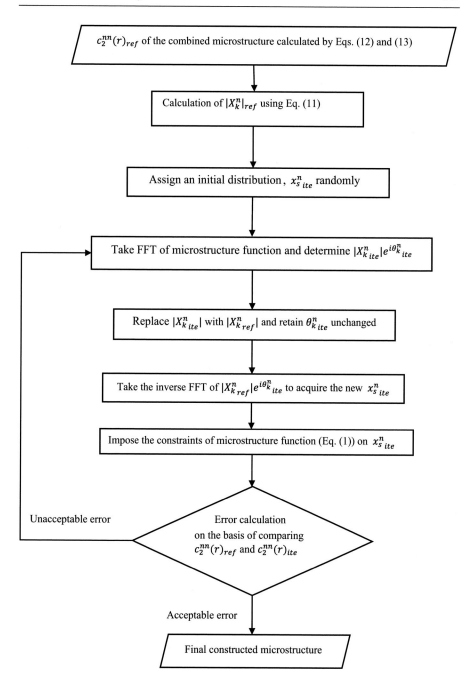

Figure 7.18 Flowchart of the phase recovery algorithm used for construction of microstructures based on assumed $c_2^{nn}(r)_{\text{ref}}$ [13].

7.3.1.3.1 Combining two isotropic microstructures

In the first example, two isotropic microstructures are considered to demonstrate how to implement the procedure illustrated in Fig. 7.18. A $150 \times 150 \times 150$ cubic microstructure with two phases (void and solid) is considered as the first microstructure [Fig. 7.19A]. The volume fraction of the void phase is set to $\nu_{void} = 0.33$. This microstructure is originally constructed by a hybrid stochastic simulation methodology based on the colony and kinetic algorithms and the Monte Carlo method [13,27].

The second microstructure (Fig. 7.19B) has the same size and volume fraction, but it is constructed by the phase recovery algorithm using a hypothetical exponentially decaying sinusoidal autocovariance function [120]. As shown in these figures the void passages of the first microstructure are relatively larger than those of the second one [13].

Fig. 7.20 illustrates the gradual change of the combined microstructure by incrementally increasing the combination factor. The form and distribution of the void and solid phases for intermediate steps are a mixture of the initial and final microstructures with a contribution determined by the combination factor [13].

In the next step, effective thermal conductivity of the combined microstructure was evaluated. It should be noted that electrical and thermal conductivity problems and diffusion properties of a specific phase (where other phases are impermeable to what passes through that specific phase) are all mathematically equivalent [19]. Therefore by determining the thermal conductivity a wide range of properties for various applications can be characterized [13].

By assuming the thermal conductivity for a totally solid or void cube, $k = 1$, the effective thermal conductivity for each phase in all three directions can be determined using Fourier's law. In these simulations, for each direction the opposing faces normal to that direction are used as a set of constant temperature boundary conditions; meanwhile, all other faces are assumed to be perfect insulators. Using

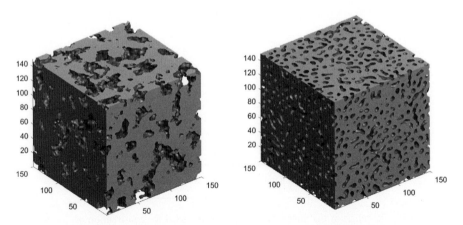

Figure 7.19 The (A) first microstructure with the size of $150 \times 150 \times 150$ voxel and $\nu_{void} = 0.33$ and (B) the second microstructure with the same size and volume fraction of the first one. The transparent region is the void phase for all figures of this section [13].

finite volume analysis the heat flux was determined to obtain the effective thermal conductivity from [13]

$$k_{\text{eff}} = -\frac{q}{(T_2 - T_1)}. \tag{7.25}$$

Since the microstructure is deemed isotropic, the average of three directions was taken to obtain k_{eff} for each phase. Fig. 7.21 illustrates the effective thermal conductivity

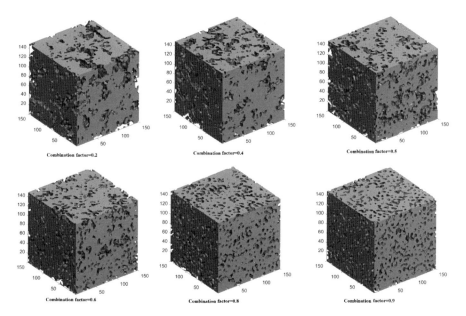

Figure 7.20 Combined microstructures for various combination factors. The intermediate microstructures are a mixture of the initial and final microstructures [13].

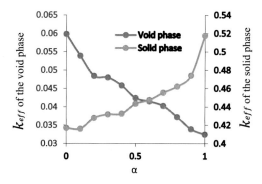

Figure 7.21 Effective thermal conductivity of the void and solid phases for isotropic combining [13].

obtained for the combined microstructures. The first microstructure compared with the second one has a relatively larger conductivity for the void phase, $k_{\text{eff}} = 0.06$, and a lower conductivity for the solid one, $k_{\text{eff}} = 0.42$. In contrast the second microstructure compared with the first one has a lower thermal conductivity for the void phase, $k_{\text{eff}} = 0.032$, and a larger thermal conductivity for the solid one, $k_{\text{eff}} = 0.52$. As shown the thermal conductivity of the void and solid phases for the combined microstructure gradually changes between these limits [13].

7.3.1.3.2 Combining two anisotropic microstructures

In this section, two $100 \times 100 \times 100$ voxel anisotropic transversely isotropic microstructures with $\nu_{\text{void}} = 0.4$ are combined. The first microstructure has enlarged void phase passages in the horizontal direction. The second microstructure is just like the first one, except that the elongation of the void phase is selected to be in the vertical direction. As shown in Fig. 7.22 by increasing the combination factor a quick visual inspection reveals that the elongation direction is gradually changed from horizontal to vertical [13].

Fig. 7.23A and B depict the void and solid thermal conductivities of the combined microstructures in horizontal and vertical directions, and it can be observed that the properties for the horizontal and vertical directions gradually switch [13].

To investigate the variation of mechanical properties, it is assumed that the two phases depicted in Fig. 7.23 are both solid with volume fractions of $\nu_1 = 0.4$ and

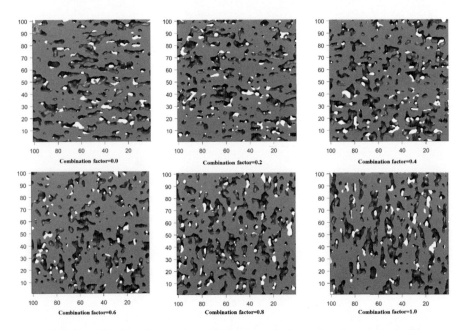

Figure 7.22 Combination of two anisotropic transversely isotropic microstructures. The elongation of the void phase gradually changes from the horizontal direction to the vertical one [13].

Microstructure hull and design

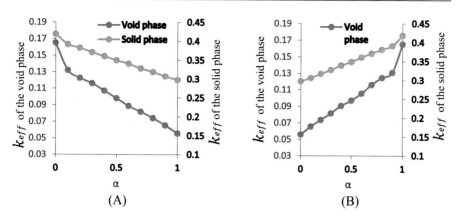

Figure 7.23 Effective thermal conductivity (A) void phases and (B) solid phases [13].

Table 7.11 Effective Young's moduli in three directions, calculated by FEM (C.F. is the abbreviation for combination factor) [13].

	C.F. = 0	C.F. = 0.3	C.F. = 0.7	C.F. = 1
E_h/E_2	3.130	2.847	2.615	2.253
E_v/E_2	2.253	2.615	2.846	3.130
E_t/E_2	2.254	2.459	2.461	2.254

$\nu_2 = 0.6$ (colored by green), respectively. Both phases are assumed to be isotropic with a relative Young's modulus of $E_1/E_2 = 10$ and a Poisson's ratio equal to 0.3 [13].

The first and second microstructures are both transversely isotropic structures. It can be seen visually, and it is also proven by finite element method (FEM) simulation that the combined microstructures have orthotropic structures [13].

The 6×6 stiffness and compliance matrix have nine and five independent elastic constants for orthotropic and transversely isotropic structures, respectively [121]. The effective Young's moduli in the vertical direction, horizontal direction, and the third one are denominated by E_v, E_h, and E_t, respectively.

Results of FEM simulations are summarized in Table 7.11. Fig. 7.24A and B respectively depict normal strain and stress field in the horizontal direction for an effective normal strain equal to 0.005 in the horizontal direction and a combination factor of 0.7 [13].

7.3.1.3.3 Combining two microstructures with unequal volume fractions

The previous combined microstructures had unique volume fractions. Also, in almost all optimization methods the volume fraction is usually considered as a fixed value [16,17]. In the next example, two $150 \times 150 \times 150$ voxel microstructures with unequal volume fractions, $\nu_{\text{void}} = 0.32$ and $\nu_{\text{void}} = 0.57$, and completely different morphologies are combined. The gradual evolution of the combined microstructure is depicted in Fig. 7.25. Also, Fig. 7.26A illustrates the thermal conductivity of

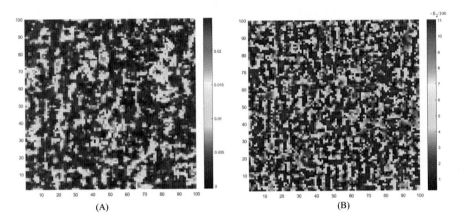

Figure 7.24 (A) Strain and (B) stress field in the horizontal direction for a combination factor of 0.7 and an effective horizontal normal strain equal to 0.005 [13].

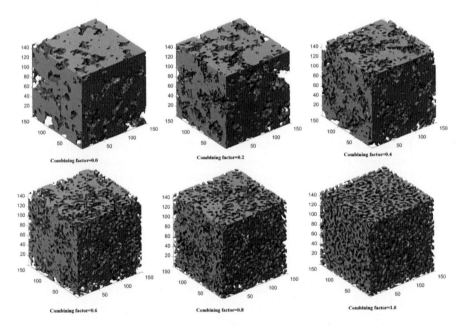

Figure 7.25 Combining two microstructures with unequal volume fractions [13].

the intermediate microstructures for void and solid phases. The thermal conductivity of the void phase of the second microstructure is almost six times larger than that of the first one, but its solid phase thermal conductivity is almost one third. The conductivities of combined microstructures are between these limits [13].

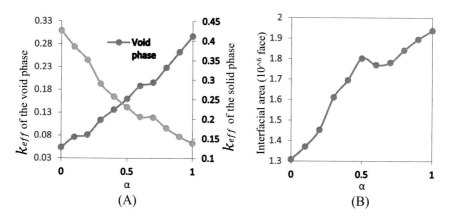

Figure 7.26 (A) Effective thermal conductivity and (B) interfacial area of combinations of two microstructures with unequal volume fractions [13].

Fig. 7.26B shows the interfacial area of the void and solid phases. To obtain the interface areas the number of interfacial faces between these phases is counted for all voxels. To avoid recounting a face, on each plane that is searched, only the right and up voxels are examined. For each of the three directions and for all planes perpendicular to each direction the described counting operation is performed. The extent of interface area is an important quantity specially for some porous microstructures; for instance the interface constitutes the sites for chemical reaction in the SOFC cathode [13,111].

The interfacial area is increased rapidly for $\nu_{void} < 0.5$ because the difference of the volume fractions of the solid and void phases is decreased and the contribution of the second microstructure that has more interfacial area is increased. In contrast, for $\nu_{void} > 0.5$ the difference between the volume fractions is increased, so the rate of increasing the interfacial area is decreased [13].

7.3.1.4 Discussion and error analysis

By investigating effective properties of the resolved cases, it is observed that the properties of the two initial microstructures are upper and lower bounds for the combined microstructures, but it cannot be said that this is true for all types of geometrical arrangement and properties. The definitive issue that can be stated is that the properties of combined structures gradually and smoothly change from the first microstructure to the second one because TPCFs are smooth and well-behaved functions [44] and there is linkage between the statistical correlation function of the microstructure and its properties such as effective thermal and stiffness tensor. For optimization purposes the combined microstructures interpolate properties of their parent microstructures, but it is possible that they include properties larger or smaller than properties of initial microstructures [13].

Using Eq. (7.24) the error of the proposed procedure for about 40 combined microstructures was investigated. The average of the obtained error was about 2.8% with the maximum amount equal to 5% as illustrated in Fig. 7.27. Because of the

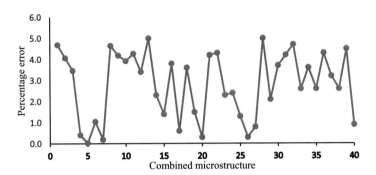

Figure 7.27 Percentage error for 40 combined microstructures [13].

closed relationship between TPCFs and effective properties, generally that error can also be considered qualitatively for the effective properties, but the inherent differences between properties cause the error values to vary numerically for different properties. It should also be noted that the only information available about the combined microstructure is its TPCFs. Therefore the reconstruction error can be considered as a qualitative indicator of the proposed method [13].

7.3.2 A framework for optimal microstructural design of random heterogeneous materials

A novel methodology was presented for microstructural design of multifunctional two-phase composites with tailored effective properties. The method utilizes efficient FFT methods for the computational reconstruction of microstructures with a given full-set TPCF and thermal and elastic homogenization analyses. The phase recovery method was used to reconstruct a given hypothetical, full-set, parametric TPCF capable of representing a wide range of isotropic and anisotropic microstructures. The accuracy and efficiency of the method were compared to those of the standard finite element models, and it was observed that for a comparable accuracy the method yields on average 1 order of magnitude improvement in the computational expense. An optimization procedure was proposed and employed to design isotropic and anisotropic microstructures with desired thermal conductivity or elastic moduli. The advantages of the presented methodology are as follows [53]:

1. applicability to multiphase, multifunctional composite materials as well as other random heterogeneous materials;
2. simplicity in the representation and capturing of microstructural complexities observed in nature and engineered composite materials such as bones, metal matrix composites, and porous metals;
3. direct linking between microstructure and effective properties;
4. optimization of properties for a wide range of microstructures with a few design variables; and therefore
5. a lower computational cost due to reliance on accelerated FFT methods.

Microstructure hull and design

7.3.2.1 Numerical homogenization

In this section, numerical homogenization of structural and thermal properties of composite materials were formulated using FFT methods. The developed methods rely on the accelerated Eyre−Milton FFT algorithm [122] to ensure robustness, efficiency, and ease of implementation [53].

7.3.2.1.1 Linear elastic properties

The homogenization problem for a periodic RVE under linear elastostatic conditions is described by [53]

$$
\begin{cases}
\varepsilon(x) = \dfrac{1}{2}\left(\nabla u(x) + \nabla^t u(x)\right), \\[2mm]
\sigma(x) = \mathbb{C}(x){:}\varepsilon(x), \\
\nabla.\sigma(x) = 0, \\
u(x) - \overline{E}.x \quad \text{periodic}, \\
\sigma(x).n \quad \text{antiperiodic},
\end{cases}
\tag{7.26}
$$

where $\mathbb{C}(x)$ is the local elastic modulus of each phase, $\overline{E}$ is a predefined strain tensor, and n is the normal vector for the faces of RVE. Following [123] the problem is cast into an iterative formulation such that [53]

$$
\varepsilon^{i+1}(x) = \varepsilon^i(x) + 2(\mathbb{C}(x) + \mathbb{C}_0)^{-1}{:}\mathbb{C}_0{:}\left[\Gamma^0(x) * \left(\mathbb{C}_0{:}\varepsilon^i(x) - \sigma^i(x)\right) - \varepsilon^i(x) + \overline{E}\right],
\tag{7.27}
$$

where the symbol $*$ denotes the convolution product, $\mathbb{C}_0$ is the elasticity modulus of the reference medium, and its associated Green operator in the Fourier space $\hat{\Gamma}^0_{khij}$ given by [53,124]

$$
\hat{\Gamma}^0_{khij}(\xi) = \frac{1}{4\mu^0|\xi|^2}\left(\delta_{ki}\xi_h\xi_j + \delta_{hi}\xi_k\xi_j + \delta_{kj}\xi_h\xi_i + \delta_{hj}\xi_k\xi_i\right) - \frac{\xi_i\xi_j\xi_k\xi_h}{|\xi|^4}\frac{\lambda^0 + \mu^0}{\mu^0\left(\lambda^0 + 2\mu^0\right)}.
\tag{7.28}
$$

where ξ is the frequency and λ^0 and μ^0 are Lamé constants of the reference medium. The following is used for the reference medium which ensures a faster convergence rate [53,123]:

$$
\mu^0 = \sqrt{\mu_{\min} \times \mu_{\max}} \text{ and } K^0 = \sqrt{K_{\min} \times K_{\max}},
\tag{7.29}
$$

where K is the bulk modulus ($K = \lambda + 2\mu/3$) and subscripts min and max denote the minimum and maximum of the composite constituents properties, respectively. The effective (homogenzied) elasticity tensor is then given by [53]

$$
\langle\sigma(x)\rangle = \mathbb{C}_{\text{eff}}{:}\langle\varepsilon(x)\rangle.
\tag{7.30}
$$

The components of $\mathbb{C}_{eff}$ were obtained by applying six individual unit strains to RVE for each of the corresponding components of the strain tensor [125]. The solution procedure for the problem is further detailed in Algorithm 7.1. This algorithm summarizes the procedure used to solve the elastic problem. Starting with a voxel-based representation a uniform strain is first applied to all voxels, and the objective of the method is to obtain the equilibrium state of the microstructure. In Algorithm 7.1, η_1 and η_2 are convergence criteria (in order to achieve less than 0.1% variance in homogenized properties, selected to be 10^{-6} and 10^{-3} in the current study, respectively).

Algorithm 7.1: *Accelerated Eyre–Milton elastic analysis algorithm.*

iteration $i = 0$:

$$\varepsilon^0(x) = \overline{E}, \quad \forall x \in \Omega_0$$

$$\sigma^0(x) = \mathbb{C}(x){:}\varepsilon^0(x), \quad \forall x \in \Omega_0$$

where x is the coordinate of each voxel, $\overline{E}$ is a predefined strain tensor, and Ω_0 is the RVE domain.

while (true)
1: $\tau^i(x) = \mathbb{C}_0{:}\varepsilon^i(x) - \sigma^i(x)$
2: $\hat{\tau}^i(\xi) = \text{FFT}[\tau^i(x)]$
3: $\hat{\varepsilon}^i_c(\xi) = \hat{\Gamma}^0(\xi){:}\hat{\tau}^i(\xi) \quad \forall \xi \neq 0$ and $\hat{\varepsilon}^i_c(0) = \overline{E}$
4: $\varepsilon^i_c(x) = \text{FFT}^{-1}[\hat{\varepsilon}^i_c(\xi)]$
5: $\varepsilon^{i+1}(x) = \varepsilon^i(x) + 2(\mathbb{C}(x) + \mathbb{C}_0)^{-1}{:}\mathbb{C}_0{:}(\varepsilon^i_c(x) - \varepsilon^i(x))$
6: $\sigma(x)^{i+1} = \mathbb{C}(x){:}\varepsilon^{i+1}(x)$
7: $\text{err}^i_1 = \frac{<\|\sigma(x)^{i+1} - \sigma(x)^i\|>}{\|\mathbb{C}_0{:}\overline{E}\|}, \text{err}^i_2 = \frac{\sqrt{<\|\xi.\hat{\sigma}^i(\xi)\|>}}{\|\hat{\sigma}^i(0)\|}$ where $\hat{\sigma}^i(\xi) = \text{FFT}[\sigma(x)^i]$
8: **if** $(\text{err}^i_1 < \eta_1$ and $\text{err}^i_2 < \eta_2)$
    break
else
    $i = i + 1$

7.3.2.1.2 Thermal properties
Similarly, for a periodic RVE the mathematical thermal homogenization problem is described by

$$\begin{cases} e(x) = \nabla T(x). \\ q(x) = K(x)\, e(x), \\ \nabla.q(x) = 0, \\ T(x) - \overline{e}.x \quad \text{periodic.} \\ q(x).n \quad \text{antiperiodic.} \end{cases} \tag{7.31}$$

where $T.e$ and q denote the temperature, its gradient, and the heat flux, respectively. Moreover, $K(x)$ and $\overline{e}$ denote the local thermal conductivity and a uniform

Microstructure hull and design 347

predefined temperature gradient tensor, respectively. Similarly, this problem can be cast into Eyre–Milton iterative formulation such that [53]

$$e^{i+1}(x) = e^i(x) + 2(K(x)+K_0)^{-1}K_0\big[G^0(x) * \big(K_0e^i(x) - q^i(x)\big) - e^i(x) + \bar{e}\big].$$

(7.32)

In Eq. (7.32) the K_0 denotes the thermal conductivity of the reference medium, and in the case of the isotropic reference medium, $K_0 = k_0I$, where I is the second-order identity tensor and k_0 is a scalar thermal conductivity. The $\hat{G}^0$ is the Green operator associated with thermal conductivity of the reference medium defined, in Fourier space, by [53,126]

$$\hat{G}^0_{ij}(\xi) = \frac{\xi_i\xi_j}{\sum_{m,n} K_{0mn}\xi_m\xi_n}.$$

(7.33)

The algorithm developed to solve Eq. (7.32) is presented in Algorithm 7.2 with guaranteed convergence [123] subject when properties for the reference medium are selected such that [53]

$$k_0 > \sqrt{k_{\min} \times k_{\max}}.$$

(7.34)

where $k_{\min}$ and $k_{\max}$ are the minimum and maximum thermal conductivities of the composite constituent, respectively. The effective thermal conductivity tensor is given by [53]

$$\langle q(x)\rangle = K_{\text{eff}}\langle e(x)\rangle$$

(7.35)

Algorithm 7.2 illustrates the procedure used for the thermal homogenization analysis. Similarly an accelerated Eyre–Milton algorithm is employed with the same convergence criteria as presented for elastic analysis. In Algorithm 7.2, η_1 and η_2 are convergence criteria (in order to achieve less than 0.1% variance in homogenized properties, selected to be 10^{-6} and 10^{-3} in the current).

Algorithm 7.2: *Accelerated FFT thermal analysis algorithm [53].*

Iteration $i = 0$:

$$e^0(x) = \bar{e} \quad \forall x \in \Omega_0$$

$$q^0(x) = K(x)e^0(x). \quad \forall x \in \Omega_0$$

where x is the coordinate of each voxel, $\bar{e}$ is predefined temperature gradient vector, and Ω_0 is the RVE domain
   while (true)
     1: $\tau^i(x) = K_0e^i(x) - q^i(x)$
     2: $\hat{\tau}^i(\xi) = \text{FFT}[\tau^i(x)]$
     3: $\hat{e}^i_c(\xi) = \hat{G}^0(\xi)\hat{\tau}^i(\xi) \quad \forall \xi \neq 0$ and $\hat{e}^i_c(0) = \bar{e}$

4: $e_c^i(x) = \mathrm{FFT}^{-1}[\hat{e}_c^i(\xi)]$

5: $e^{i+1}(x) = e^i(x) + 2(K(x) + K_0)K_0(e_c^i(x) - e^i(x))$

6: $q(x)^{i+1} = K(x)e^{i+1}(x)$

7: $\mathrm{err}_1^i = \frac{< \|q(x)^{i+1} - q(x)^i\| >}{\|Ke\|}$. $\quad \mathrm{err}_2^i = \frac{\sqrt{< \|\xi.\hat{q}^i(\xi)\| >}}{\|\hat{q}^i(0)\|}$ where $\hat{q}^i(\xi) = \mathrm{FFT}[q(x)^i]$

8: **if** ($\mathrm{err}_1^i < \eta_1$ and $\mathrm{err}_2^i < \eta_2$)

    break

  else

    $i = i + 1$

7.3.2.2 Reconstruction

7.3.2.2.1 Phase recovery algorithm

Classical microstructural reconstruction algorithms rely on optimization methods such as simulated annealing (SA) [49,127] and GA [31]. More recently, more efficient phase recovery methods [26,128] are proposed. The presented methodology relies on phase recovery for the reconstruction step as explained in Section 7.4.1.1.

Algorithm 7.3: *Reconstruction algorithm [53].*

Iteration $j = 0$:

  a: $\overline{\hat{C}_2^{nn}}(\xi) = \mathrm{FFT}[\overline{C_2^{nn}}(t)]$, $\overline{C_2^{nn}}(t)$ is input TPCF and is input of the algorithm.

  b: $\overline{\hat{C}_2^{nn}}(\xi) = \frac{1}{N}\|\overline{\hat{\kappa}^n}(\xi)\|^2 \rightarrow \|\overline{\hat{\kappa}^n}(\xi)\| = \sqrt{N\overline{\hat{C}_2^{nn}}(\xi)}$

  c: Generate a random microstructure $\kappa_0^n(x)$

while (true)

  1: $\hat{\kappa}_j^n(\xi) = \mathrm{FFT}[\kappa_j^n(x)] = \|\hat{\kappa}_j^n(\xi)\|e^{i\theta_j^n(\xi)}$

  2: $\hat{\omega}_j^n(\xi) = \|\overline{\hat{\kappa}^n}(\xi)\|e^{i\theta_j^n(\xi)}$,

  3: $\omega_j^n(x) = \mathrm{FFT}^{-1}[\hat{\omega}_j^n(\xi)]$

  4: **if** $j < n_1$

    if $\omega_j^n(x) \notin [01]$

      $\omega_j^n(x) = \kappa_j^n(x) - \beta\omega_j^n(x)$

    if $\omega_j^n(x) < 0 \ \forall x \in \Omega_0$

      $\omega_j^n(x) = 0$

    if $\omega_j^n(x) > 1 \ \forall x \in \Omega_0$

      $\omega_j^n(x) = 1$

  5: $\kappa_{j+1}^n(x) = \omega_j^n(x)$

  6: $\hat{\kappa}_{j+1}^n(\xi) = \mathrm{FFT}[\kappa_{j+1}^n(x)] \rightarrow \hat{C}_2^{nn}j+1(\xi) = \frac{1}{N}\|\hat{\kappa}_{j+1}^n(\xi)\|^2 \rightarrow$

    $C_2^{nn}j+1(t) = \mathrm{FFT}^{-1}(\hat{C}_2^{nn}j+1(\xi))$

  7: **if** $< \|C_2^{nn}j+1(t) - \overline{C_2^{nn}}(t)\| > < \eta$

    if $\kappa_{j+1}^n(x) < 0 \ \forall x \in \Omega_0$

      $\kappa_{j+1}^n(x) = 0$

Microstructure hull and design

if $\kappa_{j+1}^n(x) > 1 \ \forall x \in \Omega_0$

$\quad \kappa_{j+1}^n(x) = 1$

$\quad \kappa_o^n(x) = \text{round}[\kappa_{j+1}^n(x)]$, where $\kappa_o^n(x)$ is the reconstructed microstructure

break

else

$\quad j = j + 1$

7.3.2.3 Microstructure representation

TPCF functions are easily assessed experimentally. The methodology developed in the current work is insensitive to the complexities of the microstructures and allows for a hypothetical TPCF. A hypothetical TPCF can be guessed based on a priori knowledge using a few mathematical constraints. Torquato [129] presents the necessary condition. The same author introduces the scaled autocovariance function [53]

$$f(t) = \frac{C_2^{nn}(t) - \varphi_n^2}{\varphi_n(1 - \varphi_n)} = \exp\left(-\frac{\|t\|}{a}\right) \frac{\sin(q.\|t\|)}{q.\|t\|}, \tag{7.36}$$

where φ_n is the volume fraction of phase n and a and q are constants. Eq. (7.36) is known to capture a wide range of complex two-phase microstructures with uniform distribution of the phases [127].

A full-set TPCF, hypothetical or experimental, is needed to apply the PR. Therefore in a digital representation of Eq. (7.36), t is a vector that starts at origin (0.0.0) and ends at any voxel within the RVE. In full-set TPCF the value of each voxel is the TPCF value for a vector which ends in that voxel. For the purpose of visualization, in the current work, full-set TPCFs are translated to the center using a *fftshift* [53,58].

Generally, for anisotropic media, TPCFs are also dependent on the orientation of t. In the current work, anisotropy is captured by a scaling t. To generate an anisotropic microstructure with preferential orientation in X_1, we use $\bar{t} = \left(\frac{t_1}{\alpha}.t_2.t_3\right)$, and α is a constant controlling the extent of anisotropy in the direction. To generate phases elongated in two directions, we use an algebraic combination [53]

$$f(t) = 0.5 * f\left(\overline{t^1}\right) + 0.5 * f\left(\overline{t^2}\right), \tag{7.37}$$

where $\overline{t^1} = \left(\frac{t_1}{\alpha}.t_2.t_3\right)$ and $\overline{t^2} = \left(t_1.\frac{t_2}{\beta}.t_3\right)$ are two scaled vectors.

7.3.2.4 Design algorithm

Here the goal of the design is to find parameters of the TPCF for a microstructure with tailored properties; thus the parameters of Eq. (7.36) including volume fraction (φ_n) and scaled vector parameters (α and β) are design variables. We select the effective stiffness, thermal conductivity, and a combination of these as design objectives. In the second case study the design objective is to achieve a desired value for elastic modulus ($|E_{11} - E|$) and thermal conductivity ($|k_{11} - k|$) in one direction, where E and k are the desired elastic modulus and thermal conductivity, respectively. In the third case study,

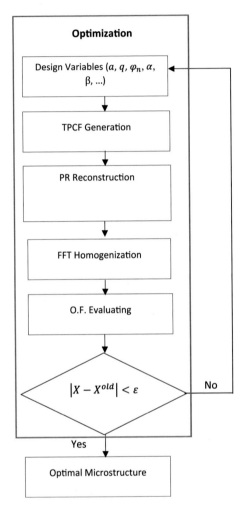

Figure 7.28 Proposed design procedure where X denotes variables used in convergence check [53].

we select the objective to be $(\frac{1}{2}|\frac{k_{11}}{k_{33}} - c_1| + \frac{1}{2}|\frac{E_{22}}{E_{33}} - c_2|)$, a combination of thermal and structural properties, for desired values of c_1 and c_2. Fig. 7.28 presents the flowchart of the design procedure [53].

7.3.2.5 Results and discussion

7.3.2.5.1 Verification study

To verify the implementation and to benchmark its performance the FFT homogenization analysis was first compared to equivalent finite element analysis. For this purpose, four sets of RVEs with volume fractions of 0.1, 0.2, 0.3, and 0.5 and

Microstructure hull and design

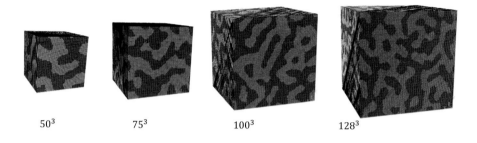

50³   75³   100³   128³

Figure 7.29 A voxel-based representative volume element used for FE and FFT performance analyses [53].

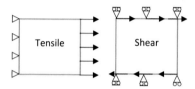

Figure 7.30 Boundary condition for tensile and shear loading in FEM [53].

different RVE sizes are reconstructed and the effective elastic modulus ($E = \frac{E_{11} + E_{22} + E_{33}}{3}$) is calculated using FEM and FFT methods. Fig. 7.29 shows three two-phase models used with 50³, 75³, 100³, and 128³ voxels for a volume fraction of $\varphi_n = 0.5$. A python script is developed for automatic modeling and homogenizing elastic properties of the reconstructed RVE in Abaqus. The finite element model was discretized to regular cubic linear mesh C3D8R (the same as the FFT model); then based on the RVE file, one element set for each phase was created, and then appropriate mechanical properties were assigned to each element set. The iterative implicit solver is selected for this problem, and six steps are considered for six loading cases [53].

Fig. 7.30 illustrates the boundary conditions used for FE study. Effective properties were obtained using three tensile and three shear loadings. The elastic moduli for the white and green phases were selected to be 1 and 100 GPa, respectively, with a Poisson's ratio of 0.3 for both phases. For FFT homogenization the stopping criteria η_1 and η_2 (see Algorithm 7.1) were selected to be 10^{-6} and 10^{-3}, respectively; it should be noted that these values are selected via a sensitivity analysis to ensure that the homogenized results do not vary more than 0.1% by reducing the η_1 and η_2. All FFT and FE analyses were carried out on a high-end workstation with AMD Opteron 6174 CPU and 64 GB RAM. Table 7.12 compares homogenized values for the elastic modulus components for FE and FFT analyses. The results show that for volume fractions of the stiffer phase more than 50% the FEM and FFT results match very well (less than 0.65% difference) and for volume fractions less than 50%, it is observed that by increasing the RVE size the difference between FFT and FEM

Table 7.12 Comparison between FEM and FFT homogenization analyses (diff $= \frac{100 \times |E^{FEM} - E^{FFT}|}{E^{FEM}}$) [53].

	$\varphi_n = 0.1$				$\varphi_n = 0.2$				$\varphi_n = 0.3$				$\varphi_n = 0.5$			
	50^3	75^3	100^3	128^3	50^3	75^3	100^3	128^3	50^3	75^3	100^3	128^3	50^3	75^3	100^3	128^3
E^{FEM}	1.3	1.29	1.29	1.28	1.91	1.86	1.82	1.81	3.94	3.83	3.82	3.79	19.8	19.79	19.84	19.45
E^{FFT}	1.27	1.26	1.26	1.26	1.8	1.8	1.76	1.75	3.73	3.67	3.66	3.62	19.75	19.92	19.71	19.33
diff	2.30	2.32	2.32	1.56	5.75	3.22	3.29	3.31	5.32	4.17	3.97	4.27	0.25	0.65	0.65	0.61

Microstructure hull and design 353

Table 7.13 Comparison between required RAM and CPU for FFT and FEM [53].

RVE size	50^3		75^3		100^3		128^3	
Required resources	**FEM**	**FFT**	**FEM**	**FFT**	**FEM**	**FFT**	**FEM**	**FFT**
RAM [MB]	1687	50	4489	160	12822	390	26014	790
CPU time [min]	24	1.5	48	7	150	23	270	32

models remains almost unchanged and the maximum of this difference is 4.27% for a volume fraction of 30%. As is found in the literature, for low contrast of phases the FFT results match very well with other methods [130]; however, for high contrasts, even for RVEs with a simple inclusion shape, there is a difference between FEM and FFT results [131] that it seems that this difference for such a complex geometry arises from numerical errors such as singular points where numerical methods such as FEM and FFT cannot handle it appropriately even if larger RVE is selected. It should be noted that the maximum difference between two methods can be considered acceptable for engineering applications. Table 7.13 shows the CPU time and required RAM for both FFT and FEM analyses. It is observed that for a comparable accuracy, FFT analyses are on average 1 order of magnitude faster and required less memory, and the difference between FEM and FFT methods drastically increased as the voxel number increased in RVE [53].

7.3.2.5.2 Anisotropic representative volume element

In this section, reconstruction of isotropic RVEs, RVEs with one preferred reconstruction direction, and a combination of the two TPCFs with two preferred directions is presented. Using TPCF presented in Eq. (7.36) and setting $\varphi_n = 0.3, a = 25$, and $q = .9$, three 151^3 RVEs were generated. The TPCF presented by Eq. (7.36) is shown in Fig. 7.31A as it is shown that the TPCF for all vectors with same dimension in respect to its orientation is the same [53].

Similarly in the reconstructed RVE based on this TPCF (Fig. 7.31D) the yellow phase is distributed randomly in RVE. To reconstruct RVE with one preferred reconstruction direction the TPCF with the scaled vector is considered (see Section 7.3.2). Fig. 7.31B shows TPCF presented by Eq. (7.36) and considering the scaled vector $\left(\frac{t_1}{\alpha}.t_2.t_3\right)$, whereas it is clear that the TPCF is elongated in the direction of axis 1. The reconstructed RVE based on the TPCF with the scaled vector is shown in Fig. 7.31E, where the yellow phase is elongated in the direction of axis 1. It should be noted that parameter α in the scaled vector can be used to tailor the elongation of the phase in direction 1. To reconstruct RVE with two preferred directions the idea of a combination of the TPCFs with two preferred directions is used (see Section 7.3.2). For this purpose the TPCF presented in the combination of the two TPCFs presented by Eq. (7.36) and two scaled vectors in two different directions is considered (See Fig. 7.31C). The reconstructed RVE based on the combination of the TPCFs with two scaled vectors in two different directions is shown in Fig. 7.31F, where the yellow phase is elongated in the direction of axis 1 and axis 2 [53].

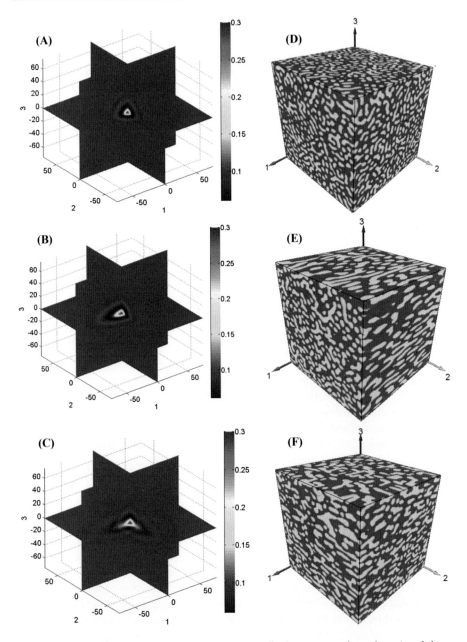

Figure 7.31 Three full-set TPCFs (only three perpendicular cross-sections shown) and the corresponding microstructures reconstructed using the presented PR methodology. An isotropic TPCF (A), a TPCF with the scaled vector in X_1 (B), and a TPCF with scaled vectors in both X_1 and X_2 (C). Corresponding microstructures reconstructed based on each TPCF (D, E, and F) [53].

7.3.2.5.3 Case study 1

A set of 151^3 voxel RVEs were generated to evaluate the thermomechanical properties of the reconstructed microstructures with volume fractions of 0.1 to 0.9 for phase n (yellow phase in Fig. 7.32). All microstructures were reconstructed based on the TPCF presented by Eq. (7.36) [53].

For all RVEs, we set $a = 25$, and the reconstructions were carried out for $q = 0.3$, 0.5, and 0.9. The elastic modulus of the yellow and second phases, the second phase is not shown in Fig. 7.32, was considered to be 100 and 1 GPa, respectively, with the Poisson's ratio of 0.3. Similarly, the thermal conductivities

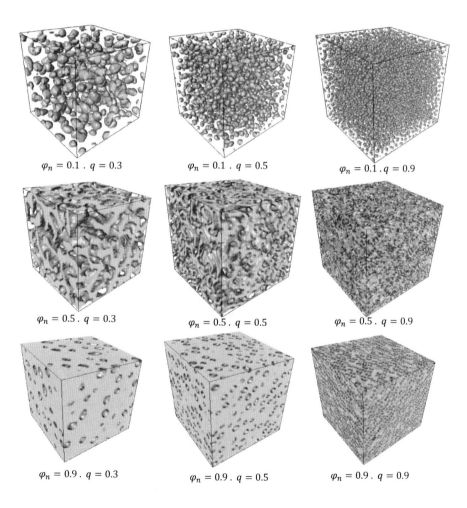

Figure 7.32 Example microstructures (only one phase shown) reconstructed for different values of volume fraction φ_n and q [53].

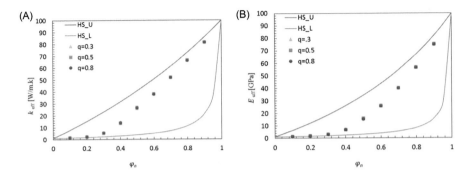

Figure 7.33 Effective thermal conductivity (A) and elastic modulus (B) for reconstructed microstructures with different volume fractions and q values [53].

for the phases were set to 100 and 1 W/(m.K), respectively. Fig. 7.32 shows the RVEs reconstructed for the volume fractions of 0.1, 0.5, and 0.9 with different values for q. It is observed that the length scale for the inclusion phase is directly related to the value of q. Effective thermal conductivities and elastic moduli for the reconstructed RVEs for different values of volume fractions and q are shown in Fig. 7.33A and B. Rigorous bounds on effective properties predicted by Hashin–Shtrikman are plotted in the same figure. It is observed that for a given volume fraction the effective thermal conductivity or elastic modulus is insensitive to the length-scale parameter q [53].

To check the dependency of the mechanical properties of the reconstructed RVE based on the same TPCF considering $a = 25$ and $q = 0.5$ for each volume fraction, 0.1, 0.3, 0.5, and 0.9, three RVEs are reconstructed and the elastic modulus in three directions is calculated. As results show (see Table 7.14) the difference between the elastic moduli of the reconstructed RVEs is dependent on the volume fraction; however, for these three RVEs the maximum difference in elastic modulus is about 1.25%, which shows that the mechanical properties of the different reconstructed RVEs based on the same TPCF are almost constant [53].

To introduce anisotropy the scaled vector $\bar{t} = (\frac{t_1}{\alpha}.\ t_2.\ t_3)$ was used. Fig. 7.34 shows the full set of TPCFs and the reconstructed RVEs obtained for a volume fraction of 0.3 and different values of α. It is observed that as α increases the anisotropy level increases in the RVE. Table 7.15 lists the directional thermal conductivities for the different values of α and volume fractions. As is expected, by increasing anisotropy in direction 1 the thermal conductivity increases in this direction and decreases in the other two directions. A detailed comparison shows that for the same value of α the $k_{11}/(k_{33} + k_{22})$ (or $E_{11}/(E_{33} + E_{22})$) ratio first increases by the volume fraction, reaches its peak value around the percolation threshold (about 25%), and then decreases [53].

Table 7.16 lists elastic moduli in three directions for different anisotropies and volume fractions; as expected, by increasing the anisotropy in direction 1 the elastic modulus increases in this direction.

Table 7.14 Elastic moduli of the three reconstructed RVEs based on the same TPCF. diff$_{max}$ is the percent of maximum difference of elastic moduli with respect to the average value for three RVEs [53].

	$\varphi_n = 0.1$			$\varphi_n = 0.3$			$\varphi_n = 0.5$			$\varphi_n = 0.9$		
	E_{11}	E_{22}	E_{33}	E_{11}	E_{22}	E_{33}	E_{11}	E_{22}	E_{33}	E_{11}	E_{22}	E_{33}
RVE1	1.251	1.253	1.254	3.767	3.744	3.685	19.535	19.487	19.243	80.207	80.303	80.276
RVE2	1.251	1.251	1.251	3.772	3.752	3.750	19.421	19.344	19.310	80.251	80.321	80.280
RVE3	1.252	1.253	1.251	3.841	3.767	3.731	19.432	19.114	19.23	80.303	80.191	80.247
E_{avr}	1.251	1.252	1.252	3.793	3.754	3.722	19.462	19.315	19.261	80.253	80.271	80.267
diff$_{max}$	0.053	0.10	0.15	1.25	0.33	0.99	0.37	1.04	0.25	0.06	0.10	0.025

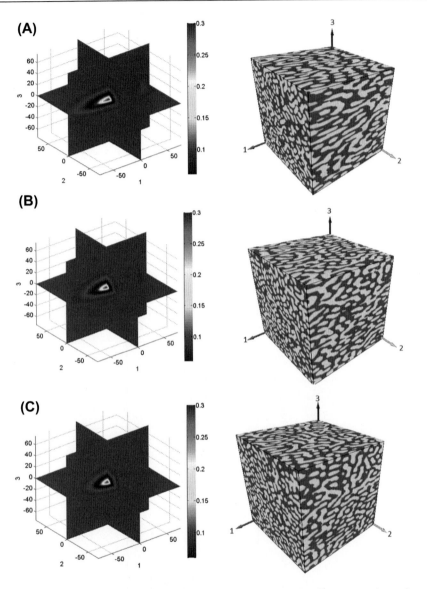

Figure 7.34 Representation of the TPCF and equivalent reconstructed representative volume elements obtained for different values of α, the parameter in scaled vector $\bar{t} = (t_1/\alpha.t_2.t_3)$, for (A) $\alpha = 3.15$, (B) $\alpha = 2.2$, and (C) $\alpha = 1.8$ [53].

7.3.2.5.4 Case study 2

This case study investigates the application of the proposed method for the reconstruction of an anisotropic RVE with desired values of elastic modulus or thermal

Table 7.15 Effective thermal conductivity for different representative volume elements reconstructed based on the different values of α in different volume fractions [53].

φ_n	$\alpha = 3.15$			$\alpha = 2.2$			$\alpha = 1.8$			k_{HS_U}	k_{HS_L}
	k_{11}	k_{22}	k_{33}	k_{11}	k_{22}	k_{33}	k_{11}	k_{22}	k_{33}		
0.1	2.86	1.38	1.39	2.31	1.42	1.45	2.02	1.47	1.48	7.85	1.32
0.3	14.62	4.69	4.73	11.9	5.61	5.53	11.26	6.12	6.15	23.08	2.23
0.5	37.14	17.61	17.49	33.66	19.42	20.02	31.59	20.97	20.57	40.71	3.82
0.7	61.3	41.47	41.83	58.58	44.41	43.96	56.86	45.61	45.52	61.37	7.35
0.9	85.6	76.3	76.31	84.51	77.56	77.21	83.67	78.2	78.15	85.91	21.72

Table 7.16 Effective elasticity modulus and thermal conductivity for different representative volume elements reconstructed based on the different values of α in different volume fractions [53].

φ_n	$\alpha = 3.15$			$\alpha = 2.2$			$\alpha = 1.8$			E_{HS_U}	E_{HS_L}
	E_{11}	E_{22}	E_{33}	E_{11}	E_{22}	E_{33}	E_{11}	E_{22}	E_{33}		
0.1	1.96	1.25	1.25	1.65	1.26	1.26	1.52	1.27	1.27	6.26	1.22
0.3	8.76	2.56	2.58	6.53	2.84	2.81	5.95	3.08	3.08	18.59	1.84
0.5	28.58	8.63	8.85	24.13	9.83	10.46	21.50	10.92	10.88	34.18	2.94
0.7	54.70	27.35	28.05	50.66	30.61	30.07	47.36	32.04	31.89	54.50	5.41
0.9	82.74	66.96	66.79	80.57	68.65	68.09	79.02	69.59	69.55	82.11	16.10

conductivity in the X_1 direction. Microstructures were reconstructed using Eq. (7.36) with the scaled vector $\bar{t} = (t_1/\alpha.t_2.t_3)$. Other parameters used in this study are $\varphi_n = 0.3$, $a = 15$, and $q = 0.9$ with 101^3 RVEs and the same material properties assumed as in the previous case study. The desired elastic modulus of $E_{11} = 15$ GPa and a thermal conductivity equal to $k_{22} = 15$ W/(m.K) were selected, so [53]

$$\text{O.F.}_1 = |E_{11} - 15 \times 10^9| \text{ for the structural problem, and} \qquad (7.38)$$

$$\text{O.F.}_2 = |k_{22} - 15| \text{ for the thermal problem.} \qquad (7.39)$$

Due to the noisy nature of the objective functions, a gradient-free optimization algorithm (COBYLA) [132] was used. The optimization algorithm was implemented using *NLopt* library [133]. Moreover, for both problems, the constraint was selected to be [53]

$$0.7 \le \alpha \le 32 \qquad (7.40)$$

A stopping criterion $|\alpha_i - \alpha_{i-1}| < 0.02$ was used, where i is the optimization iteration. The optimization procedure was converged after 17 iterations (see Fig. 7.35C) with the optimal design variables ($\alpha = 6.01.\text{OF}_1 = 0.12 \times 10^9$) for the elasticity problem with elastic moduli of $E_{11} = 15.12$ GPa, $E_{22} = 2.36$ GPa, and $E_{33} = 2.40$ GPa, respectively. Similarly the thermal optimization procedure was converged in 13 iterations (see Fig. 7.36C) with optimal parameters ($\alpha = 2.86.\text{OF}_2 = 0.21$), and the components of the thermal conductivity tensor are $k_{11} = 5.39$ W /(m.K), $k_{22} = 15.21$ W /(m. K), and $k_{33} = 5.40$ W /(m.K) [53].

Fig. 7.35A and 7.36A show optimized microstructures obtained for the structural and thermal problems, respectively. As is expected the phase with a higher elastic modulus and thermal conductivity is elongated in the related directions. Fig. 7.35B shows σ_{11} distribution in the optimized microstructure after applying $\varepsilon_{11} = 0.001$ (all other strain components set to 0). Similarly, Fig. 7.36B shows the distribution of heat flux (q_{22}) in the thermally optimized microstructure for the temperature gradient $e_{22} = 0.1$ (all other components were set to 0). It should be noted that because we find TPCF of the optimal microstructure and then reconstruct the RVE and also considering the point that reconstructed RVE based on the TPCF is not unique, based on this method a unique optimum microstructure does not exist; however, all optimum microstructures have the same TPCF [53].

7.3.2.5.5 Case study 3

This case study demonstrates the application of the presented method in the design of a multifunctional composite material microstructure. The phase with higher values of thermal conductivity and elastic modulus is elongated in the $(X_1.X_2)$-plane where the elastic modulus in X_1 is 3 times larger than that of X_3, and simultaneously the thermal conductivity in X_2 is 3 times larger than X_3. The RVE size, TPCF parameters, and materials properties were selected to be identical to the second case study. As explained in Section 7.3.2 the combination of the two TPCFs with scaled vectors was used for

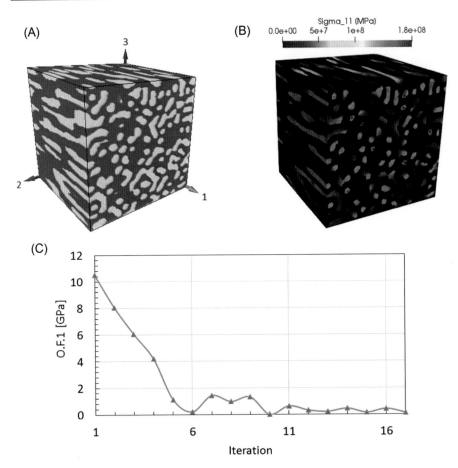

Figure 7.35 Optimized microstructures obtained for the structural problem (A), σ_{11} for the optimized microstructure for $\varepsilon_{11} = 0.001$ (B), and the convergence history for the optimization procedure (C) [53].

the reconstruction. A single objective function with the same weight factor for the elasticity modulus and thermal conductivity components was used with [53]

$$\text{O.F.} = \frac{1}{2}\left|\frac{E_{11}}{E_{33}} - 3\right| + \frac{1}{2}\left|\frac{k_{22}}{k_{33}} - 3\right|. \tag{7.41}$$

Like the second case study a gradient-free optimization algorithm was used to find optimal α and β values with constraints $0.7 < \alpha$ and $\beta < 32$. The optimal microstructure was obtained after 31 (see Fig. 7.37B) iterations with ($\alpha = 9.27$. $\beta = 7.91$. O.F. $= 0.218$) with optimal components of the thermal conductivity tensor of $k_{11} = 11.21$ W /(m.K), $k_{22} = 13.41$W /(m.K), and $k_{33} = 4.22$ W /(m.K). Moreover, the effective elastic moduli

Microstructure hull and design 363

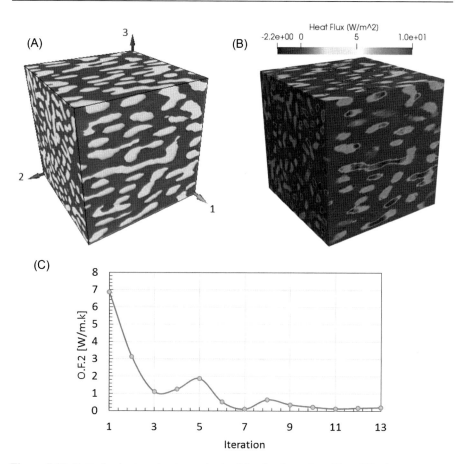

Figure 7.36 Optimized microstructures obtained for the thermal problem (A), the distribution of heat flux q_{22} in optimized representative volume element for the temperature gradient $e_{22} = 0.1$ (B), and the convergence history for the optimization procedure (C), [53].

in three directions were $E_{11} = 7.21$ GPa, $E_{22} = 9.81$ GPa, and $E_{33} = 2.63$ GPa, respectively. Fig. 7.37A shows the optimized microstructure with the blue phase elongated in two directions to minimize the objective function given by Eq. (7.41) [53].

7.4 Practical approach for materials design of periodic heterogeneous materials

7.4.1 3D-PUC design using density-based topology optimization

In materials design of porous and periodic structures, void (empty) and occupied (filled) regions of the 3D-PUC need to be identified for the numerical homogenization

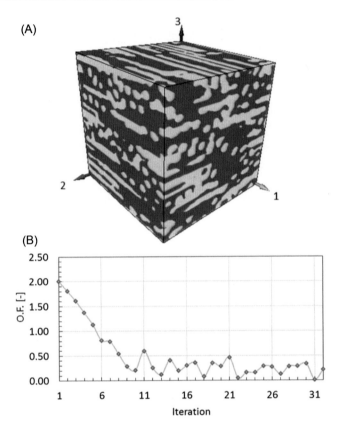

Figure 7.37 Optimized multifunctional microstructure based on the objective function given by Eq. (7.41) (A) and optimization history plot (B) [53].

procedure. In the finite element simulation of 3D-PUCs with a total of N elements, each element e is assigned with a characterization parameter $X_e \in \{0, 1\}$, which is in fact the design variable indicating the presence or absence of the solid material in that element. These design variables are utilized for evaluating the effective (homogenized) stiffness tensor C_{ijkl}^H. There is also the volume constraint $\sum_{e=1}^{N} V_e X_e / V \leq \vartheta$ which must be satisfied, where V_e is the volume of the element e and ϑ is the target material-occupied volume fraction (VF) of 3D-PUC and the final microstructure. Denoting the vector of design variables by $\mathbf{X} = \{X_1, X_2, \ldots, X_N\}$ the optimization problem can be stated as [54]

$$\text{Minimize}: f\left(C_{ijkl}^H(\mathbf{X})\right)$$
$$\text{Subject to}: \left\{ \begin{array}{l} 3DPBCs \\ \sum_{e=1}^{N} V_e X_e / V \leq \vartheta \\ X_e \in \{0, 1\} \end{array} \right\} \quad (7.42)$$

in which $f(C^H_{ijkl}(\mathbf{X}))$ is the scalar objective function defined on the components of the homogenized stiffness tensor. Effective Poisson's ratio and bulk modulus are examples of function f that are commonly used in most research studies. Regarding the imposition of 3D PBCs, they are applied on the nodes lying on the PUC faces, similar to the procedure implemented by Andreassen et al. [134] for 2D RVEs [54].

Expressions more or less similar to the problem (2.1) are employed in the relevant studies, although their optimization tools may differ remarkably. The optimization Eq. (7.42) is basically a specific case of the integer programming problems with binary design variables. Although there are specifically developed codes and methods for such problems, in materials design applications the final (optimal) result should present an acceptable topology with a reasonable continuity in Von Neumann neighborhood of the solid regions [135] together with an acceptable smoothness to fulfill the manufacturability. It is therefore not merely a question of which elements in the unit cell are empty or occupied but discontinuous topologies (checkerboards) are inacceptable (Fig. 7.38). Furthermore, bearing in mind that solving such constrained discrete variable problems is iterative in nature the iterative method of OCA takes care of finding the optimal topology of 3D-PUC in the methodology of this work. The OCA approach is suitable for such constrained discrete variable problems with a large number of design variables [136]. However, there is a possibility that the discretized topology in successive iterations does not converge to an acceptable topology and instead alternates between some topologies. In recent years, some methods have been developed to avoid such a drawback. These methods usually replace the binary design variables X_e with the corresponding continuous ones $0 \leq \rho_e \leq 1$, known as density design variables. They are initially assigned sort of randomly to the elements, and by their gradual variation toward 0 or 1

Figure 7.38 An example of a discontinuous topology: gray and white cubes show occupied and empty regions, respectively [54].

a "black-and-white" pattern for the topology is obtained, which satisfies the volumetric and periodic constraints of the optimization problem. The modified SIMP method [85] is one of the successful methods that have proved to be efficient in taking care of driving ρ_e's to either of 0 or 1 limit values [54].

Using the SIMP method and replacing the vector $\mathbf{X}$ by the vector of continuous design variables $\rho = \{\rho_1, \rho_2, ..., \rho_N\}$ the optimization Eq. (7.42) is restated as follows [54]:

$$\text{Minimize}: f\left(C_{ijkl}^H(\rho)\right)$$

$$\text{Subject to}: \left\{ \begin{array}{l} 3DPBCs \\ \sum_{e=1}^{N} V_e \rho_e / V \leq \vartheta \\ 0 \leq \rho_e \leq 1 \end{array} \right\} \tag{7.43}$$

As a major advantage of the current methodology, it is free from the imposition of additional constraints (such as isotropy constraint, considered by some authors—see Radman et al. [71] and Andreassen et al. [134] for example) in optimization formulation. It should be pointed out that for some particular cases, mathematical constraints may be added to the optimization equation [Eq. (7.43)]. For example, in section 6.2 the desired moduli ratios in the principal directions are considered as constraints for orthotropic microstructure design. The constraint can be stated as [54]:

$$\text{Subject to}: |E_i/E_j - r| \leq \varepsilon \tag{7.44}$$

where E_i and E_j are the Young's moduli in the ith and jth principal directions, respectively, r is the desired ratio, and ε is a small positive number, and this constraint can be written for as many ratios as desired [54].

The optimization Eq. (7.43) begins by initializing the elements' densities such that the volume fraction constraint $\sum V_e \rho_e / V \approx \vartheta$ is satisfied. As a common practice the initial microstructural topology to begin the optimization with is generated by concentrating the lighter phase at the corners of the unit cell, as explained and implemented in Amstutz et al. [137] and Xia and Breitkopf [138]. The commonly used practice of initialization consists of 1/8 of a sphere whose radius is approximately 1/3 of the PUC edge, as illustrated in Fig. 7.39. In this way, it is possible to perform an analysis of sensitivity to the densities during the optimization process, while the PBCs are being applied [54].

After initialization the modified SIMP method is called to calculate and update the equivalent Young's modulus of each element, E_e, in the intermediate topologies. According to the modified SIMP method, E_e is expressed as follows [54]:

$$E_e(\rho_e) = E_{\min} + \rho_e^p(E^0 - E_{\min}) \tag{7.45}$$

in which E_e is the Young's modulus of the solid phase and $E_{\min}$ is the modulus of the void phase, taken to be a small positive number to avoid singularity problems in

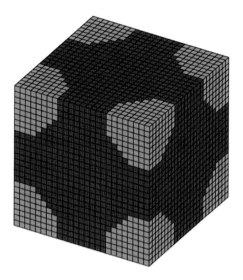

Figure 7.39 Illustration of the initial guess for a 3D unit cell with 25 × 25 × 25 elements [54].

the numerical calculation of the stiffness tensor. The exponent p serves as a penalty parameter that pushes ρ_e's toward either 0 or 1. The penalty factor p is chosen based on the following inequality, which is developed from Hashin−Shtrikman (H-S) bounds [115] for an isotropic material [54,139]:

$$p \geq \max\left\{15\frac{1-\nu^0}{7-5\nu^0}, \frac{3}{2}\frac{1-\nu^0}{1-2\nu^0}\right\} \qquad (7.46)$$

Here, ν^0 denotes the Poisson's ratio of the base material (solid phase).

In the next step the effective stiffness tensor of the 3D-PUC is calculated using FEM, followed by the sensitivity analysis of the objective function $f(C_{ijkl}^H(\rho))$, which is accompanied by its modification using a filtering algorithm. Finally an updating scheme based on OCA, which is elaborated further below in Section 7.4, is used to update the density of all elements. All these steps are repeated iteratively until the convergence criterion, which is the smallness of density variations in successive iterations, being realized. The flowchart of Fig. 7.40 provides an illustrative summary of the steps involved. Further details about each step are provided in the following sections [54].

7.4.1.1 SIMP-based numerical homogenization of 3D-PUC

To evaluate C_{ijkl}^H of periodic microstructures the usual procedure is to use FEM in which the 3D-PUC is discretized to eight-node, linear cubic elements. In this

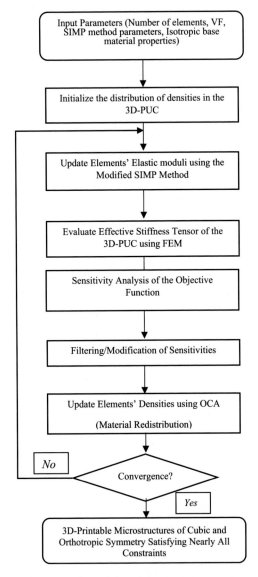

Figure 7.40 The flowchart of the SIMP-OCA methodology for optimal design of 3D-printable microstructures [54].

section, we generalized the numerical homogenization approach deployed by Andreassen and Andreasen [134] for 2D RVEs to the effective stiffness estimation of 3D-PUCs based on the density design variables. The key point of the numerical homogenization lies in developing the 3D FEM formulation and considering

Microstructure hull and design

3D-PBCs. Based on the 3D numerical homogenization scheme and following the SIMP method the homogenized stiffness tensor expressed in terms of density vectors $C_{ijkl}^H(\rho)$ corresponds to the volumetric average of total strain energies of the elements of 3D-PUC [54]:

$$C_{ijkl}^H(\rho) = \frac{1}{V} \sum_{e=1}^{N} \int_{V_e} \left(\mathbf{u}_e^{(ij)0} - \mathbf{u}_e^{(ij)}\right)^T \mathbf{k}_e(\rho_e)\left(\mathbf{u}_e^{(kl)0} - \mathbf{u}_e^{(kl)}\right) dV_e \tag{7.47}$$

in which V is the volume of the 3D-PUC, $\mathbf{k}_e(\rho_e)$ are 3D element stiffness matrices expressed in terms of element densities [140], and $\mathbf{u}_e^0$ represents the six displacement fields corresponding to the six unit test strains given by [54]

$$\varepsilon^1 = (1, 0, 0, 0, 0, 0)^T, \ \varepsilon^2 = (0, 1, 0, 0, 0, 0)^T, \ \ldots, \ \varepsilon^6 = (0, 0, 0, 0, 0, 1)^T \tag{7.48}$$

The elements' unknown displacements $\mathbf{u}_e$ are induced by the elements' initial displacement fields $\mathbf{u}_e$. $\mathbf{u}_e$ is an array of six columns corresponding to the six displacement vectors resulting from the application of six unit strains of Eq. (7.48). Because all elements of the PUC are identical, $\mathbf{u}_e$ that is found accordingly applies to all elements. In fact, $\mathbf{u}_e$ is the solution to the element's characteristic Eq. (7.49), in which sufficient degrees of freedom are constrained to avoid the singularity of the element stiffness matrix. The displacement vectors are obtained for six load cases [54]:

$$\mathbf{K}(\rho)\mathbf{u}^i = \mathbf{F}^i(\rho) \quad i = 1, \ldots, 6 \tag{7.49}$$

where the displacement vectors $\mathbf{u}^i$ are assumed to be periodic. The above equation can be recast as follows, in which the modified SIMP relation Eq. (7.45) is incorporated [54]:

$$\mathbf{K}(\rho) = \sum_{e=1}^{N} \mathbf{k}_e(\rho_e) = \sum_{e=1}^{N} E_e(\rho_e)\mathbf{k}_{le} \tag{7.50}$$

Here, $\mathbf{k}_{le}$ denotes the element stiffness matrices corresponding to unit Young's modulus in FEM formulation. Likewise the force term on the right-hand side of Eq. (7.49) is reshaped as follows [54]:

$$\mathbf{F}^i(\rho) = \sum_{e=1}^{N} \mathbf{f}_e^i(\rho_e) = \sum_{e=1}^{N} E_e(\rho_e)\mathbf{f}_{le}^i \tag{7.51}$$

where $\mathbf{f}_{le}$ are element loading (internal force) matrices corresponding to unit Young's modulus in FEM formulation [54].

7.4.1.2 OCA-type optimizer

The OCA methodology is implemented by taking the density constraints $0 \le \rho_e \le 1$ in problem Eq. (7.43) as inactive and only keeping the VF constraint into account. Thus the first-order optimality criterion for the combined Lagrange function of the problem can be defined as $B_e = 1$, where [54]

$$B_e = -\frac{\frac{\partial \hat{f}}{\partial \rho_e}}{\lambda \frac{\partial V}{\partial \rho_e}} \qquad (7.52)$$

considers the filtered sensitivity of the objective function f and the volume constraint with respect to the densities. Because all volume elements of the PUC are identical, $\frac{\partial V}{\partial \rho_e} = 1$. The VF constraint is enforced through the Lagrange multiplier, λ, which in turn is found by means of a root-finding numerical tool such as the bisection method [54].

The OCA methodology requires the sensitivity analysis of the objective function to the densities, which are stated by the first-order derivatives of f appearing in the numerator of the fraction in Eq. (7.52). Given that f is a function of the homogenized stiffness's $C_{ijkl}^H(\rho)$ the numerator in Eq. (7.52) is calculated using the chain rule. Next the derivatives of the effective moduli with respect to the densities are required, which are determined by means of the adjoint method [14], given by [54]

$$\frac{\partial C_{ijkl}^H(\rho)}{\partial \rho_e} = \frac{1}{V} p \rho_e^{p-1} (E^0 - E_{\min}) \int_{V_e} \left(\mathbf{u}_e^{(ij)0} - \mathbf{u}_e^{(ij)} \right)^T \mathbf{k}_{le} \left(\mathbf{u}_e^{(kl)0} - \mathbf{u}_e^{(kl)} \right) dV_e \qquad (7.53)$$

In this study, sensitivity analysis is modified by the original mesh-independent sensitivity filter [60,88]. Implementation of this filtering technique is inspired by the formulation of Andreassen et al. [84]. To ensure mesh independency the filtration is provided with an adjustable radius $R_{\min}$ around the considered element, which defines the neighboring elements that influence its modification [54].

As such the sensitivity filter modifies the sensitivities $\partial f / \partial \rho_e$ as follows:

$$\frac{\partial \hat{f}}{\partial \rho_e} = \frac{1}{\max(\alpha, \rho_e) \sum_{i \in N_e} W_{ei}} \sum_{i \in N_e} W_{ei} \rho_i \frac{\partial f}{\partial \rho_i} \qquad (7.54)$$

where N_e denotes the set of elements i whose center to center distance $\delta(e, i)$ with element e is smaller than the filter radius $R_{\min}$ and W_{ei} is a weight factor or a convolution operator defined as [54]

$$W_{ei} = \max(0, R_{\min} - \delta(e, i)) \qquad (7.55)$$

Also the term α in Eq. (7.54) is a small positive number, say 10^{-3}, which is used to avoid division by 0 [54].

The filtered sensitivity of the objective function with respect to the design variables (densities) is used to update the densities, moving them toward 0 or 1, while

Microstructure hull and design

satisfying the volume constraint. The updating process based on Bendsøe's heuristic scheme [82] is given by

$$\rho_e^{new} = \begin{cases} \max(0, \rho_e - \text{move}) & \text{if } \rho_e B_e^\Lambda \le \max(0, \rho_e - \text{move}) \\ \rho_e B_e^\Lambda & \text{if } \max(0, \rho_e - \text{move}) \le \rho_e B_e^\Lambda < \min(1, \rho_e + \text{move}) \\ \min(1, \rho_e + \text{move}) & \text{if } \rho_e B_e^\Lambda \ge \min(1, \rho_e + \text{move}) \end{cases}$$

(7.56)

in which "move" is an allowable positive move limit for the densities which prevents their fluctuation with a high amplitude from one iteration to another, and Λ serves as a damping coefficient. These parameters take care of the changes that take place during the microstructure updating and can be adjusted to control the performance of the method. The updating procedure is continued until the convergence of all elements densities. At this point the optimized topology in a nearly "black-and-white" form is obtained (i.e., the densities become either 0 or 1, with an allowable tolerance). Finally the stopping criteria are set as a maximum for the number of iterations or [54]

$$\rho_e^{new} = \begin{cases} \max(0, \rho_e - \text{move}) & \text{if } \rho_e B_e^\Lambda \le \max(0, \rho_e - \text{move}) \\ \rho_e B_e^\Lambda & \text{if } \max(0, \rho_e - \text{move}) \le \rho_e B_e^\Lambda < \min(1, \rho_e + \text{move}) \\ \min(1, \rho_e + \text{move}) & \text{if } \rho_e B_e^\Lambda \ge \min(1, \rho_e + \text{move}) \end{cases}$$

(7.57)

(the tolerance $\Delta\rho$ is a relatively small positive number), whichever happens first.

7.4.1.3 Numerical implementation

In FEM simulations the PUCs are discretized into $25 \times 25 \times 25$ cubic elements. The Young's modulus and Poisson's ratio of the solid phase are chosen as 100 units and 0.3, respectively. An exceedingly small value, say 10^{-9}, is considered for the elastic modulus of the void phase. As recommended by Xia and Breitkopf [86] the OCA updating parameters are set as move $= 0.2$ and $\Lambda = 0.5$. Also the optimal topologies of the PUCs are obtained through an iterative process with different penalization factors ($p \ge 3$) and filter radii ($1 < R_{min} \le 0.1 \times 25$). The convergence criterion is satisfied when the normalized difference between densities, $\Delta\rho$, of two successive iterations is less than 0.01. It should be noted that each simulation takes an average of 1 hour on Intel Core i7$-$6700K CPU @ 4.00 GHz and with 16GB RAM. However, in OCA-based topology optimization the convergence to the optimal topology strongly depends on the adjustable parameters. Finding the right set of parameters is achieved by trial and error, which might be more or less challenging [54].

7.4.1.4 Numerical examples and experimental evaluations

In this section the SIMP-OCA methodology is employed for designing 3D-manufacturable microstructures of cubic and orthotropic symmetry. In the objective functions that follow the following reduced 6 by 6 homogenized stiffness matrix $\mathbf{C}^H$ is utilized [54]:

$$
\mathbf{C}^H = \begin{bmatrix} C_{11}^H & C_{12}^H & C_{13}^H & C_{14}^H & C_{15}^H & C_{16}^H \\ & C_{22}^H & C_{23}^H & C_{24}^H & C_{25}^H & C_{26}^H \\ & & C_{33}^H & C_{34}^H & C_{35}^H & C_{36}^H \\ & \text{Sym} & & C_{44}^H & C_{45}^H & C_{46}^H \\ & & & & C_{55}^H & C_{56}^H \\ & & & & & C_{66}^H \end{bmatrix} \tag{7.58}
$$

It is known that for maximizing the objective functions of bulk modulus, shear modulus, and negative Poisson's ratio a cubic symmetry is obtained for the final optimal microstructure (Suresh, 2014) [54].

7.4.1.4.1 PUCs of cubic symmetry

Bulk modulus maximization The objective function for this case is expressed as follows [54,78]:

$$
f = -\sum_{i,j=1}^{3} C_{ij}^H \tag{7.59}
$$

Four values of VFs are considered: %65, %50, %35, and %25. Table 7.17 shows the resulting PUCs along with the corresponding stiffness matrix $\mathbf{C}^H$ and the number of iterations [54].

Fig. 7.41 shows the resulting bulk moduli K along with the H-S upper bounds given by [54,115]

$$
\frac{K}{K^0} \leq \frac{4G^0 VF}{3K^0(1 - VF) + 4G^0}, \tag{7.60}
$$

where K^0 and G^0 are the bulk and shear moduli of the isotropic base material, respectively. It is seen that the optimized results are close to but not beyond this limit Eq. (7.61), as they should be [54].

Shear modulus maximization The associated objective function is defined as [54,78]

$$
f = -\sum_{i=4}^{6} C_{ii}^H \tag{7.61}
$$

Here the same VFs as the previous case are considered. Table 7.18 shows the resulting PUCs along with the corresponding stiffness matrix $\mathbf{C}^H$ and the number of iterations. It should be stated that the optimal topologies in Table 7.18 are comparable to those presented in similar works (see Andreassen et al. [134]. and Gao et al. [78] for example) [54].

Table 7.17 Optimal results for 3D-printable PUCs with a maximized bulk modulus [54].

Case	VF	Optimized PUC	C^H	Iteration
1	%65		$\begin{bmatrix} 52.04 & 20.03 & 20.03 & 0 & 0 & 0 \\ & 52.04 & 20.03 & 0 & 0 & 0 \\ & & 52.04 & 0 & 0 & 0 \\ & & & 16.49 & 0 & 0 \\ & \text{Sym} & & & 16.49 & 0 \\ & & & & & 16.49 \end{bmatrix}$	63
2	%50		$\begin{bmatrix} 33.18 & 13.12 & 13.12 & 0 & 0 & 0 \\ & 33.18 & 13.12 & 0 & 0 & 0 \\ & & 33.18 & 0 & 0 & 0 \\ & & & 10.25 & 0 & 0 \\ & \text{Sym} & & & 10.25 & 0 \\ & & & & & 10.25 \end{bmatrix}$	83
3	%35		$\begin{bmatrix} 16.96 & 7.16 & 7.16 & 0 & 0 & 0 \\ & 16.96 & 7.16 & 0 & 0 & 0 \\ & & 16.96 & 0 & 0 & 0 \\ & & & 5.29 & 0 & 0 \\ & \text{Sym} & & & 5.29 & 0 \\ & & & & & 5.29 \end{bmatrix}$	42
4	%25		$\begin{bmatrix} 8.42 & 4.37 & 4.37 & 0 & 0 & 0 \\ & 8.42 & 4.37 & 0 & 0 & 0 \\ & & 8.42 & 0 & 0 & 0 \\ & & & 2.68 & 0 & 0 \\ & \text{Sym} & & & 2.68 & 0 \\ & & & & & 2.68 \end{bmatrix}$	54

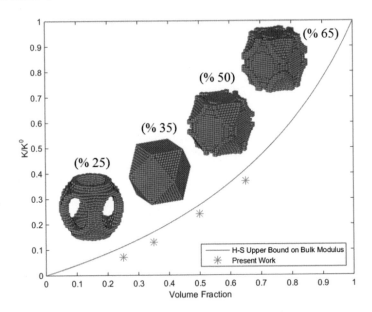

Figure 7.41 Comparison of the bulk modulus of optimized PUCs and the H-S upper bound for different VFs [54].

Negative Poisson's ratio maximization The objective function of this optimization problem is expressed as follows:

$$f = -\sum_{i,j=1, i\neq j}^{3} C_{ij}^{H} \tag{7.62}$$

With a VF of %50 an optimized NPR of -0.0235 was achieved. A further attempt led to a better NPR of -0.19 at a VF of %43.3.

Table 7.19 shows the optimized topology, the corresponding stiffness matrix $\mathbf{C}^{H}$, and the number of iterations.

Maximization of axial stiffness in cubic symmetry In this case the objective function is defined as [54]

$$f = -\sum_{i=1}^{3} C_{ii}^{H} \tag{7.63}$$

For this case a VF = %50 of the solid phase is considered. The resulting optimum microstructure is presented in Table 7.20 and is consistent with the designed topology reported in Zhang et al. [69]. In the table the effective elastic stiffness tensor $\mathbf{C}^{H}$ and the number of iterations of the optimization process are also reported.

Table 7.18 Optimal results for 3D-printable PUCs with a maximized shear modulus [54].

Case	VF	Optimized PUC	C^H	Iteration
1	%65		$\begin{bmatrix} 50.92 & 20.2 & 20.2 & 0 & 0 & 0 \\ & 50.92 & 20.2 & 0 & 0 & 0 \\ & & 50.92 & 0 & 0 & 0 \\ & & & 17.7 & 0 & 0 \\ & \text{Sym} & & & 17.7 & 0 \\ & & & & & 17.7 \end{bmatrix}$	41
2	%50		$\begin{bmatrix} 31.34 & 13.07 & 13.07 & 0 & 0 & 0 \\ & 31.34 & 13.07 & 0 & 0 & 0 \\ & & 31.34 & 0 & 0 & 0 \\ & & & 11.27 & 0 & 0 \\ & \text{Sym} & & & 11.27 & 0 \\ & & & & & 11.27 \end{bmatrix}$	95
3	%35		$\begin{bmatrix} 14.78 & 7.61 & 7.61 & 0 & 0 & 0 \\ & 14.78 & 7.61 & 0 & 0 & 0 \\ & & 14.78 & 0 & 0 & 0 \\ & & & 6.69 & 0 & 0 \\ & \text{Sym} & & & 6.69 & 0 \\ & & & & & 6.69 \end{bmatrix}$	185
4	%25		$\begin{bmatrix} 7.65 & 3.71 & 3.71 & 0 & 0 & 0 \\ & 7.65 & 3.71 & 0 & 0 & 0 \\ & & 7.65 & 0 & 0 & 0 \\ & & & 3.67 & 0 & 0 \\ & \text{Sym} & & & 3.67 & 0 \\ & & & & & 3.67 \end{bmatrix}$	95

Table 7.19 Optimal results for 3D-printable PUC with maximized NPR [54].

VF	Optimized PUC	$\mathbf{C}^H$						Iteration
%43.3		$\begin{bmatrix} 5.8 & -0.93 & -0.93 & 0 & 0 & 0 \\ & 5.8 & -0.93 & 0 & 0 & 0 \\ & & 5.8 & 0 & 0 & 0 \\ & & & 0.86 & 0 & 0 \\ & \text{Sym} & & & 0.86 & 0 \\ & & & & & 0.86 \end{bmatrix}$						56

Table 7.20 Optimal results for 3D-printable PUC with maximized axial stiffness [54].

VF	Optimized PUC	$\mathbf{C}^H$						Iteration
%50		$\begin{bmatrix} 32.66 & 11.18 & 11.18 & 0 & 0 & 0 \\ & 32.66 & 11.18 & 0 & 0 & 0 \\ & & 32.66 & 0 & 0 & 0 \\ & & & 9.1 & 0 & 0 \\ & \text{Sym} & & & 9.1 & 0 \\ & & & & & 9.1 \end{bmatrix}$						137

7.4.1.4.2 PUCs of orthotropic symmetry

We also tested the applicability of the methodology to generate 3D-printable orthotropic microstructures. Although there are more material constants to be investigated, an interesting and practical case would be to have different axial stiffnesses in three directions. This is done by defining the objective function as the weighted sum of the axial stiffness, taking the unknown weight factors as additional design variables and desired moduli ratios as constraints [see Eq. (7.44)]. Thus the objective function is stated as [54]

$$f = -(A_1 C_{11}^H + A_2 C_{22}^H + A_3 C_{33}^H), \tag{7.64}$$

where A_1, A_2, and A_3 are the weight factors. This problem is solved for several cases, and the VFs, weight factors, resulting PUCs, effective elastic stiffness tensors $\mathbf{C}^H$, and Young's moduli ratios are presented in Table 7.21 [54].

In this research an effective optimum design procedure was proposed for 3D-printable PUC of heterogeneous materials using SIMP-OCA methodology for microstructures of cubic and orthotropic symmetry. The major highlights of the method are capability in designing 3D microstructures with prescribed elastic properties and manufacturability of the generated microstructures with properties close to those of the optimized structure. It should be stated that the 3D-printable orthotropic microstructure

Table 7.21 Optimal results for 3D-printable PUCs of orthotropic symmetry [54].

Case	Weight factors	VF	Optimized PUC	$\mathbf{C}^H$	Young's moduli ratios
1	$A_1 = 1$ $A_2 = 2$ $A_3 = 1$	%57	 (Top View)	$\begin{bmatrix} 37.26 & 13.19 & 6.7 & 0 & 0 & 0 \\ & 58.8 & 13.19 & 0 & 0 & 0 \\ & & 37.26 & 0 & 0 & 0 \\ & & & 13.12 & 0 & 0 \\ & \text{Sym} & & & 13.12 & 0 \\ & & & & & 1.97 \end{bmatrix}$	$\frac{E_2}{E_1} = 1.5 \ \frac{E_3}{E_1} = 1$
2	$A_1 = 0.75$ $A_2 = 2$ $A_3 = 1$	%42	 (Top View)	$\begin{bmatrix} 14.12 & 4.89 & 2.19 & 0 & 0 & 0 \\ & 38.59 & 9.51 & 0 & 0 & 0 \\ & & 29.52 & 0 & 0 & 0 \\ & & & 5.02 & 0 & 0 \\ & \text{Sym} & & & 10.42 & 0 \\ & & & & & 0.36 \end{bmatrix}$	$\frac{E_2}{E_1} = 2.5 \ \frac{E_3}{E_1} = 2$

(Continued)

Table 7.21 (Continued)

Case	Weight factors	VF	Optimized PUC	$\mathbf{C}^H$						Young's moduli ratios
3	$A_1 = 0.5$ $A_2 = 2$ $A_3 = 1$	%55.7		$\begin{bmatrix} 17.58 & 6.46 & 3.95 & 0 & 0 & 0 \\ & 57.44 & 16.31 & 0 & 0 & 0 \\ & & 50.40 & 0 & 0 & 0 \\ & & & 6.27 & 0 & 0 \\ & \text{Sym} & & & 17.47 & 0 \\ & & & & & 0.48 \end{bmatrix}$						$\frac{E_2}{E_1} = 3 \ \frac{E_3}{E_1} = 2.7$

design is a remarkable ability of the proposed methodology, which is rarely considered in the literature. In future parts of this work, we plan to generalize the proposed methodology to the design of multiphase microstructures composed of two or more solid phases with probably other physical properties set as the requirements of the topology optimization. As an interesting extension to the current study, we are working on the design of microstructures satisfying coupled thermomechanical properties [54].

7.4.2 3D-PUC design using topology optimization and innovative building block

In this section a novel method is developed to design a structure with a tunable effective Young's modulus. In this method a gap is generated in the middle of each unit cell in the structure. Compression loading was used as a stimulus to change the effective Young's modulus. Upon applying a compressive load the gap was closed, and an extra resistance was added to the cell, so the effective Young's modulus was increased. This stiffness change mostly depends on the material which is used and size of the gap. Also, other geometrical parameters' effects were studied. The material selection depends on the gap's distance. A bigger gap's distance requires a material with a lower Young's modulus. A small gap's distance requires more accurate methods to print the sample, and this increases the cost. Numerical simulations and experimental tests have been performed on a periodic structure to prove the effectiveness of the design. The effective Young's modulus increases by 142% using this method. GA was used to design a structure with desired stress and strain curves. This method was based on defining geometric parameters which have the most effect in changing the effective Young's modulus. There are some advantages for this design. First, many different structures with different ranges of changes in effective Young's modulus could be designed. Second, these structures can response to the changes fast. Third, this method is simple and cheap [89].

7.4.2.1 Proposed building block

A unit cell or building block with a gap in it has been designed (Fig. 7.42A). A compressive stress applied on the outer surfaces perpendicular to the gap's direction, and

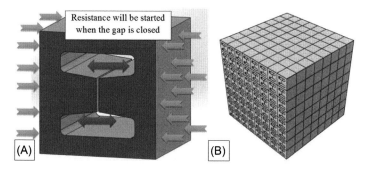

Figure 7.42 (A) Method to increase the effective Young's modulus in the building block; (B) one of the structures which were made by repeating the flat surface contact cell [89].

the corresponding strain due to this stress has been calculated. An effective Young's modulus is evaluated by dividing this stress to the obtained strain. This result is considered as stress loading effective Young's modulus. At the beginning, there was no change of effective Young's modulus and the stress—strain curve was linear. When the gap's surfaces reach each other, an extra resistance was added to the structure. This contact does not allow the structure to deform like a linear material, so the stress—strain curve of the structure becomes nonlinear [89].

Later a periodic structure was constructed by repeating the unit cell in all three directions (Fig. 7.42B). This structure can alter its effective Young's modulus under external load. The amount of change depends on many factors like geometry, material, and loading. There are many geometric factors which are important to alter the effective Young's modulus. In this section, some of the most effective geometrical parameters are studied [89].

Obviously, one of the most effective factors to change the effective Young's modulus is the gap's shape which is inside of the cell. Five different shapes for the gap have been considered to determine which one is most effective.

A building block with the dimension of 250 mm was considered. Fig. 7.43 shows five different contact surfaces which are considered in the middle of the cell. These surfaces are called contact surfaces [89].

Fig. 7.44 illustrates the stress—strain curve for five different shapes of contact surfaces. As shown by the results, the cell with the flat contact surfaces has the most changes in effective Young's modulus and nonlinearity in stress—strain curve.

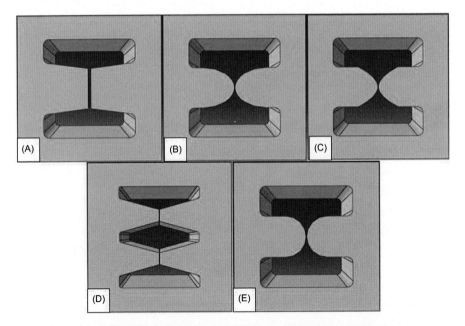

Figure 7.43 Different contact surfaces were studied. (A) Flat surface, (B) oval surfaces, (C) parabola surface, (D) two parts of the flat surface, and (E) semicircle surface [89].

Local maximum stress in cells with curve-like contact surfaces exceeds the allowed stress because of stress concentration in these cells. Hence because of entering the plastic deformation region, all samples except flat surfaces do not have stress–strain curves at strains higher than 0.008.

The gap's distance and surface areas of cases A and D are the same, and both shapes create the same void volume fraction in the structure; that is why stress–strain curves are almost the same for these two cases. For the same reason, cases B, C, and E have similar elastic responses [89].

The second important parameter is the gap's distance. As shown in Fig. 7.45, five different values were selected to study the effect of the gap's distance. As was expected the process of alteration of the effective Young's modulus speeds up by decreasing the gap's distance. In other words, cells with a smaller gap's distance become stiffer and reach the maximum Von Mises stress in lower strain compared with the one with a larger gap's distance. Therefore the gap's distance is important for designing a material with tunable stiffness, and depending on specific applications, this parameter regulates stiffness change. All cells had the same effective Young's modulus after gap closure [89].

Length of the contact area is another important parameter. Fig. 7.46 illustrates this parameter for a flat contact surface. To determine the effect of this parameter,

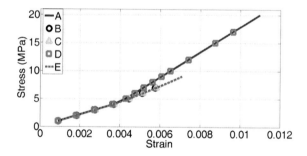

Figure 7.44 Stress–strain curves for five different shapes for contact surfaces. Absolute values of stress and strain were considered because the loading was compression [89].

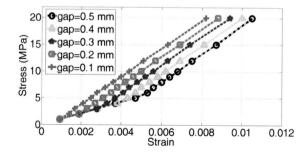

Figure 7.45 Stress–strain curves of five samples with different gap distances. Absolute values of stress and strain were considered because the loading was compression [89].

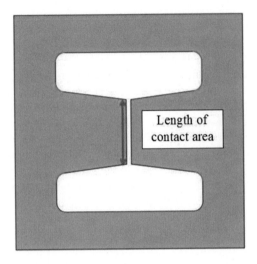

Figure 7.46 Length of the contact area for a flat surface contact area sample [89].

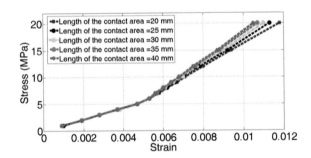

Figure 7.47 Stress–strain curves of five different samples with different lengths of the contact area (length of the contact area). Absolute values of stress and strain were considered because the loading was compression [89].

five different lengths have been chosen. As shown in Fig. 7.47, by increasing the contact length, maximum stress decreases and the effective Young's modulus increases [89].

There are also many other parameters and shapes for both the inner and outer boundaries of the cell, and some of them will be discussed in the following sections. In this section the focus was on finding the effective geometrical parameters on effective Young's modulus. Therefore selecting the proper material has not been one of the concerns. In Section 7.2.1, poly(methyl methacrylate) (PMMA) was considered as the material for all simulations. PMMA has the Young's modulus of 3 GPa [141] and a Poisson' ratio of 0.4 [142]. For the purpose of this research a material with a lower Young's modulus was required, so PLA has been chosen. Its mechanical properties will be discussed in Section 7.3.2. All simulations in Section 7.2.2 and after that were done with PLA properties [89].

7.4.2.2 Designing representative volume elements

To determine the geometry and minimum size of the building block, some factors should be considered. Many different shapes and patterns with at least one gap can be designed. The most important parameter was the size of the building block. The geometry of the building block is complicated, so the structure has been fabricated using advanced additive manufacturing techniques such as 3D printing. Also the selected material should satisfy the required initial Young's modulus. If the structure's material has a high Young's modulus, too much external strain is required to close the gap of the cell. This may create high stress in the cell, and this stress may exceed allowable stress and causes plastic deformation. Meanwhile the cell's material should not be so soft that deforms easily and reaches the plastic strain region. A commonly used polymer in 3D printing, polylactic acid (PLA), was chosen as a filament which has a proper initial Young's modulus for the purpose of the design [89].

Fig. 7.48 and Table 7.22 illustrate dimensions of building blocks. In Fig. 7.49 a building block is shown after applying a compressive load of 1.5 MPa. To estimate RVE size, different sizes of microstructures (building blocks) are generated and effective properties are calculated to get size-independent properties. Because of discontinuity in the microstructure (Figs. 7.50 and 7.51), the Young's modulus was obtained, dividing mean stress by strain [143,144]. To validate if the selected

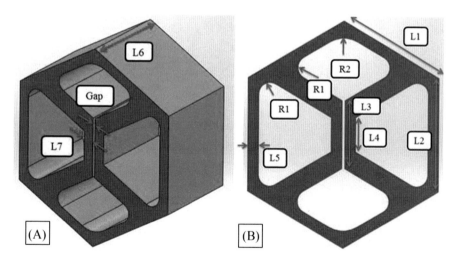

Figure 7.48 Geometrical parameters of the building block. (A) Isometric view and (B) front view [89].

Table 7.22 Dimensions of the geometric parameters of the building block in mm [89].

L1	L2	L3	L4	L5	L6	L7	Gap	R1	R2
20	20	12	6	2	2	20	0.5	2	5

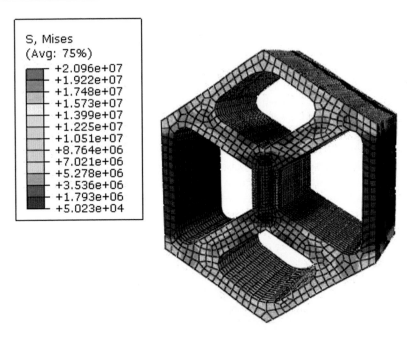

Figure 7.49 Von-Mises stress contour of the building block after applying 1.5 MPa compression. All stresses in Pa [89].

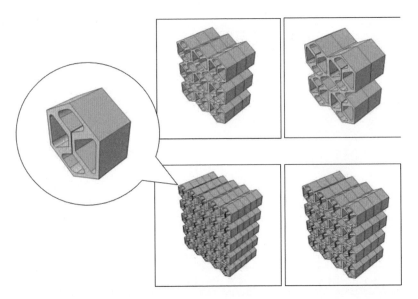

Figure 7.50 Five samples with different numbers of cells were considered and simulated [89].

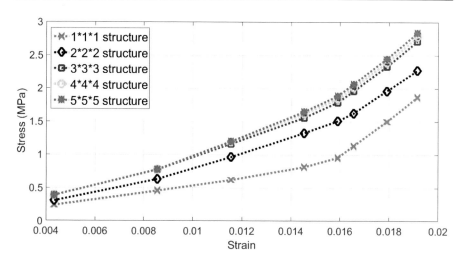

Figure 7.51 Comparing stress—strain curves of five samples with different numbers of cells has been done. The 3 × 3 × 3 structure result is the same as the results of 4 × 4 × 4 and 5 × 5 × 5 structures [89].

realization is RVE, two separate loadings were applied (Fig. 7.52). First, compression load was applied to the outer surface of the cell, and then the strain response was calculated. An effective Young's modulus was evaluated by dividing the stress to the obtained strain. This result is considered as stress loading effective Young's modulus. Second a displacement was applied to the outer surface of the unit cell and the mean stress on the outer surfaces was calculated [89,145—147].

7.4.2.3 Optimization procedure

For a desired overall response of the bulk material, geometrical parameters of the RVE should be evaluated using the optimization procedure. Table 7.23 shows the desired stress—strain curve's data. GA was utilized as an optimization method [148]. This optimization process is performed to find five effective geometric parameters of a 2D cell. These parameters are the most effective parameters to change the effective Young's modulus of the cell. Fig. 7.53 shows these five geometric parameters. A desired stress—strain curve has been chosen, and GA was used to determine these five parameters such that the stress—strain curve of the cell fits the desired curve. A 2D cell was used to speed up the optimization process. Fig. 7.54 illustrates the optimization process. GA uses the objective function to determine the criterion. In order to compare two stress—strain curves, 15 points from the desired stress—strain curve were selected. In each step a script code is run to calculate the new 15 desired stress—strain data. These data are compared with the desired data. Since strains of the obtained data are same as desired data the difference in stresses is considered. The objective function uses a script code and a MATLAB code to calculate the results. This method can determine a

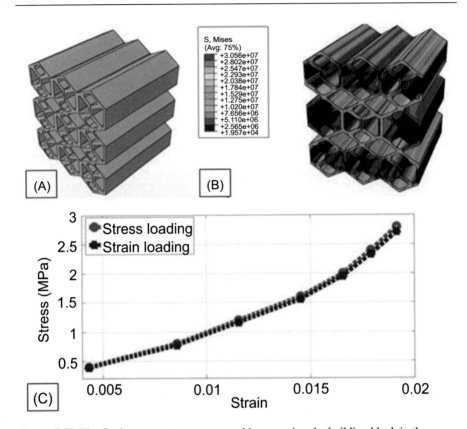

Figure 7.52 The final structure was constructed by repeating the building block in three directions. (A) Structure and (B) Von-Mises stress contour of the structure after applying 1.0 MPa compression. All stresses in Pa. (C) Stress–strain curves of the final structure. Absolute values of stress and strain were considered because the loading was compression [89].

Table 7.23 Desired values for stress and strain [89].

Strain × 10^{-3}	4.4	5.1	5.9	6.6	7.4	8.1	8.8	9.6	10.3	11	11.8	12.5	13.2	14	14.7
Stress (MPa)	0.62	0.73	0.83	0.93	1.04	1.14	1.25	1.41	1.75	2.14	2.55	2.98	3.41	3.86	4.3

specific cell with the desired stress–strain curve. The objective function is considered as shown in Eq. (7.65) [89]

$$\text{Objective function} = \int |\sigma_2 - \sigma_1| d\varepsilon \qquad (7.65)$$

where σ_2 is the stress of the obtained data in each iteration and σ_1 is stress of desired values. To achieve the desired stress–strain curve the objective function

Microstructure hull and design

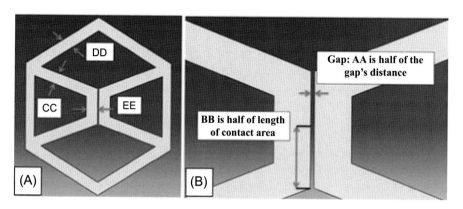

Figure 7.53 Geometrical parameters that are used in the optimization process. (A) Front view and (B) zoomed view of the gap [89].

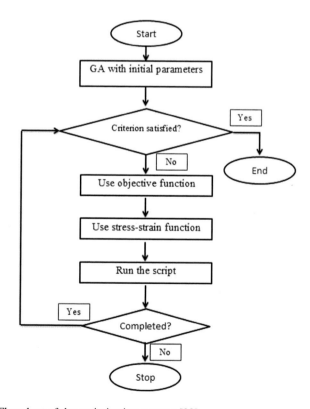

Figure 7.54 Flowchart of the optimization process [89].

should be near 0. This means that the area between the target curve and the current obtained curve is 0. The MATLAB code uses the trapezoidal integration rule to calculate the objective function. The objective function final value for the first optimization was 0.0073. It is possible for a desired curve that the code could not find proper answers or there would be more than one answer [89].

Different values for geometric parameters have been tested, and the result showed that the stress−strain response is more sensitive to half of the gap's distance AA and thickness of the outer boundary of cell DD. As shown in Fig. 7.55, GA found the results with six generations. Geometric parameters which have been calculated by GA are shown (Table 7.24) [89].

Fig. 7.56 shows the comparison between the target stress−strain curve and optimized stress−strain curve estimated with GA [89].

As shown in Fig. 7.57 the fminsearch function from the optimization toolbox in MATLAB software was used to optimize the problem. This function uses a derivative-free method. The optimization problem is nonlinear, so the solutions are only converging locally. An initial guess was made to begin the algorithm. This guess was close enough to the answer, so the optimization process was ended after 15 iterations. Table 7.25 shows the results of optimization by using the fminsearch function [89].

7.4.2.4 Results

7.4.2.4.1 Designing a new structure

The optimization method explained in section 2.3 was used to construct a periodic structure. This structure is 2D. Fig. 7.58A shows the unit cell of this structure. Two

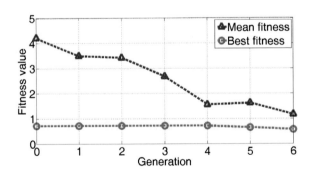

Figure 7.55 GA's convergence [89].

Table 7.24 Geometric parameters which were calculated by GA. All data are in mm [89].

	AA	BB	CC	DD	EE
Upper bound	0.20	5.50	4.00	3.50	3.00
Lower bound	0.10	3.00	2.50	2.50	2.00
Desired	0.15	3.75	2.83	2.73	2.18
Calculated by GA	0.16	3.50	3.41	2.89	2.11

Microstructure hull and design

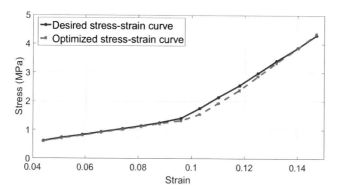

Figure 7.56 Comparison between the desired stress−strain curve and optimized stress−strain curve with GA. GA results are shown in Table 7.24 [89].

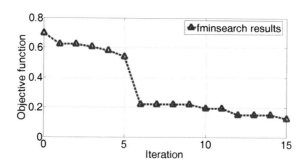

Figure 7.57 Results of optimization using the fminsearch algorithm [89].

Table 7.25 Geometric parameters which were calculated by the fminsearch algorithm. All data are in mm [89].

	AA	BB	CC	DD	EE
Upper bound	0.20000	5.50	4.00	3.50	3.00
Lower bound	0.10000	3.00	2.50	2.50	2.00
Desired	0.15000	3.75	2.83	2.73	2.18
Calculated by fminsearch	0.15337	4.24	2.67	2.74	2.29
Initial guess	0.15000	4.00	3.00	2.80	2.20

different gaps were created in this cell. All other geometrical parameters were constant. Fig. 7.58B illustrates the periodic structure. A desired stress−strain curve is selected. The stress−strain curve of the structure alters by changing these two gaps. Gap 1 was 0.4 mm, and gap 2 was 0.2 mm. As shown in Table 7.26, GA determined the required parameters for constructing the structure [89].

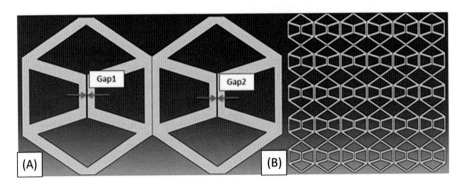

Figure 7.58 2D structure and its cell which were used in the second optimization. (A) Cell and (B) periodic structure [89].

Table 7.26 Gap1 and Gap2 are two parameters which were utilized in the second optimization [89].

	Gap1	Gap2
Upper bound	0.40	0.40
Lower bound	0.20	0.20
Desired	0.30	0.20
Calculated by GA	0.28	0.20

7.4.2.4.2 Experimental verification

As mentioned earlier, PLA has been chosen as a filament for 3D printing the designed structure. The sample was printed with the fused deposition modeling (FDM) technique. Printing causes material degradation and changes the mechanical properties of the final product [149], so a compression test has been done on a printed sample in order to reveal the final mechanical response. The Young's modulus of the printed sample was calculated to be 1725.52 MPa using these data. The Poisson's ratio of the FDM PLA was considered 0.36 [149,150]. All designs have been simulated with these properties except for Section 7.2.1 [89].

Compression test has been done on the periodic structure to verify the results of numerical simulations (Fig. 7.59). This structure is the final structure which was simulated in Section 7.3.1. The speed of the pressing was 0.5 mm/min, and the displacement was from 0 to 1.8 mm. This test has been repeated five times to verify the results [89].

Fig. 7.60 shows how the effective modulus of the structure changes under strain. The effective stiffness varies from 85 to 206 MPa. The effective Young's modulus was increased by 142% using the proposed method in this paper. As shown in Fig. 7.61 the force−displacement and stress−strain curves of the test are both nonlinear [89].

Numerical and experimental stress−strain responses are compared in Fig. 7.62, and they are in good agreement. Both these simulations have begun from −0.4 MPa, but experiments started from 0 loading [89].

Microstructure hull and design

Figure 7.59 The final structure is being compressed [89].

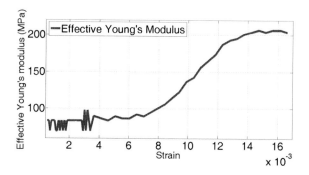

Figure 7.60 Change of effective Young's modulus of the structure [89].

7.4.3 *Directional elastic modulus of the TPMS structures and a novel hybridization method to control anisotropy*

Unlike random cellular materials the distribution of the ligament size and orientation is constant in periodic structures. This can lead to the directional dependency of the mechanical properties in these structures. It is shown that TPMS structures are characterized by a significantly directional stiffness; that is, there are strong and weak directions in these structures, especially in the low volume fractions of the solid phase [90].

Generally, anisotropy can be considered as an undesirable property when a structure can be exposed to a load with an unknown direction, especially in energy-absorbing applications [151]. It is shown that at least in plate-based [152,153] and

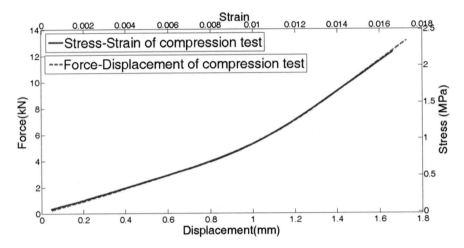

Figure 7.61 (-) Force–displacement curve of compression test of the final structure. Absolute values of force and displacement were considered because the loading was compression. (—) Stress–strain curve of compression test of the final structure. Absolute values of stress [89].

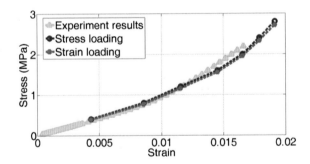

Figure 7.62 Comparison of experimental results with simulation results. The final structure's stress–strain curves. Absolute values of stress and strain were considered because the loading was compression [89].

truss-based unit cells [151] a uniform directional elastic modulus can be achieved via a simple superposition of the face-centered cubic and body-centered cubic structures. While it is not possible to superpose the TPMS structures to generate a new type of structure, it is shown that a hybrid structure with a smooth transition between two TPMS structures [154,155] can be used to fabricate laminated or matrix-spherical inclusion hybrid structures of TPMS [156]. Strong and weak directions of the TPMS-based structures depend on the TPMS type. Moreover, their anisotropy index depends on the volume fraction of the solid phase. In this section the directional elastic modulus of seven TPMS structures (i.e., Schwarz-P, IWP, Gyroid, diamond, FKS, FRD, and Neovius) was investigated. The possibility of

Microstructure hull and design

combining Schwarz-P and Neovius with each of the five other structures to form laminate or spherical-type hybrid structures aiming to achieve a more uniform directional elastic modulus was explored. The study indicated that such hybrid structures can effectively decrease the anisotropy. It also revealed that there is an optimal combination ratio of the parent structures that can lead to the lowest state of anisotropy of the elastic modulus in a hybrid structure [90].

7.4.3.1 TPMS-based structures

The seven TPMS structures (i.e., Schwarz-P, IWP, Diamond, Gyroid, Neovius, FRD, and FKS) investigated here are considered as solid networks (or skeletal structures). The equations of these surfaces are given in Eqs. (7.66)−(7.72) [157,158].

IWP:
$$f_I(x.y.z) = [\cos(2x) + \cos(2y) + \cos(2z)] - 2[\cos(x)\cos(y) + \cos(y)\cos(z) + \cos(z)\cos(x)], \tag{7.66}$$

Schwarz-P:
$$f_S(x.y.z) = \cos(x) + \cos(y) + \cos(z), \tag{7.67}$$

Gyroid:
$$f_G(x.y.z) = \cos(x)\sin(y) + \cos(y)\sin(z) + \cos(z)\sin(x), \tag{7.68}$$

Diamond:
$$f_D(x.y.z) = \sin(x)\sin(y)\sin(z) + \sin(x)\cos(y)\cos(z) + \cos(x)\sin(y)\cos(z) + \cos(x)\cos(y)\sin(z), \tag{7.69}$$

FRD:
$$f_{FRD}(x.y.z) = 4[\cos(x)\cos(y)\cos(z)] - [\cos(2x)\cos(2y) + \cos(2y)\cos(2z) + \cos(2z)\cos(2x)], \tag{7.70}$$

FKS:
$$f_{FKS}(x.y.z) = \cos(2x)\sin(y)\cos(z) + \cos(x)\cos(2y)\sin(z) + \sin(x)\cos(y)\cos(2z), \tag{7.71}$$

and Neovius:
$$f_N(x.y.z) = 3[\cos(x) + \cos(y) + \cos(z)] + 4\cos(x)\cos(y)\cos(z), \tag{7.72}$$

where $f_i \leq t$ gives the equation of the skeletal structures and t is the surface level parameter that indicates the isosurface level of the TPMS and specifies the volume fraction of the solid phase in the unit cell. Fig. 7.63 shows the relationship between volume fraction ρ^* (ratio of the solid-phase volume to the unit cell volume) and the surface level parameter t for different skeletal structures, which confirms the results of the previous studies (e.g., [159]). It should be noted that the solid phase and void phase are not connected for the entire range of the surface level parameter. The critical surface level parameters (i.e., t_1 and t_2) and associated densities (i.e., ρ_1^* and ρ_2^*) are defined as the limits where solid and void phases are connected, respectively. Fig. 7.64 shows the limits for different structures where solid and void phases are shown as red and blue, respectively. Furthermore the values for critical volume fraction and surface level parameter for each structure are shown in Table 7.27.

7.4.3.2 Hybrid structures

Hybrid structures can be easily generated using a simple combination of the two TPMS structures as shown in Eq. (7.73) [154]:

$$f_h(x.y.z) = s(x.y.z)f_1(x.y.z) + (1 - s(x.y.z))f_2(x.y.z), \tag{7.73}$$

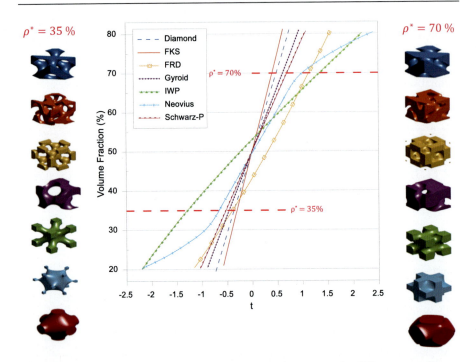

Figure 7.63 Relation between density and surface level parameter for different type of TPMS structures [90].

where f_h is the new hybridized structure function, f_1 and f_2 are the parent structure functions, and $s(x, y, z)$ is a weighting function that controls the transition between two structures. Similarly the surface level parameter of the hybrid structure should be considered as

$$t_h(x.y.z) = s(x.y.z)t_1(x.y.z) + (1 - s(x.y.z))t_2(x.y.z), \qquad (7.74)$$

where t_h, t_1, and t_2 are the associated surface level parameters of f_h, f_1, and f_2, respectively. The sigmoid function Eq. (7.75) is an appropriate function to define a smooth transition between two structures given by [154]

$$s(x.y.z) = \frac{1}{1 + \exp[\alpha g(x.y.z)]}. \qquad (7.75)$$

The g function determines what the boundary between the two parent structures looks like, and the α coefficient determines how fast the transition between the two structures occurs. If g is the equation of a sphere ($g = x^2 + y^2 + z^2 - R^2$), the boundary between the two parent structures will be a sphere. Assume that the transition between the two structures only happens in the z-direction the function $g(x.y.z)$ simply reduces to $g(z)$, and by setting the function g as $g(z) = z - a$ the transition boundary will be placed around the $z = a$ plane. In here, hybridization using two g

Microstructure hull and design 395

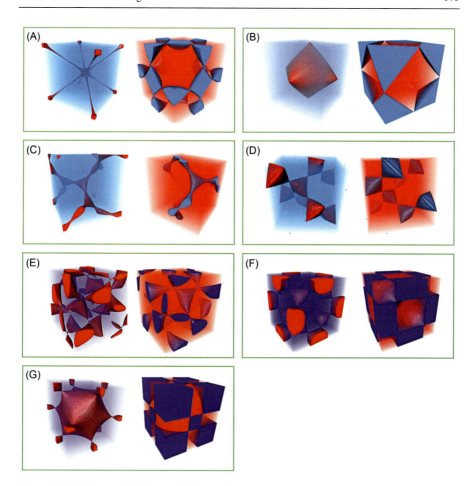

Figure 7.64 The two limiting conditions where solid and void phases remain connected. The red and blue colors indicate solid and void phases, respectively, for A: IWP, B: Schwarz-P, C: Gyroid, D: diamond, E: FKS, F: FRD, and G: Neovius structures [90].

Table 7.27 Critical densities and surface level parameters for different types of TPMS structures [90].

Structure	$t_1[-]$	$\rho_1^*[\%]$	$t_2[-]$	$\rho_2^*[\%]$
IWP	−2.99	0.78	2.99	89.83
Schwarz-P	−1	21.31	1	78.68
Gyroid	−1.4	2.2	1.4	97.76
diamond	−0.99	8.5	0.99	91.5
FRD	−1.1	21.42	1.0	66.7
FKS	−0.75	11.48	0.75	88.52
Neovius	0.75	33.26	0.75	66.16

functions ($g = x^2 + y^2 + z^2 - R^2$ and $g = z - a$) will be investigated and they will be referred to as spherical hybrids and laminated hybrids, respectively.

It should be noted that it is necessary to carefully set the relative position of two structures in order to achieve a smooth transition between two structures.

7.4.3.3 Finite element model

An integrated program was developed for automatic voxel-based structure generation for structural analysis of TPMS structures inside commercial finite element software ABAQUS and also to calculate the effective stiffness tensor. At the first step, using the program a cubic structure with dimensions of $L_x = 2\pi n_x l_x$, $L_y = 2\pi n_y l_y$, and $L_z = 2\pi n_z l_z$ is created, where n_i and $2\pi l_i$ are the number of unit cells and length of the unit cells, respectively, which are discretized into N_x, N_y, and N_z voxels in each direction. Each voxel can be represented by its voxel count as (i,j,k), where $0 \leq i.j.k \leq N_i - 1$. Then the coordinate of the center of each voxel (which should be used in Eqs. (7.66)–(7.72) as $(x.y.z)$) can be presented as $(\alpha_x(i+0.5).\alpha_y(j+0.5).\alpha_z(k+0.5))$, where $\alpha_i = L_i/N_i$.

In the finite element model, each voxel was treated as a trilinear brick (C3D8R) element. Two element sets for solid and void phases were created, and appropriate mechanical properties were assigned to each element set. To avoid numerical difficulties an airy material with an elastic modulus of 0.1 Pa was assigned to the void phase and a linear elastic model with $E = 72$GPa and $\nu = 0.33$ was used to represent the solid phase. The iterative implicit solver was selected for this problem, and six loading cases (three axial and three shear loadings) were used to obtain independent constants of macroscopic stiffness tensor. Fig. 7.65 illustrates a set of loading and boundary conditions applied in each case. In each step, one of the loading cases is applied to the structure and the mean stress and strain were computed using [160]

$$\bar{\sigma} = \frac{1}{V} \int_V \sigma dV,$$

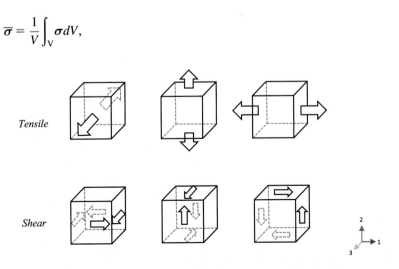

Figure 7.65 Boundary conditions for tensile and shear loadings in FEM [90].

$$\bar{\varepsilon} = \frac{1}{V} \int_V \varepsilon dV, \tag{7.76}$$

where V is the volume of the RVE. Then the effective stiffness tensor of the structures ($\mathbb{C}^*$) was computed using [160]

$$\bar{\sigma} = \mathbb{C}^* : \bar{\varepsilon}. \tag{7.77}$$

Considering the unit vector d the effective elastic modulus of the structure in the direction of the vector d is given by [161,162]

$$\frac{1}{E*(d)} = (d \otimes d) : \mathbb{S}^* : (d \otimes d), \tag{7.78}$$

where $E*(d)$ and S^*_{ijkl} are the effective directional elastic modulus and effective compliance tensor of the structure ($\mathbb{S} = \mathbb{C}^{-1}$), respectively. Further the effective shear modulus and Poisson's ratio in a plane with normal vector d and in the direction of vector n are given by [161,162]

$$\frac{1}{2G*(d.n)} = M : \mathbb{S}^* : M, \qquad M = \frac{\sqrt{2}}{2}(n \otimes d + d \otimes n), \tag{7.79}$$

$$\frac{-\nu^*(d,n)}{E^*(d)} = (d \otimes d) : \mathbb{S}^* : (n \otimes n), \tag{7.80}$$

7.4.3.4 Cell tessellation and mesh sensitivity analysis

The number of tessellated cells in a structure will affect its mechanical properties. As the number of cells increases, the effect of boundaries can be ignored and therefore the mechanical properties practically do not change. Maskeri et al. [163] showed that structures with $4 \times 4 \times 4$ tessellated cells can give a very good estimate of the elastic behavior of lattice TPMS structures. Therefore in this research, $4 \times 4 \times 4$ lattice arrangement has been used. Also, based on the mesh sensitivity analysis (Fig. 7.66), it was found that the use of 128 voxels in each direction ($\sim 2{,}100{,}000$ elements in total) is sufficient, and in addition to the convergence of the properties, it also has a reasonable computational cost.

7.4.3.5 Results and discussion

7.4.3.5.1 Directional elastic modulus

Effective directional properties of the TPMS-based structures were computed following the methodology described in the previous section. It is well known that the deformation of the cellular materials is a combination of the bending and stretching (or compressing) of their ligaments [164]. Therefore the structure's stiffness may vary in different directions based on the governing deformation mode in that

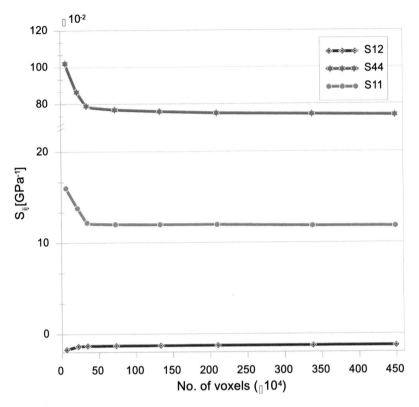

Figure 7.66 Sensitivity analysis of the compliance tensor components with respect to the total number of voxels in the RVE [90].

direction. For the materials with a random structure due to the random distribution of the ligament size and orientation, an almost uniform distribution of the stiffness is expected. In contrast, in TPMS structures the stiffness of the structure is dependent on the load direction.

Analysis of the results shows that with a good approximation the effective stiffness tensor of the seven TPMS-based structures has cubic symmetry as reported elsewhere [165]. The stiffness tensor has six symmetry transformations, including 90 degree reflection and rotation about all three orthogonal planes. These groups of symmetries lead to only three independent components of the stiffness tensor, which are summarized in Eq. (7.81) using Voigt notation [166]

$$C = \begin{bmatrix} C_{11} & C_{12} & C_{12} & 0 & 0 & 0 \\ & C_{11} & C_{12} & 0 & 0 & 0 \\ & & C_{11} & 0 & 0 & 0 \\ & & & C_{44} & 0 & 0 \\ & \text{sym} & & & C_{44} & 0 \\ & & & & & C_{44} \end{bmatrix} \quad (7.81)$$

Fig. 7.67A shows the elastic modulus of the Schwarz-P structure. As is shown, there is a considerable difference between the minimum and maximum directional elastic moduli for all ranges of solid-phase volume fractions. Fig. 7.68A shows the surface plot of directional elastic modulus of the Schwarz-P structure for 22%–78% volume fraction of the solid phase. As is clear the strong directions of this structure are aligned with coordinate axis directions (i.e., $[\pm 1.0.0].[0. \pm 1.0]$ and $[0.0. \pm 1]$). In contrast the eight weak directions of the structure are aligned with diagonal directions (i.e., $[\pm 1. \pm 1. \pm 1]$). For other directions the elastic modulus of the structure varies between these two extrema. The power function fitted for the highest and the lowest elastic modulus is also plotted in Fig. 7.67A. These functions closely approximate the extrema values of elastic modulus of the Schwarz-P

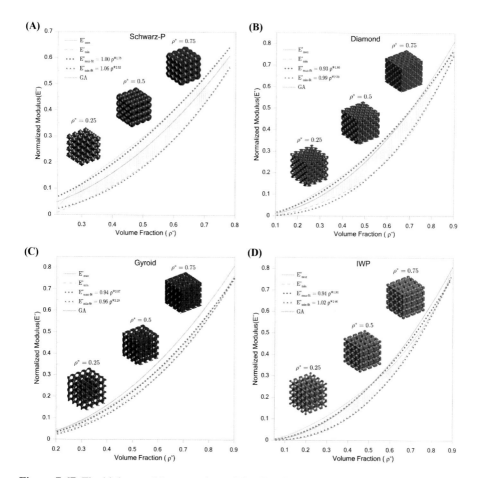

Figure 7.67 The highest and lowest values of the directional elastic modulus for (A) Schwarz-P, (B) diamond, (C) Gyroid, (D) IWP, (E) FKS, (F) FRD, and (G) Neovius. Moreover the power fitted function to lower and higher bounds and GA approximation is plotted for each structure [90].

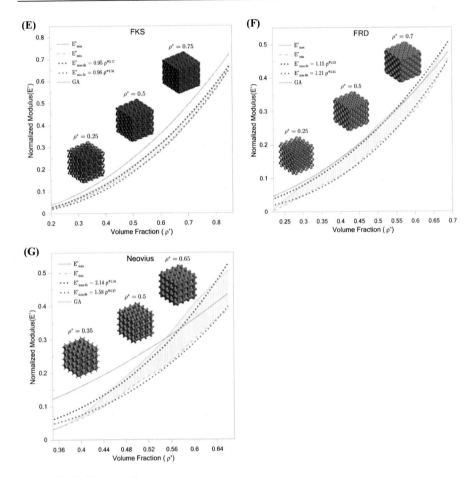

Figure 7.67 (Continued).

structure. The classical Gibson–Ashby (GA) power equation ($E* = \rho^{*2}$) [167], also shown in the figure, predicts a value between the maximum and minimum elastic moduli of the Schwarz-P structure.

Fig. 7.67B shows the maximum and minimum bounds of the directional elastic modulus of the diamond structure. Unlike the Schwarz-P structure, in the diamond structure, as shown in Fig. 7.68B, the strong and weak directions are diagonal and orthogonal, respectively. The ratio between maximum and minimum directional elastic moduli decreases as the volume fraction of the solid phase increases; however, compared to other structures, shown in Fig. 7.69A, this ratio is higher for solid-phase volume fractions of more than 50%. Further the GA approximation is close to the maximum bound for this structure. The directional elastic modulus of the gyroid structure is shown in Fig. 7.67C. Similar to the diamond structure, as shown in Fig. 7.68C, the strong and weak directions of this structure are diagonal and orthogonal, respectively. The IWP structure shows a strong directional dependency of the elastic

Microstructure hull and design 401

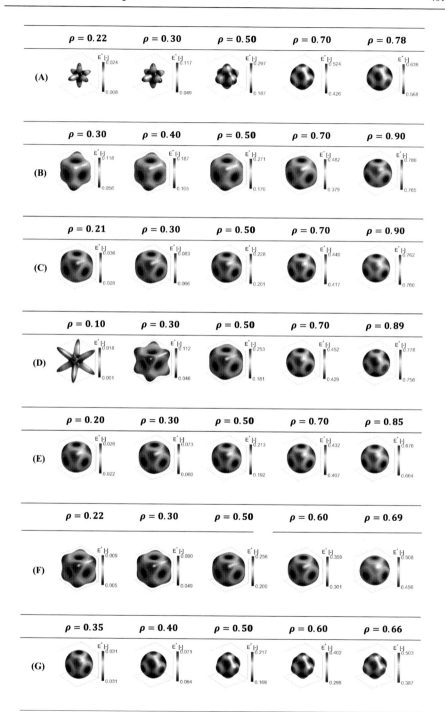

Figure 7.68 Surface plot of the normalized directional elastic moduli at specified volume fractions for (A) Schwarz-P, (B) diamond, (C) Gyroid, (D) IWP, (E) FKS, (F) FRD, and (G) Neovius [90].

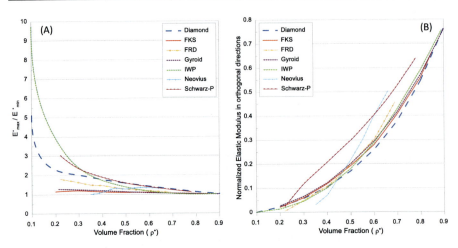

Figure 7.69 (A) The ratio of the maximum to minimum directional elastic moduli and (B) normalized elastic modulus in orthogonal directions for different TPMS structures [90].

modulus for the low volume fractions of the solid phase as shown in Fig. 7.69A, where the ratio of the maximum to minimum directional elastic moduli is about 10 for structures with 10% solid phase. The maximum and minimum bounds and the power fitted functions of the directional elastic modulus are shown in Fig. 7.67D, where similar to the diamond and gyroid structures the strong and weak directions are diagonal and orthogonal, respectively (c. f. Fig. 7.68D). As an interesting observation, for the high volume fractions of the IWP (i.e., more than 80%), unlike lower volume fractions, the strong and weak directions change and are in the orthogonal and diagonal directions, respectively. However, in such a high volume fraction the difference between the minimum and maximum elastic moduli is negligible. The directional elastic modulus of the FKS structure is shown in Fig. 7.68E. In addition to its geometric resemblance to the gyroid, FKS has a similar elastic modulus behavior. The directional elastic modulus is almost uniform, and the ratio of the maximum to minimum moduli is almost constant at different volume fractions. However, as shown in Fig. 7.69A, this value slightly decreases with increasing volume fraction. Also, in this structure the strong and weak directions are diagonal and orthogonal, respectively. The FRD structure, as shown in Fig. 7.68F, has a diagonal strong direction. As shown in Fig. 7.67F the Gibson equation is very close to the upper limit of the elastic modulus, and it predicts the behavior well, especially around the middle volume fractions. The Neovius structure behaves different from other structures (c. f. Figs. 7.67G and 7.68G). This structure has a sphere-like geometry in low volume fractions, and with increasing volume fraction and growth of its ligaments the anisotropy of the structure increases. However, this trend changes in about 60% volume fraction, and with increasing volume fraction the anisotropy decreases as shown in Fig. 7.69A. Like the Schwarz-P structure, this structure has a larger modulus in the orthogonal direction and a smaller modulus in the diagonal direction. However, according to Fig. 7.68G, this seems to be the opposite in volume fractions less than 35%.

Fig. 7.69B shows the elastic modulus of TPMS structures in orthogonal directions. As shown in the figure, Schwarz and Neovius structures (where their stiffness in the orthogonal direction is maximum) have more moduli in a wide range of volume fractions than other structures. On the other hand, the diamond structure in volume fractions of more than 50% has the lowest elastic modulus among the structures. This has also been observed in [156], where three structures of Schwarz, diamond, and IWP in the volume fraction range of 15%−55% are studied. The diagram presented in the paper shows that the Schwarz structure has the highest elastic modulus in [97] the direction for the entire range of volume fractions and the IWP and diamond structures have the lowest modulus in volume fractions less than 45% and more than 45%, respectively. Also, Yang et al. [168], by examining the gyroid structure, concluded that this structure shows the highest and lowest elastic moduli in the [108] and [001] directions, respectively, which is consistent with our findings.

The 2D and 3D representations of the shear modulus and the Poisson's ratio of some of the structures are given in Fig. 7.70. As shown in Eqs. (7.79)−(7.80), two orthogonal vectors are required to determine the shear modulus and Poisson's ratio. For 2D and 3D representations of the Poisson's ratio and shear modulus, only the maximum and minimum values are displayed for each direction. Also, to distinguish between positive minimum and negative minimum values of the Poisson's ratio in 2D plots, negative values are drawn with a different color. In IWP and FRD structures with low volume fractions (volume fractions less than 40% for IWP and less than 25% for FRD), a negative Poisson's ratio was observed.

7.4.3.5.2 Hybridization

As shown in the previous part the directional elastic stiffness tensor of all structures has almost a cubic symmetry. Moreover the strong and weak directions of the Schwarz-P and Neovius structures are laid in axial and diagonal directions, respectively. In contrast, in five other structures the weak and strong directions are in the reverse order. Therefore it is possible to combine the Schwarz-P/Neovius structure with any of the other five structures to achieve a structure with a more uniform directional elastic modulus. Relying on a smooth transition between two TPMS structures a laminate-type and a matrix spherical inclusion-type hybrid structure are created.

As mentioned earlier in section 6.2, for a smooth transition between structures the relative position of them must be determined. Considering different structures with the same unit cell dimensions, Table 7.28 shows our recommended parameters for both spherical and laminate-type hybridization that can be used to generate hybrid structures with a smooth transition. Here, in order to reduce the transition region a large value of alpha is considered ($\alpha = 10$). Fig. 7.71 shows the laminated and spherical hybrid structures of Schwarz-P/Neovius-Gyroid (S/N-G), Schwarz-P/Neovius-Diamond (S/N-D), Schwarz-P/Neovius-IWP (S/N-I), Schwarz-P/Neovius-FRD (S/N-FRD), and Schwarz-P/Neovius-FKS (S/N-FKS) using parameters presented in Table 7.28.

In studying the mechanical properties of hybrid structures in this section the modeling of the transition region and the modeling of the geometry of the parent structures are neglected, and instead the stiffness tensors obtained in the previous section for each structure and in different volume fractions are used. By assigning

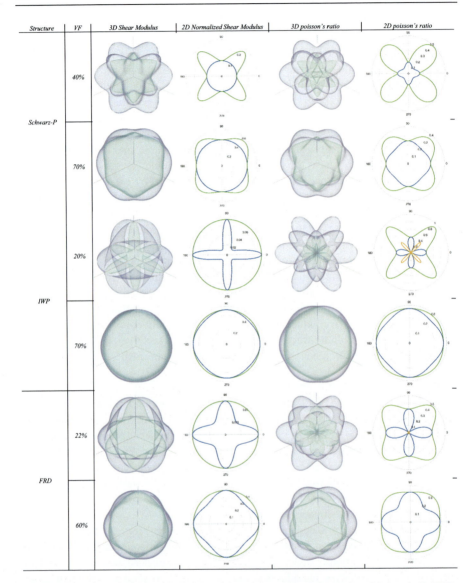

Figure 7.70 3D representation of the shear modulus and Poisson's ratio of Schwarz-P, IWP, and FRD and their projections on the XY-plane. The blue, yellow, and green curves represent the positive minimum, negative minimum, and maximum, respectively [90].

the stiffness tensors to a layered structure that is stacked in one direction a laminated hybrid structure is created (see Fig. 7.72A). The spherical hybrids are modeled by assigning the tensors to spheres within the RVE (see Fig. 7.72B). More details on modeling both types of hybrids are given in the following.

In order to calculate the effective elastic modulus of the laminate-type structures a cubic RVE was modeled and was divided into n sections in the z-direction as illustrated

Table 7.28 Recommended parameters for hybrid structures (f_1 is related to Neovius or Schwarz-P structures, and f_2 is one of the five remaining structures) [90].

| Spherical hybrid equation | $f_{sph-hyb} = s(z)f_1(x.y.z) + [1 - s(z)]f_2(x - x_c.y - y_c.z - z_c); s(z) = \frac{1}{1 + \exp(\alpha*(x^2 + y^2 + z^2 - r^2))}$ | | | | | |
| Laminated hybrid equation | $f_{lam-hyb} = s(z)f_1(x.y.z) + [1 - s(z)]f_2(x - x_c.y - y_c.z); s(z) = \frac{1}{1 + \exp(\alpha z)}$ | | | | | |

Hybrid structure	x_c	y_c	Hybrid structure	x_c	y_c	z_c
Lam-S-G	$-\pi/10$	0	Sph-S-G	$\pi/10$	$\pi/10$	$\pi/10$
Lam-S-I	$4\pi/25$	$4\pi/25$	Sph-S-I	0	0	0
Lam-S-D	$9\pi/8$	$9\pi/8$	Sph-S-D	$3\pi/8$	$3\pi/8$	$3\pi/8$
Lam-S-FRD	$4\pi/25$	0	Sph-S-FRD	0	0	0
Lam-S-FKS	$2\pi/25$	0	Sph-S-FKS	$\pi/10$	$\pi/10$	$\pi/10$
Lam-N-G	0	0	Sph-N-G	0	0	0
Lam-N-I	$2\pi/25$	$2\pi/25$	Sph-N-I	$\pi/7$	$\pi/7$	$\pi/7$
Lam-N-D	0	0	Sph-N-D	$\pi/6$	$\pi/6$	$\pi/6$
Lam-N-FRD	0	0	Sph-N-FRD	$\pi/4$	$\pi/4$	$\pi/4$
Lam-N-FKS	0	0	Sph-N-FKS	0	0	0

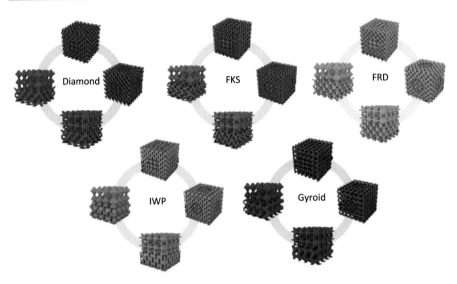

Figure 7.71 Laminated and spherical hybrid structures. In each structure the right image shows a spherical hybrid with Schwarz-P, and the top image shows a spherical hybrid with Neovius. The left image shows a laminated hybrid with Schwarz-P, and the bottom image shows a laminated hybrid with Neovius [90].

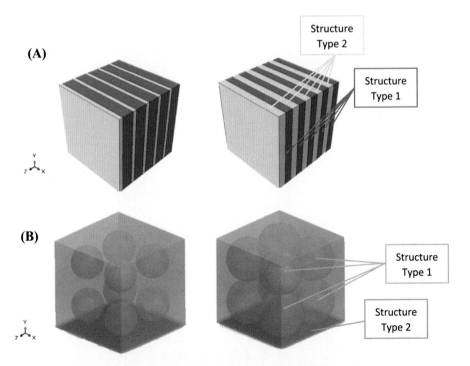

Figure 7.72 Hybrid structures with equal contribution of the parent structures (right) and unequal contribution of parent structures (left) for (A) laminate-type and (B) spherical-type hybrid structures [90].

Microstructure hull and design

in Fig. 7.72A. Then the homogenized stiffness tensors of Schwarz-P (or Neovius) and one of the other five structures were assigned to each two successive layers.

For spherical hybrid structures a cubic RVE was modeled consisting of m^3 smaller equal cubes (each edge of the cube was divided into m equal parts) with a sphere at the center of each one as illustrated in Fig. 7.72B. The homogenized stiffness tensor of the Schwarz-P (or Neovius) structure was assigned to the spheres, and the stiffness tensor of one of the other five structures was assigned to the remaining space inside the cube.

Homogenized stiffness tensors of the hybrid structures were calculated based on the procedure introduced in section 6.3. Fig. 7.72 shows spherical and laminate-type hybrid structures with different combination ratios of the parent structures.

A sensitivity analysis was performed to ensure that a sufficient number of layers (for laminate-type hybridization) and spheres (for spherical-type) are selected. Fig. 7.73 shows the variation of the different components of the stiffness tensor versus the number of layers/spheres of the hybrid structure for S-G with 50% of solid-phase volume fraction of the parent structures with equal contribution of each structure ($50\%S_{50\%}-50\%G_{50\%}$). As shown in Fig. 7.73, by increasing the number of layers from 10 to 12 and the number of spheres (in each direction) from 2 to 4 the relative variation of all stiffness components was less than 0.2%. Thus for the rest of this study, laminate-type structures with 12 layers and matrix-spherical inclusion structures with 2 spheres in each direction are considered.

Universal anisotropy index (AU) is used to express the anisotropy of the structures in here. Universal anisotropy index is a measure to quantify the anisotropy of all classes of crystals, and it is defined by [169,170]

$$A^U = \mathbb{C}^V : \mathbb{S}^R - 6 = 5\frac{G^V}{G^R} + \frac{K^V}{K^R} - 6 \geq 0, \qquad (7.82)$$

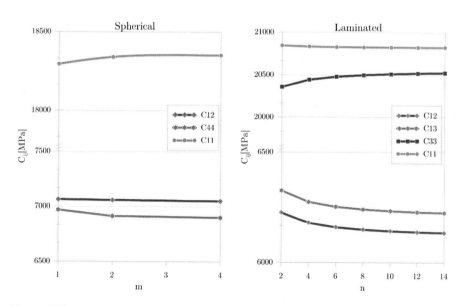

Figure 7.73 Sensitivity analysis of the stiffness tensor components with respect to the number of layers (n) for laminate structures and the number of cubes (m) in each direction for spherical-type hybrid structures [90].

where $\mathbb{C}$, $\mathbb{S}$, G, and K are the stiffness tensor, compliance tensor, shear modulus, and bulk modulus, respectively, and the superscripts V and R denote the Voigt and Reuss estimates. For isotropic crystals the A^U is 0. A larger value of A^U represents a higher state of anisotropy for the corresponding crystal.

Fig. 7.74 shows the 3D representation of the Young modulus for several matrix-spherical inclusion and laminated hybrid structures with different combination ratios of the parent structures. As observed in the figure, hybridized structures (both

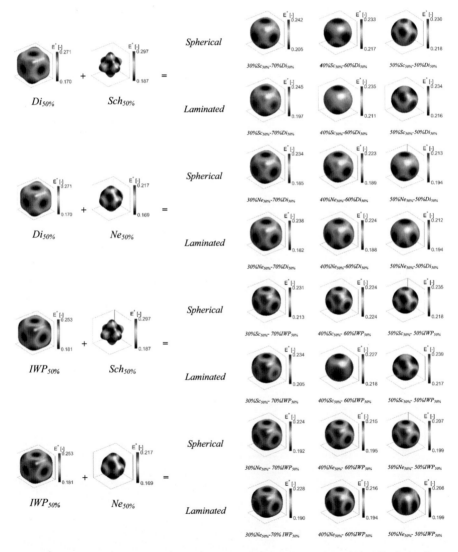

Figure 7.74 Surface plot of the directional elastic modulus for different combination ratios of the parent structures [90].

spherical and laminated) lead to reduced anisotropy and more uniform directional elastic modulus values. In laminated-type hybrids the two structures were combined in series in the z-direction and in parallel in the x- and y-directions. For this reason the properties in the z-direction are different from that of x- and y-directions. However, in the spherical type, because the hybridization of the structures was the same in all directions, more uniform properties were obtained. From this point of view, this is an advantage for this type of hybridization. As an example the universal anisotropy index for Neovius and diamond at 40% solid-phase volume fraction was 0.02 and 0.44, respectively, while this index for the laminated hybrid structures with combination ratios of 30%Sc−70%Di, 40%Sc−60%Di, and 50%Sc−50%Di is 0.14076, 0.128, and 0.086, respectively, and for the spherical-type, it is 0.17286, 0.10969, and 0.060719, respectively. This means that hybrid structures have a more uniform directional elastic modulus distribution than their parent structure. However, the most uniform distribution can be achieved by selecting the appropriate ratio of the combination of the structures.

Fig. 7.75 shows the effect of the combination ratio on the anisotropy of the hybrid structures for different volume fractions of the parent structures. For this purpose, hybrid structures with 30%, 40%, and 50% of the solid phase in the parent structures were considered, and the effect of the combination ratio of parent

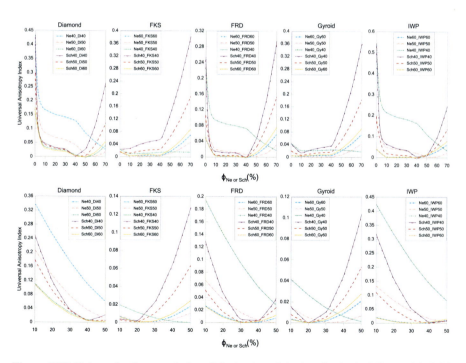

Figure 7.75 Universal anisotropy index of the laminated (top) and spherical (bottom) hybrid structures for different combination ratios of the parent structures. (Φ defines the contribution of the Schwarz or Neovius in the hybridized structure) [90].

structures was investigated. By increasing the contribution of the Schwarz-P/ Neovius in the hybrid structure, first the anisotropy of the hybridized structure decreases, and then after reaching a minimum value, it starts to increase. This minimum contribution of the Schwarz-P/Neovius structure can be considered an optimum point where the hybrid structure has the minimum anisotropy. As shown by the results the optimum combination ratio of a hybrid structure depends on the type of the parent structures and their solid-phase volume fraction.

7.5 Conclusion and remarks

In this chapter, we have introduced a variety of numerical techniques for the optimal design of periodic and nonperiodic heterogeneous materials.

For periodic microstructures, we resort to two methods; first an effective optimum design procedure was proposed for 3D-printable PUC of heterogeneous materials using SIMP-OCA methodology. The main advantage of the method is a capability in designing 3D microstructures with prescribed elastic properties and manufacturability of the generated microstructures with properties close to those of the optimized structure; second, topology optimization was exploited to design a structure with tunable properties. It should be stated that the 3D-printable microstructure design is a remarkable ability of these proposed methodologies. The proposed methodologies can be implemented to the design of multiphase microstructures composed of two or more solid phases. Designing hybrid TPMS microstructures has been explained, and useful sample microstructures have been realized.

For nonperiodic microstructures, FFT methods have been developed for 3D reconstruction, homogenization, and materials design of heterogeneous materials based on the full-set TPCF. The methodology can be applied to multiphase, multifunctional complex random heterogeneous materials. Due to the few design variables the method is very efficient and has a good computational performance and a low computational cost.

References

[1] Li H, Li H, Xiao M, Zhang Y, Fu J, Gao L. Robust topology optimization of thermoelastic metamaterials considering hybrid uncertainties of material property. Compos Struct 2020;248:112477. Available from: https://doi.org/10.1016/j.compstruct.2020.112477.

[2] Peng Y, Wei K, Mei M, Yang X, Fang D. Simultaneously program thermal expansion and Poisson's ratio in three dimensional mechanical metamaterial. Compos Struct 2020;113365. Available from: https://doi.org/10.1016/j.compstruct.2020.113365.

[3] Zhang J, Yanagimoto J. Topology optimization of microlattice dome with enhanced stiffness and energy absorption for additive manufacturing. Compos Struct 2021;255:112889. Available from: https://doi.org/10.1016/j.compstruct.2020.112889.

[4] Wang F, Sigmund O. Extreme 3D architected isotropic mater tunable stiffness buckling strength, 2020.

[5] Yang H, Wang B, Ma L. Designing hierarchical metamaterials by topology analysis with tailored Poisson's ratio and Young's modulus. Compos Struct 2019;214:359−78. Available from: https://doi.org/10.1016/j.compstruct.2019.01.076.

[6] Giraldo-Londoño O, Paulino GH. Fractional topology optimization of periodic multi-material viscoelastic microstructures with tailored energy dissipation. Comput Methods Appl Mech Eng 2020;372:113307. Available from: https://doi.org/10.1016/j.cma.2020.113307.

[7] Saber Hashemi M, Safdari M, Sheidaei A. A supervised machine learning approach for accelerating the design of particulate. Compos Appl Therm Conduct 2020;.

[8] Shi Z, Tsymbalov E, Dao M, Suresh S, Shapeev A, Li J. Deep elastic strain engineering of bandgap through machine learning. Proc Natl Acad Sci 2019;116(10):4117−22. Available from: https://doi.org/10.1073/pnas.1818555116.

[9] Bessa MA, Glowacki P, Houlder M. Bayesian machine learning in metamaterial design: fragile becomes supercompressible. Adv Mater 2019;31(48):1904845. Available from: https://doi.org/10.1002/adma.201904845.

[10] Gu GX, Chen C-T, Richmond DJ, Buehler MJ. Bioinspired hierarchical composite design using machine learning: simulation, additive manufacturing, and experiment. Mater Horiz 2018;5(5):939−45. Available from: https://doi.org/10.1039/C8MH00653A.

[11] Riazat M, Tafazoli M, Baniassadi M, Safdari M, Faraji G, Garmestani H. Investigation of the property hull for solid oxide fuel cell microstructures. Computat Mater Sci 2017;127:1−7.

[12] Tafazoli M, Shakeri M, Baniassadi M, Babaei A. Investigation of the geometric property hull for infiltrated solid oxide fuel cell electrodes. Int J Energy Res 2017;41(14):2318−31.

[13] Hasanabadi A, Baniassadi M, Abrinia K, Safdari M, Garmestani H. Optimal Combining of Microstructures Using Statistical Correlation Functions. Int J Solids Struct 2019;160:177−86.

[14] Bendsøe MP, Sigmund O. Topology optimization theory, methods, and applications. 2 ed. Springer-Verlag Berlin Heidelberg; 2004.

[15] Bendsøe MP, Kikuchi N. Generating optimal topologies in structural design using a homogenization method. Comput Methods Appl Mech Eng 1988;71(2):197−224. Available from: https://doi.org/10.1016/0045-7825(88)90086-2.

[16] Torquato S. Optimal design of heterogeneous materials. In: Clarke DR, Ruhle M, Zok F, editors. Annual review of materials research, Vol 40. Annual Review of Materials Research; 2010. p. 101−29.

[17] Guest JK, Prévost JH. Optimizing multifunctional materials: design of microstructures for maximized stiffness and fluid permeability. Int J Solids Struct 2006;43 (22−23):7028−47. Available from: https://doi.org/10.1016/j.ijsolstr.2006.03.001.

[18] Adams B.L., Kalidindi S.R., Fullwood, David T. Microstructure-sensitive design for performance optimization. Waltham, MA: Butterworth-Heinemann; 2013.

[19] Torquato S. Random heterogeneous materials: microstructure and macroscopic properties. New York: Springer-Verlag; 2002.

[20] Fullwood DT, Niezgoda SR, Adams BL, Kalidindi SR. Microstructure sensitive design for performance optimization. Prog Mater Sci 2010;55(6):477−562.

[21] Lu B, Torquato S. Lineal-path function for random heterogeneous materials. Phys Rev A 1992;45(2):922.

[22] Matheron G. Random sets and integral geometry. New York: John Wiley & Sons; 1975.

[23] Jiao Y, Stillinger FH, Torquato S. A superior descriptor of random textures and its predictive capacity. Proc Natl Acad Sci U S A 2009;106(42):17634−9. Available from: https://doi.org/10.1073/pnas.0905919106.

[24] Torquato S, Stell G. Microstructure of Two-Phase Random Media. I. The n-Point Probability Functions. J Chem Phys 1982;77:2071−7.

[25] Jiao Y, Stillinger FH, Torquato S. Modeling heterogeneous materials via two-point correlation functions: basic principles. Phys Rev E Stat Nonlin Soft Matter Phys 2007;76 (3 Pt 1):031110. Available from: https://doi.org/10.1103/PhysRevE.76.031110.

[26] Hasanabadi A, Baniassadi M, Abrinia K, Safdari M, Garmestani H. 3D microstructural reconstruction of heterogeneous materials from 2D cross sections: A modified phase-recovery algorithm. Computat Mater Sci 2016;111:107−15. Available from: https://doi.org/10.1016/j.commatsci.2015.09.015.

[27] Baniassadi M, Garmestani H, Li D, Ahzi S, Khaleel M, Sun X. Three-phase solid oxide fuel cell anode microstructure realization using two-point correlation functions. Acta Mater 2011;59(1):30−43.

[28] Hadwiger H. Vorlesungen über inhalt, oberfläche und isoperimetrie. Berlin: Springer-Verlag Berlin Heidelberg; 1975.

[29] Arns CH, Mecke J, Mecke K, Stoyan D. Second-order analysis by variograms for curvature measures of two-phase structures. Eur Phys J B 2005;47(3):397−409. Available from: https://doi.org/10.1140/epjb/e2005-00338-5.

[30] Mecke KR. Additivity, convexity, and beyond: applications of minkowski functionals in statistical physics. In: Mecke KR, Stoyan D, editors. Statistical physics and spatial statistics. Berlin, Heidelberg: Springer Berlin Heidelberg; 2000. p. 111−84.

[31] Matheron G. Eléments pour une théorie des milieux poreux. Paris: Masson; 1967.

[32] Stoyan D, Mecke K. The Boolean Model: from Matheron till Today. Space, structure and randomness. Springer; 2005. p. 151−81.

[33] Arns CH, Knackstedt MA, Mecke KR. Boolean reconstructions of complex materials: integral geometric approach. Phys Rev E Stat Nonlin Soft Matter Phys 2009;80(5 Pt 1):051303. Available from: https://doi.org/10.1103/PhysRevE.80.051303.

[34] Schroder-Turk GE, Kapfer S, Breidenbach B, Beisbart C, Mecke K. Tensorial Minkowski functionals and anisotropy measures for planar patterns. J Microsc 2010;238(1):57−74. Available from: https://doi.org/10.1111/j.1365-2818.2009.03331.x.

[35] Panchal JH, Kalidindi SR, McDowell DL. Key computational modeling issues in integrated computational materials engineering. Comput Des 2013;45(1):4−25. Available from: https://doi.org/10.1016/j.cad0.2012.06.006.

[36] Ruggles TJ, Rampton TM, Rose SA, Fullwood DT. Reducing the microstructure design space of 2nd order homogenization techniques using discrete Fourier Transforms. Mech Mater 2013;59:14−23. Available from: https://doi.org/10.1016/j.mechmat.2012.11.007.

[37] Safdari M, Baniassadi M, Garmestani H, Al-Haik MS. A modified strong-contrast expansion for estimating the effective thermal conductivity of multiphase heterogeneous materials. J Appl Phys 2012;112(11):114318.

[38] Jeulin D. Random structures in physics. Space, structure and randomness. Springer; 2005. p. 183−219.

[39] Beran MJ. Statistical continuum theories, monographs in statistical physics and thermodynamics. New York: Interscience; 1968.

[40] Torquato S. Effective stiffness tensor of composite media − I. Exact series expansion. J Mech Phys Solids 1997;45(9):1421−48.

[41] Sheidaei A, Baniassadi M, Banu M, Askeland P, Pahlavanpour M, Kuuttila N, et al. 3-D microstructure reconstruction of polymer nano-composite using FIB−SEM and statistical correlation function. Compos Sci Technol 2013;80:47−54.

[42] Hasanabadi A, Baniassadi M, Abrinia K, Safdari M, Garmestani H. Efficient three-phase reconstruction of heterogeneous material from 2D cross-sections via phase-recovery

algorithm. J Microsc 2016;264(3):384−93. Available from: https://doi.org/10.1111/jmi.12454.

[43] Hasanabadi A, Baniassadi M, Abrinia K, Safdari M, Garmestani H. Optimization of solid oxide fuel cell cathodes using two-point correlation functions. Computat Mater Sci 2016;123:268−76. Available from: https://doi.org/10.1016/j.commatsci.2016.07.004.

[44] Niezgoda SR, Turner DM, Fullwood DT, Kalidindi SR. Optimized structure based representative volume element sets reflecting the ensemble-averaged 2-point statistics. Acta Mater 2010;58(13):4432−45. Available from: https://doi.org/10.1016/j.actamat.2010.04.041.

[45] Fullwood DT, Niezgoda SR, Kalidindi SR. Microstructure reconstructions from 2-point statistics using phase recovery algorithms. Acta Mater 2008;52:942−8.

[46] Jiao Y, Stillinger FH, Torquato S. Modelling Heterogeneous materials via two-point correlation functions. II. Algorithmic details and applications. Phys Rev E 2008;77:031135.

[47] Cule D, Torquato S. Generating random media from limited microstructural information via stochastic optimization. J Appl Phys 1999;86(6):3428−37.

[48] Liebscher A, Jeulin D, Lantuejoul C. Stereological reconstruction of polycrystalline materials. J Microsc 2015;258(3):190−9. Available from: https://doi.org/10.1111/jmi.12232.

[49] Yeong CLY, Torquato S. Reconstructing random media. II. Three-dimensional media from two-dimensional cuts. Phys Rev E 1998;58(1):224.

[50] Blumenfeld R, Torquato S. Coarse-graining procedure to generate and analyze heterogeneous materials: Theory. Phys Rev E Stat Phys Plasmas Fluids Relat Interdiscip Top 1993;48(6):4492−500. Available from: https://doi.org/10.1103/physreve.48.4492.

[51] Fullwood DT, Kalidindi SR, Niezgoda SR, Fast A, Hampson N. Gradient-based microstructure reconstructions from distributions using fast Fourier transforms. Mater Sci Eng A 2008;494:68−72.

[52] Gerchberg RW, Saxton WO. A practical Algorithm for the determination of phase from image and diffraction plane pictures. OPTIK. 1972;35(2):237−46.

[53] Nosouhi Dehnavi F, Safdari M, Abrinia K, Hasanabadi A, Baniassadi M. A framework for optimal microstructural design of random heterogeneous materials. Computat Mech 2020;1−17.

[54] Moshki A, Ghazavizadeh A, Atai AA, Baghani M, Baniassadi M. 3D-printable unit cell design for cubic and orthotropic porous microstructures using topology optimization based on optimality criteria algorithm. Int J Appl Mech 2018;10(06) 1850060.

[55] Montazerian H, Davoodi E, Asadi-Eydivand M, Kadkhodapour J, Solati-Hashjin M. Porous scaffold internal architecture design based on minimal surfaces: a compromise between permeability and elastic properties. Mater Des 2017;126:98−114.

[56] Kaur M, Yun TG, Han SM, Thomas EL, Kim WS. 3D printed stretching-dominated micro-trusses. Mater Des 2017;134:272−80.

[57] Wang Y, Zhang L, Daynes S, Zhang H, Feih S, Wang MY. Design of graded lattice structure with optimized mesostructures for additive manufacturing. Mater Des 2018;142:114−23.

[58] Hassani B, Hinton E. A review of homogenization and topology optimization I—homogenization theory for media with periodic structure. Comput Struct 1998;69(6):707−17.

[59] Guedes J, Kikuchi N. Preprocessing and postprocessing for materials based on the homogenization method with adaptive finite element methods. Comput Methods Appl Mech Eng 1990;83(2):143−98.

[60] Sigmund O. Design of material structures using topology optimization. Techn Univ Den Lyngby 1994.

[61] Hassani B, Hinton E. A review of homogenization and topology opimization II—analytical and numerical solution of homogenization equations. Comput Struct 1998; 69(6):719−38.

[62] Sigmund O. Tailoring materials with prescribed elastic properties. Mech Mater 1995;20 (4):351−68.

[63] Sigmund O, Torquato S, Aksay IA. On the design of 1−3 piezocomposites using topology optimization. J Mater Res 1998;13(4):1038−48.

[64] Bendsøe MP. Optimal shape design as a material distribution problem. Struct Optim 1989;1(4):193−202.

[65] Zhou M, Rozvany G. The COC algorithm, Part II: Topological, geometrical and generalized shape optimization. Comput Methods Appl Mech Eng 1991;89(1-3):309−36.

[66] Rozvany GI, Zhou M, Birker T. Generalized shape optimization without homogenization. Struct Optim 1992;4(3-4):250−2.

[67] Rozvany G. The SIMP method in topology optimization-theoretical background, advantages and new applications. Proceedings of the eighth symposium on multidisciplinary analysis and optimization conference, 2000;4738.

[68] Svanberg K. The method of moving asymptotes—a new method for structural optimization. Int J Numer Methods Eng 1987;24(2):359−73.

[69] Zhang W, Dai G, Wang F, Sun S, Bassir H. Using strain energy-based prediction of effective elastic properties in topology optimization of material microstructures. Acta Mechan Sin 2007;23(1):77−89.

[70] Challis V, Roberts A, Wilkins A. Design of three dimensional isotropic microstructures for maximized stiffness and conductivity. Int J Solids Struct 2008;45(14-15):4130−46.

[71] Radman A, Huang X, Xie Y. Topological optimization for the design of microstructures of isotropic cellular materials. Eng Optim 2013;45(11):1331−48.

[72] Querin OM, Steven GP, Xie YM. Evolutionary structural optimisation (ESO) using a bidirectional algorithm. Eng Computat 1998.

[73] Yang X, Xie Y, Steven G, Querin O. Bidirectional evolutionary method for stiffness optimization. AIAA J 1999;37(11):1483−8.

[74] Yang X, Huang X, Rong J, Xie Y. Design of 3D orthotropic materials with prescribed ratios for effective Young's moduli. Computat Mater Sci 2013;67:229−37.

[75] Özdemir İ. Topological derivative based optimization of 3D porous elastic microstructures. Computat Mater Sci 2014;81:319−25.

[76] Suresh K. Efficient microstructural design for additive manufacturing. Int Des Eng Techn Conf Comput Inf Eng Conf Am Soc Mech Eng 2014. p. V01AT2A045.

[77] Andreassen E, Lazarov BS, Sigmund O. Design of manufacturable 3D extremal elastic microstructure. Mech Mater 2014;69(1):1−10.

[78] Gao J, Li H, Gao L, Xiao M. Topological shape optimization of 3D micro-structured materials using energy-based homogenization method. Adv Eng Softw 2018;116:89−102.

[79] Chen D, Skouras M, Zhu B, Matusik W. Computational discovery of extremal microstructure families. Sci Adv 2018;4(1). Available from: https://doi.org/10.1126/sciadv.aao7005.

[80] Ming Z, Rozvany G. DCOC N. An optimality criteria method for large system, Part I: theory [J]. Struct Optim 1992;5(1):12−25.

[81] Zhou M, Rozvany G. DCOC: an optimality criteria method for large systems Part II: algorithm. Struct Optim 1993;6(4):250−62.

[82] Bendsøe MP, Sigmund O. Optimization of structural topology, shape, and material. Springer; 1995.

[83] Sigmund O. A 99 line topology optimization code written in MATLAB. Struct Multidiscip Optim 2001;21(2):120−7.

[84] Andreassen E, Clausen A, Schevenels M, Lazarov BS, Sigmund O. Efficient topology optimization in MATLAB using 88 lines of code. Struct Multidiscip Optim 2011;43(1):1−16.

[85] Sigmund O. Morphology-based black and white filters for topology optimization. Struct Multidiscip Optim 2007;33(4-5):401–24.

[86] Xia L, Breitkopf P. Recent advances on topology optimization of multiscale nonlinear structures. Arch Computat Methods Eng 2017;24(2):227–49.

[87] Kalpakjian S, Schmid S. Manufacturing, engineering and technology SI 6th Edition-Serope Kalpakjian and Stephen Schmid: manufacturing, engineering and technology. Digital Des 2006.

[88] Sigmund O. On the design of compliant mechanisms using topology optimization. J Struct Mech 1997;25(4):493–524.

[89] Hajighasemi MR, Safarabadi M, Sheidaei A, Baghani M, Baniassadi M. Design and Manufacture of a smart macro-structure with changeable effective stiffness. Int J Appl Mech 2020;12(01):2050001.

[90] Khaleghi S, Dehnavi FN, Baghani M, Safdari M, Wang K, Baniassadi M. On the directional elastic modulus of the TPMS structures and a novel hybridization method to control anisotropy. Mater Des 2021;210:110074. Available from: https://doi.org/10.1016/j.matdes.2021.110074.

[91] Baniassadi M, Ahzi S, Garmestani H, Ruch D, Remond Y. New approximate solution for N-point correlation functions for heterogeneous materials. J Mech Phys Solids 2012;60(1):104–19.

[92] Janardhanan VM, Heuveline V, Deutschmann O. Three-phase boundary length in solid-oxide fuel cells: a mathematical model. J Power Sources 2008;178(1):368–72.

[93] Sebdani MM, Baniassadi M, Jamali J, Ahadiparast M, Abrinia K, Safdari M. Designing an optimal 3D microstructure for three-phase solid oxide fuel cell anodes with maximal active triple phase boundary length (TPBL. Int J Hydrog Energy 2015;40(45):15585–96. Available from: https://doi.org/10.1016/j.ijhydene.2015.09.086.

[94] Fleig J, Maier J. The polarization of mixed conducting SOFC cathodes: effects of surface reaction coefficient, ionic conductivity and geometry. J Eur Ceram Soc 2004;24(6):1343–7.

[95] Anselmi-Tamburini U, Chiodelli G, Arimondi M, Maglia F, Spinolo G, Munir Z. Electrical properties of Ni/YSZ cermets obtained through combustion synthesis. Solid State Ion 1998;110(1):35–43.

[96] Zheng K, Ni M. Reconstruction of solid oxide fuel cell electrode microstructure and analysis of its effective conductivity. Sci Bull 2016;61(1):78–85.

[97] Rhazaoui K, Cai Q, Kishimoto M, Tariq F, Somalu MR, Adjiman C, et al. Towards the 3D modelling of the effective conductivity of solid oxide fuel cell electrodes—validation against experimental measurements and prediction of electrochemical performance. Electrochim Acta 2015;168:139–47.

[98] Chen X, Khor K, Chan S, Yu L. Influence of microstructure on the ionic conductivity of yttria-stabilized zirconia electrolyte. Mater Sci Eng A 2002;335(1):246–52.

[99] He W, Lv W, Dickerson J. Gas transport in solid oxide fuel cells. Springer; 2014.

[100] Zhao F, Armstrong TJ, Virkar AV. Measurement of O 2 N 2 effective diffusivity in porous media at high temperatures using an electrochemical cell. J Electrochem Soc 2003;150(3):A249–56.

[101] Adams BL, Kalidindi S, Fullwood DT. Microstructure-sensitive design for performance optimization. Butterworth-Heinemann; 2013.

[102] Tanner CW, Fung KZ, Virkar AV. The effect of porous composite electrode structure on solid oxide fuel cell performance I. Theoretical analysis. J Electrochem Soc 1997;144(1):21–30.

[103] Kishimoto M, Lomberg M, Ruiz-Trejo E, Brandon NP. Towards the Microstructural Optimization of SOFC Electrodes Using Nano Particle Infiltration. ECS Trans 2014;64(2):93–102.

[104] Kishimoto M, Iwai H, Saito M, Yoshida H. Quantitative evaluation of solid oxide fuel cell porous anode microstructure based on focused ion beam and scanning electron microscope technique and prediction of anode overpotentials. J Power Sources 2011;196(10):4555−63.

[105] Bertei A, Nucci B, Nicolella C. Microstructural modeling for prediction of transport properties and electrochemical performance in SOFC composite electrodes. Chem Eng Sci 2013;101:175−90.

[106] Rüger B, Joos J, Weber A, Carraro T, Ivers-Tiffée E. 3D electrode microstructure reconstruction and modelling. ECS Trans 2009;25(2):1211−20. Available from: https://doi.org/10.1149/1.3205650.

[107] Joos J, Rüger B, Carraro T, Weber A, Ivers-Tiffée E. Electrode reconstruction by FIB/SEM and microstructure modeling. ECS Trans 2010;28(11):81−91. Available from: https://doi.org/10.1149/1.3495834.

[108] Cai Q, Adjiman CS, Brandon NP. Modelling the 3D microstructure and performance of solid oxide fuel cell electrodes: computational parameters. Electrochim Acta 2011;56(16):5804−14.

[109] Jiang SP. Nanoscale and nano-structured electrodes of solid oxide fuel cells by infiltration: advances and challenges. Int J Hydrog Energy 2012;37(1):449−70.

[110] Jiang Z, Xia C, Chen F. Nano-structured composite cathodes for intermediate-temperature solid oxide fuel cells via an infiltration/impregnation technique. Electrochim Acta 2010;55(11):3595−605.

[111] Adler SB, Lane J, Steele B. Electrode kinetics of porous mixed-conducting oxygen electrodes. J Electrochem Soc 1996;143(11):3554−64. Available from: https://doi.org/10.1149/1.1837252.

[112] Babaei A, Jiang SP, Li J. Electrocatalytic promotion of palladium nanoparticles on hydrogen oxidation on Ni/GDC anodes of SOFCs via spillover. J Electrochem Soc 2009;156(9):B1022−9. Available from: https://doi.org/10.1149/1.3156637.

[113] Kishimoto M., Lomberg M., Ruiz-Trejo E., Brandon N.P. Towards the Design-led optimization of solid oxide fuel cell electrodes. ECS conference on electrochemical energy conversion & storage with SOFC-XIV (July 26−31, 2015): ECS; 2015.

[114] Kishimoto M, Lomberg M, Ruiz-Trejo E, Brandon NP. Numerical modeling of nickel-infiltrated gadolinium-doped ceria electrodes reconstructed with focused ion beam tomography. Electrochim Acta 2016;190:178−85.

[115] Hashin Z, Shtrikman S. A variational approach to the theory of the elastic behaviour of multiphase materials. J Mech Phys Solids 1963;11(2):127−40.

[116] Gokhale AM, Tewari A, Garmestani H. Constraints on microstructural two-point correlation functions. Scr Mater 2005;53:989−93.

[117] Torquato S. Exact conditions on physically realizable correlation functions of random media. J Chem Phys 1999;111:8832−7.

[118] Torquato S. Necessary Conditions on Realizable Two-Point Correlation Functions of Random Media. Ind Eng Chem Res 2006;6923−8.

[119] Fienup JR. Reconstruction of an object from the modulus of its Fourier transform. Opt Lett 1978;3(1):27−9. Available from: https://doi.org/10.1364/ol.3.000027.

[120] Ali Hasanabadi, Majid Baniassadi, Karen Abrinia, Masoud Safdari, Garmestani H. Optimization of solid oxide fuel cell cathodes using two-point correlation functions. Computational Mater Sci, 2016.

[121] Sadd MH. Elasticity theory, applications, and numerics. third ed Oxford: Academic Press;; 2014.

[122] Eyre DJ, Milton GW. A fast numerical scheme for computing the response of composites using grid refinement. Eur Phys J App Phys 1999;6(1):41−7.

[123] Moulinec H, Silva F. Comparison of three accelerated FFT-based schemes for computing the mechanical response of composite materials. Int J Numer Methods Eng 2014;97(13):960–85.

[124] Michel JC, Moulinec H, Suquet P. Effective properties of composite materials with periodic microstructure: a computational approach. Comput Methods Appl Mech Eng 1999;172(1-4):109–43.

[125] Colabella L, Ibarra Pino AA, Ballarre J, Kowalczyk P, Cisilino AP. Calculation of cancellous bone elastic properties with the polarization-based FFT iterative scheme. Int J Numer Method Biomed Eng 2017;33(11). Available from: https://doi.org/10.1002/cnm.2879.

[126] Lemaitre S., Salnikov V., Choi D., Karamian P. Computation of thermal properties via 3D homogenization of multiphase materials using FFT-based accelerated scheme. arXiv:150407499. 2015.

[127] Jiao Y, Stillinger FH, Torquato S. Modeling heterogeneous materials via two-point correlation functions: basic principles. Phys Rev E 2007;76(3):31110.

[128] Fullwood DT, Niezgoda SR, Kalidindi SR. Microstructure reconstructions from 2-point statistics using phase-recovery algorithms. Acta Mater 2008;56(5):942–8.

[129] Torquato S. Necessary conditions on realizable two-point correlation functions of random media. Ind Eng Chem Res 2006;45(21):6923–8.

[130] Anglin BS, Lebensohn RA, Rollett AD. Validation of a numerical method based on Fast Fourier Transforms for heterogeneous thermoelastic materials by comparison with analytical solutions. Computat Mater Sci 2014;87:209–17. Available from: https://doi.org/10.1016/j.commatsci.2014.02.027.

[131] Dunant CF, Bary B, Giorla AB, Péniguel C, Sanahuja J, Toulemonde C, et al. A critical comparison of several numerical methods for computing effective properties of highly heterogeneous materials. Adv Eng Softw 2013;58:1–12.

[132] Powell MJD. A direct search optimization method that models the objective and constraint functions by linear interpolation. Springer; 1994. p. 51–67.

[133] Johnson S.G. NLopt nonlinear-optimization package, 2014.

[134] Andreassen E, Andreasen CS. How to determine composite material properties using numerical homogenization. Computat Mater Sci 2014;83:488–95.

[135] Rémond Y., Ahzi S., Baniassadi M., Garmestani H. Applied RVE reconstruction and homogenization of heterogeneous materials. Wiley Online Library; 2016.

[136] Hassani B, Hinton E. A review of homogenization and topology optimization III—topology optimization using optimality criteria. Comput Struct 1998;69(6):739–56.

[137] Amstutz S, Giusti S, Novotny A, de Souza Neto E. Topological derivative for multiscale linear elasticity models applied to the synthesis of microstructures. Int J Numer Methods Eng 2010;84(6):733–56.

[138] Xia L, Breitkopf P. Design of materials using topology optimization and energy-based homogenization approach in Matlab. Struct Multidiscip Optim 2015;52(6):1229–41.

[139] Sigmund O, BENDSOE M. Material interpolation schemes in topology optimization. Archive Appl Mech 1999;69(9/10):635–54.

[140] Liu K, Tovar A. An efficient 3D topology optimization code written in Matlab. Struct Multidiscip Optim 2014;50(6):1175–96.

[141] Boger A, Bisig A, Bohner M, Heini P, Schneider E. Variation of the mechanical properties of PMMA to suit osteoporotic cancellous bone. J Biomater Sci Polym Ed 2008;19(9):1125–42. Available from: https://doi.org/10.1163/156856208785540154.

[142] Bhushan B, Burton Z. Adhesion and friction properties of polymers in microfluidic devices. Nanotechnology 2005;16(4):467.

[143] Nemat-Nasser S, Hori M. Micromechanics: overall properties of heterogeneous materials. Elsevier; 2013.

[144] Swaminathan S, Ghosh S, Pagano N. Statistically equivalent representative volume elements for unidirectional composite microstructures: Part I-Without damage. J Compos Mater 2006;40(7):583−604.

[145] Gitman I, Askes H, Sluys L. Representative volume: existence and size determination. Eng Fract Mech 2007;74(16):2518−34.

[146] Hori M, Nemat-Nasser S. On two micromechanics theories for determining micro-−macro relations in heterogeneous solids. Mech Mater 1999;31(10):667−82.

[147] Liang B, Nagarajan A, Ahmadian H, Soghrati S. Analyzing effects of surface roughness, voids, and particle−matrix interfacial bonding on the failure response of a heterogeneous adhesive. Comput Methods Appl Mech Eng 2019;346:410−39.

[148] Mitchell M. An introduction to genetic algorithms. MIT press; 1998.

[149] Torres J, Cotelo J, Karl J, Gordon AP. Mechanical property optimization of FDM PLA in shear with multiple objectives. JOM 2015;67(5):1183−93.

[150] Kumar SA, Narayan YS. Tensile testing and evaluation of 3D-printed PLA specimens as per ASTM D638 type IV standard. Innovative Design and Development Practices in Aerospace and Automotive Engineering (I-DAD 2018). Springer; 2019. p. 79−95.

[151] Xu S, Shen J, Zhou S, Huang X, Xie YM. Design of lattice structures with controlled anisotropy. Mater Des 2016;93:443−7. Available from: https://doi.org/10.1016/j.matdes.2016.01.007.

[152] Chen Z, Xie YM, Wu X, Wang Z, Li Q, Zhou S. On hybrid cellular materials based on triply periodic minimal surfaces with extreme mechanical properties. Mater Des 2019;183:108109. Available from: https://doi.org/10.1016/j.matdes.2019.108109.

[153] Tancogne-Dejean T, Diamantopoulou M, Gorji MB, Bonatti C, Mohr D. 3D plate-lattices: an emerging class of low-density metamaterial exhibiting optimal isotropic stiffness. Adv Mater 2018;30(45):1803334. Available from: https://doi.org/10.1002/adma.201803334.

[154] Yang N, Quan Z, Zhang D, Tian Y. Multi-morphology transition hybridization CAD design of minimal surface porous structures for use in tissue engineering. Comput Des 2014;56:11−21. Available from: https://doi.org/10.1016/j.cad.2014.06.006.

[155] Yoo D-J, Kim K-H. An advanced multi-morphology porous scaffold design method using volumetric distance field and beta growth function. Int J Precis Eng Manuf 2015;16(9):2021−32. Available from: https://doi.org/10.1007/s12541-015-0263-2.

[156] Maskery I, Aremu AO, Parry L, Wildman RD, Tuck CJ, Ashcroft IA. Effective design and simulation of surface-based lattice structures featuring volume fraction and cell type grading. Mater Des 2018;155:220−32. Available from: https://doi.org/10.1016/j.matdes.2018.05.058.

[157] Lu Y, Zhao W, Cui Z, Zhu H, Wu C. The anisotropic elastic behavior of the widely-used triply-periodic minimal surface based scaffolds. J Mech Behav Biomed Mater 2019;99:56−65. Available from: https://doi.org/10.1016/j.jmbbm.2019.07.012.

[158] Yin H, Liu Z, Dai J, Wen G, Zhang C. Crushing behavior and optimization of sheet-based 3D periodic cellular structures. Compos Part B Eng 2020;182:107565. Available from: https://doi.org/10.1016/j.compositesb.2019.107565.

[159] Jiang W, Liao W, Liu T, Shi X, Wang C, Qi J, et al. A voxel-based method of multiscale mechanical property optimization for the design of graded TPMS structures. Mater Des 2021;204:109655. Available from: https://doi.org/10.1016/j.matdes.2021.109655.

[160] Drago A, Pindera M-J. Micro-macromechanical analysis of heterogeneous materials: Macroscopically homogeneous vs periodic microstructures. Compos Sci Technol 2007;67 (6):1243−63. Available from: https://doi.org/10.1016/j.compscitech.2006.02.031.

[161] Nordmann J, Aßmus M, Altenbach H. Visualising elastic anisotropy: theoretical background and computational implementation. Contin Mech Thermodyn 2018;30 (4):689−708. Available from: https://doi.org/10.1007/s00161-018-0635-9.

[162] Böhlke T, Brüggemann C. Graphical Representation of the Generalized Hooke's Law. Techn Mechanik 2001;21(2):145−58.

[163] Maskery I, Sturm L, Aremu AO, Panesar A, Williams CB, Tuck CJ, et al. Insights into the mechanical properties of several triply periodic minimal surface lattice structures made by polymer additive manufacturing. Polymer 2018;152:62−71. Available from: https://doi.org/10.1016/j.polymer.2017.11.049.

[164] Novak N, Al-Ketan O, Krstulović-Opara L, Rowshan R, Abu Al-Rub RK, Vesenjak M, et al. Quasi-static and dynamic compressive behaviour of sheet TPMS cellular structures. Compos Struct 2021;266:113801. Available from: https://doi.org/10.1016/j.compstruct.2021.113801.

[165] Bonatti C, Mohr D. Smooth-shell metamaterials of cubic symmetry: Anisotropic elasticity, yield strength and specific energy absorption. Acta Mater 2019;164:301−21. Available from: https://doi.org/10.1016/j.actamat.2018.10.034.

[166] Ganczarski AW, Egner H, Skrzypek JJ. Introduction to Mechanics of Anisotropic Materials. In: Skrzypek JJ, Ganczarski AW, editors. Mechanics of anisotropic materials. Cham: Springer International Publishing; 2015. p. 1−56.

[167] Gibson LJ, Ashby MF. Cellular solids: structure and properties. Cambridge solid state science series. 2 ed. Cambridge: Cambridge University Press;; 1997.

[168] Yang L, Yan C, Fan H, Li Z, Cai C, Chen P, et al. Investigation on the orientation dependence of elastic response in Gyroid cellular structures. J Mech Behav Biomed Mater 2019;90:73−85. Available from: https://doi.org/10.1016/j.jmbbm.2018.09.042.

[169] Ranganathan SI, Ostoja-Starzewski M. Universal elastic anisotropy index. Phys Rev Lett 2008;101(5):055504. Available from: https://doi.org/10.1103/PhysRevLett.101.055504.

[170] Healy D, Timms NE, Pearce MA. The variation and visualisation of elastic anisotropy in rock-forming minerals. Solid Earth 2020;11(2):259−86. Available from: https://doi.org/10.5194/se-11-259-2020.

Index

Note: Page numbers followed by "*f*" and "*t*" refer to figures and tables, respectively.

A

ABAQUS commercial software, 13−14, 237
Accelerated Eyre-Milton elastic analysis algorithm, 346
Accelerated FFT thermal analysis algorithm, 347
Active three-phase boundary length, 118, 119*t*, 120*f*
Additive manufacturing (AM), 304
Agglomeration, 318−320
Al-Si alloys, 222
AM. *See* Additive manufacturing
Anisotropic microstructure
  random Voronoi tessellation, homogenization and elastic percolation threshold, 244−249, 246*f*, 247*f*, 248*f*
Anisotropic microstructures
  of Voronoi-based microstructures, mechanical characterization of, 223−249
    composite structures, homogenization and elastic percolation of, 224−232, 225*f*, 226*t*
    random Voronoi tessellation, homogenization and elastic percolation threshold, 232−249
    with Voronoi tessellation, 223−249
Anisotropic models, in maximum, minimum, and average TPBL, 103*t*
ANN. *See* Artificial neural network
Approximation of TPCFs, three-phase microstructures, 109
  anisotropic reconstruction, 115*f*, 116−117
  approximation formulation, 110−112, 111*f*
  isotropic reconstruction, 114−117

microstructural properties, 116*f*, 117*f*, 118−122
  n-point correlation functions and conditional probability, 109
  number of independent sets, 110
  phase recovery algorithm, modification of, 112−114, 114*f*
Arbitrary binary model, 10
Architectured porous biomaterials, 23
Artificial neural network (ANN), 104−105, 182, 192
  generalization of limited homogenized samples with, 189, 189*f*
  liver tissue, characterization of, 192, 193*f*
  in maximum, minimum, and average TPBL, 106*t*
Asymptotic homogenization, 20−21
Axial stiffness maximization, 374−375, 376*t*
Axonal death, 146

B

Back propagation neural network (BPNN), 310−311
BESO. *See* Bidirectional evolutionary structural optimization
Bessel function of order, 4
Bidirectional evolutionary structural optimization (BESO), 305
Biodegradable composites, 253
BMD. *See* Bone mineral density
Bone cell activation, 205
Bone degradation, degraded microstructures, 255−257
Bone degradation kinetics, 284−286
Bone density loss in osteoporosis, numerical model of, 274−290
  bone microstructure data, 281

Bone density loss in osteoporosis, numerical model of (*Continued*)
  bone microstructure, 3D reconstruction of, 275–276, 276f
  microgravity, determination of bone quality in, 276–277, 277f
  model objectives, 275
  optimization algorithm, 279–281, 280f
  PSO optimization, 282–284, 283f, 283t, 284f, 285f
  rate of degradation, available bone surface on, 284–288, 286t, 287f
  rate of degradation, surface contact area on, 281–282, 282f
  sample scale, effective elastic modulus at, 288–290, 289f
  theoretical model, 277–279, 278f
Bone formation process, 256–257
Bone mechanical behavior, 195
Bone microstructure, 146
  3D reconstruction of, 275–276, 276f
  statistical reconstruction and mechanical characterization of, 194–210
    bone microstructure, statistical reconstruction of, 196–199, 198f, 199f
    developed stresses and strain energy density in, 203f, 204f
    finite element homogenization, 199–201
    mean strain energy density variation, 208t
    mechanical properties, 210
    modulus of elasticity and Poisson's ratio, 202t
    specimen preparation, 196
    statistical bone microstructures, 201
    statistical reconstruction method, 210
Bone microstructure distribution, 195
Bone mineral density (BMD), 276, 290
Boolean model, 301–302
BPNN. *See* Back propagation neural network
Brain tissue
  homogenization of, 149–151, 154f
  viscoelastic homogenization of, 149–164
    cerebral cortex tissue constituents, elastic properties of, 155t

homogenization of brain tissue, 149–151, 154f
    representative volume element and finite element modeling, 149–151
Bulk modulus maximization, 372, 373t

C

CAD. *See* Computer-aided design (CAD) data
Carbon nanosprings (CNSs), 29–30
Carbon nanotubes (CNTs), 29–30
  orientation of, 55, 57f
Cauchy-Green deformation tensor, 168
CCNT
  coil diameter, 57, 58f
  helical angle, 55–56, 57f
  tube diameter, 58, 59f
CDM. *See* Continuum damage mechanism
Cell tessellation, TPMS structures and novel hybridization method to control anisotropy, 397, 398f
Cellular automata model of corrosion, 262, 263f
Cellular automata, PLA/Mg composites, 271–274, 271f, 273f, 274f
Cellular automata (CA) techniques, 99, 254–255, 262, 265, 266f
Cement composites, 29
Cerebral cortex tissue
  elastic modulus, neurons volume fraction on, 159f
  elastic properties of, 155t
  finite element investigation of effective mechanical behavior of, 146–181
    brain tissue, viscoelastic homogenization of, 149–164
    heterogeneous brain tissue under quasistatic loading, homogenization of, 164–181
  mechanical properties of, 157t
  viscoelastic properties of, 156t
CGANs. *See* Conditional generative adversarial networks
Characteristic function (CF), of the heterogeneous microstructure, 196
Chemical vapor deposition (CVD), 29–30
CNSs. *See* Carbon nanosprings
CNTs. *See* Carbon nanotubes
Coefficient of thermal expansion (CTE), 299

Index

Cohesive interfaces, damage, nanocomposites with, 58–61, 59*f*, 60*f*
Cohesive zone model, 53*t*, 66*t*
Combined FE-optimization method, homogenization with, 184–186, 185*f*, 191–192, 191*f*, 192*f*
Complex microstructures, computational modeling of
  asymptotic homogenization, 20–21
  effective properties, 12–19
    calculation of, 13–19
  Hull space and materials design, 21–23, 22*f*
  numerical reconstruction of heterogeneous materials, 7–11, 11*f*, 12*f*
  representative volume element, 5–7, 7*f*
  statistical descriptor, 1–5
    correlation functions, 1
    N-point correlation functions, 4–5
    two-point cluster functions, 5
    two-point correlation functions, 2–4, 2*f*
Composite structures, homogenization and elastic percolation of, 224–232, 225*f*, 226*t*
  finite element modeling of microstructure, 224–225, 227*f*
  honeycomb model, 231, 232*f*
  overall elastic modulus of microstructure, 229*f*, 230*f*
  random microstructure, protocol for generation of, 225–228, 227*f*, 228*t*
  Rhombus model, 231, 233*f*
  stress contour plots, for microstructures, 232, 234*f*, 235*f*, 236*f*
Computational homogenization, 131
Computational modeling, of complex microstructures
  asymptotic homogenization, 20–21
  effective properties, 12–19
    calculation of, 13–19
  Hull space and materials design, 21–23, 22*f*
  numerical reconstruction of heterogeneous materials, 7–11, 11*f*, 12*f*
  representative volume element, 5–7, 7*f*
  statistical descriptor, 1–5
    correlation functions, 1
    N-point correlation functions, 4–5

two-point cluster functions, 5
two-point correlation functions, 2–4, 2*f*
Computed tomography (CT) scan, 181–182
Computer-aided design (CAD) data, 223
Concentric cylinder micromechanics model, 222–223
Conditional generative adversarial networks (CGANs), 8–9
Conditional probability functions, 4–5
Conditional probability, n-point correlation functions and, 109
Constitutive model, heterogeneous brain tissue under quasistatic loading, 167–169
Contact surface density of particles, 321
Continuum damage mechanism (CDM), 253
Correlation functions, 1
Cost function (CF), 10
Cubic symmetry, PUCs of, 372–375
  axial stiffness maximization, 374–375, 376*t*
  bulk modulus maximization, 372, 373*t*
  negative Poisson's ratio maximization, 374
  shear modulus maximization, 372–373, 375*t*
CVD. *See* Chemical vapor deposition

D

DAI. *See* Diffuse axonal injury
Deep convolutional network, 8–9
Degraded microstructures, numerical modeling of
  bone degradation, 255–257
  bone density loss in osteoporosis, 274–290
    bone microstructure data, 281
    bone microstructure, 3D reconstruction of, 275–276, 276*f*
    microgravity, determination of bone quality in, 276–277, 277*f*
    model objectives, 275
    optimization algorithm, 279–281, 280*f*
    PSO optimization, 282–284, 283*f*, 283*t*, 284*f*, 285*f*
    rate of degradation, available bone surface on, 284–288, 286*t*, 287*f*
    rate of degradation, surface contact area on, 281–282, 282*f*

Degraded microstructures, numerical modeling of (*Continued*)
   sample scale, effective elastic modulus at, 288−290, 289*f*
   theoretical model, 277−279, 278*f*
  PLA/Mg composites, 257−274
   cellular automata, 271−274, 271*f*, 273*f*, 274*f*
   corrosion test experimental setup, 267−270, 268*f*, 268*t*, 269*t*
   modeling corrosion, 260−267
   reconstruction, 270, 270*f*, 271*f*
   sample preparation, 258, 258*f*
   statistical reconstruction of samples, 259−260
  PLLA/magnesium composite, degradation of, 253−255
Density-based topology optimization, 363−379, 365*f*, 367*f*, 368*f*
  cubic symmetry, PUCs of, 372−375
   axial stiffness maximization, 374−375, 376*t*
   bulk modulus maximization, 372, 373*t*
   negative Poisson's ratio maximization, 374
   shear modulus maximization, 372−373, 375*t*
  3D-PUC design using, 363−379, 365*f*, 367*f*, 368*f*
  numerical examples and experimental evaluations, 371−379
  numerical implementation, 371
  OCA-type optimizer, 370−371
  orthotropic symmetry, PUCs, 376−379
  SIMP-based numerical homogenization, 367−369
Diffuse axonal injury (DAI), 145−147
Diffusion, 261, 265−267, 266*f*, 267*f*
Diffusion factor (DF), 99
Disorientation, 146
Dispersion of particles, 320
3D periodic unit cells (3D-PUCs), 304
  using density-based topology optimization, 363−379, 365*f*, 367*f*, 368*f*
3D-PUC design
  topology optimization and innovative building block, 379−390

designing representative volume elements, 383−385, 383*f*, 383*t*, 385*f*
  experimental verification, 390, 391*f*, 392*f*
  new structure, designing, 388−389, 390*f*, 390*t*
  optimization procedure, 385−388, 386*t*, 388*f*, 388*t*, 389*f*, 389*t*
  proposed building block, 379−382, 379*f*, 381*f*, 382*f*
3D reconstruction
  of microstructure, 96−97, 111−112
  using VCAT software, 45−47, 46*f*

E

Eigen microstructure, 109
Effective properties, heterogonous materials, 12
  calculation of, 13−19
  finite element modeling, 13−15
  Mori−Tanaka method, 15−17
  strong contrast method, 17−19
Elastic modulus, 172*f*
  of nanocomposites, 51*t*
  strong contrast method for, 18−19
Elastic percolation, of composite structures, 224−232, 225*f*, 226*t*
Elastic percolation threshold, random Voronoi tessellation, 232−249
  anisotropic microstructures, 244−249, 246*f*, 247*f*, 248*f*
  isotropic microstructures, 242−244, 242*f*, 243*f*, 244*f*, 245*f*
  tessellation method and finite element modeling, 235−241, 237*f*, 239*f*, 240*t*, 241*f*
Electric conductivity, 308
Electron microscopy, 36−37
Ellipsoidal inclusions, composites loaded with, 32−34, 33*f*
Energy-based homogenization method (EBHM), 305
Error function, 10
Extracellular matrix (ECM), 147−148
Eyre−Milton elastic analysis algorithm, 346
Eyre−Milton FFT algorithm, 132−133

F

Fast Fourier transform (FFT), 3, 97, 259–260, 303
FD. *See* Finite difference
FDM. *See* Fused deposition modeling
FEA. *See* Finite element analysis
FEM. *See* Finite element methods
FFT. *See* Fast Fourier transform
FFT thermal analysis algorithm, 135, 347
Filler index, 35–36
Finite difference (FD), 97
Finite element (FE), 97
Finite element analysis (FEA), 7–8
Finite element homogenization, 199–201
  of coupling thermomechanical properties, 72–84
    analytical and numerical results, 82–84, 83*f*, 84*f*
    methodology, 74–82, 76*f*, 77*f*, 78*f*, 79*f*, 80*f*, 81*f*
  of polymer nanocomposites, 47–50
    mechanical properties, percolation of interphase, 48–50, 49*f*, 50*f*
    methodology, 47–48, 48*t*
Finite element methods (FEM), 13–15, 31, 146, 253–254
  brain tissue, viscoelastic homogenization of, 149–151
  homogenization of nanocomposites with, 50–61, 51*t*
  interfacial debonding, homogenization of nanocomposites with, 50–61, 51*t*
    CCNT's coil diameter, 57, 58*f*
    CCNT's helical angle, 55–56, 57*f*
    CCNT's tube diameter, 58, 59*f*
    CNT, orientation of, 55
    cohesive interfaces, damage in, 58–61, 59*f*, 60*f*
    cohesively and perfectly bonded nanocomposites, volume fraction effect and filler type on, 54–55, 56*f*
    nanofillers and matrix, modeling interface between, 52–54, 52*f*, 53*t*
  of microstructure, 224–225, 227*f*
  modeling interface, nanofillers and matrix, 53–54, 53*f*, 54*t*, 55*f*
  nonlinear matrix properties, homogenization of nanocomposites with, 61–72

    numerical homogenization, 66–72, 67*f*, 68*f*, 69*f*, 70*f*, 71*f*, 72*f*, 73*t*
    SMP, constitutive equations for, 61–63, 61*f*
    three-dimensional modeling and numerical considerations, 64–66, 64*f*, 65*t*
  random heterogeneous materials, numerical realization and characterization of, 122–131
    elastic properties, 127–129, 127*f*, 129*f*
    microstructural aspects, 125–127, 126*f*
    thermal conductivity and thermal expansion, 129–131, 129*f*, 130*f*, 131*f*
  random Voronoi tessellation, homogenization and elastic percolation threshold, 235–241, 237*f*, 239*f*, 240*t*, 241*f*
  TPMS structures and novel hybridization method, 396–397, 396*f*
Finite element simulation, heterogeneous brain tissue under quasistatic loading, 169–171
Focused ion beam/scanning electron microscopy (FIB-SEM), 7–8
Fourier transform, 112
Fused deposition modeling (FDM), 305–306

G

GA. *See* Genetic algorithm
Gas diffusion, 309
Gas transport factor, 321–322
Gaussian filtering method, 96
Generalized Effective Media model, 222–223
Genetic algorithm (GA)
  in maximum, minimum, and average TPBL, 106*t*
Genetic algorithm (GA), 8–9
Gradient-based multi objective algorithm, 185
Graphene, 29
Graphene oxide (GO), 29
Graphite nanoplatelets (GNPs), 29
Gravity forces, 255
GUI application, 192–194, 194*f*

H

Halloysite nanotube composite, 45, 46f
Halpin—Tsai models, 12
Hanks' buffer solution, 268t
Hanks solution, 254
Heat transfer (DC3D4) elements, 14—15
Helical inclusions, 34—36, 35f, 36t
Heterogeneous brain tissue under quasistatic loading
  homogenization of, 164—181
    constitutive model and simulation, 167—169
    ECM and neurons, material properties of, 172t, 173t
    finite element simulation, 169—171
    isotropy, RVE size and irregular distribution, 171—174
    viscohyper elastic model, development, 174—181, 175t, 176f, 177f, 178f, 179f, 180f, 181f
Heterogeneous microstructure
  random segment in, 5, 6f
  TPCF vectors in, 2, 2f
Heterogeneous model, vessels and surrounding tissue, 186—187
Heterogonous materials, effective properties of, 12
  calculation of, 13—19
  finite element modeling, 13—15
  Mori—Tanaka method, 15—17
  strong contrast method, 17—19
HFIR. *See* High flux isotope reactor
High diffusion factor, microstructure properties with, 315, 316f
Higher-order correlation functions, 17
High flux isotope reactor (HFIR), 39
Holzapfel—Gasser—Ogden model, 189
Homogenization
  of brain tissue, 149—151, 154f
  random heterogeneous microstructures, 97
Homogenization
  with combined FE-optimization method, 184—186, 185f, 191—192, 191f, 192f
  of composite structures, 224—232, 225f, 226t
  of heterogeneous brain tissue under quasistatic loading, 164—181
    constitutive model and simulation, 167—169

    ECM and neurons, material properties of, 172t, 173t
    finite element simulation, 169—171
    isotropy, RVE size and irregular distribution, 171—174
    viscohyper elastic model, development, 174—181, 175t, 176f, 177f, 178f, 179f, 180f, 181f
  random Voronoi tessellation, 232—249
    anisotropic microstructures, 244—249, 246f, 247f, 248f
    isotropic microstructures, 242—244, 242f, 243f, 244f, 245f
    tessellation method and finite element modeling, 235—241, 237f, 239f, 240t, 241f
Homogenization of nanocomposites
  with interfacial debonding using FEMs, 50—61, 51t
    CCNT's coil diameter, 57, 58f
    CCNT's helical angle, 55—56, 57f
    CCNT's tube diameter, 58, 59f
    CNT, orientation of, 55
    cohesive interfaces, damage in, 58—61, 59f, 60f
    cohesively and perfectly bonded nanocomposites, volume fraction effect and filler type on, 54—55, 56f
    nanofillers and matrix, modeling interface between, 52—54, 52f, 53t
  with nonlinear matrix properties using finite element methods, 61—72
    numerical homogenization, 66—72, 67f, 68f, 69f, 70f, 71f, 72f, 73t
    SMP, constitutive equations for, 61—63, 61f
    three-dimensional modeling and numerical considerations, 64—66, 64f, 65t
Homogenization theory, for structural design and optimization, 304—305
Homogenized model, liver tissue, 188—189
Honeycomb model, 231, 232f
Hooke's law, 21
Hot extrusion, 258
Hull space, 21—23, 22f
Hull space, infiltrated SOFC microstructures, 318—332

Index

microstructure generation and characterization, 319–322
contact surface density of particles, 321
direct search and multiobjective GA method, optimum microstructures in, 330*t*
gas transport factor, 321–322
gas transport factor, microstructure property closure, 327, 327*f*
microstructure hull, 322–324, 323*f*
optimization scenarios, 330*t*
optimization scheme, 324–327, 325*f*, 325*t*, 326*f*
surface density of particles, 329, 329*f*
TPB density variation, 328–329, 328*f*
Hybridization, TPMS structures and novel hybridization method to control anisotropy, 403–410, 405*t*, 406*f*
Hybrid numerical analytic model, 222–223
Hybrid structures, TPMS structures and novel hybridization method to control anisotropy, 393–396
Hydrogen measurement system, 268*f*
Hyperelastic constitutive models, 147
Hypothetical microstructures, gas diffusion factor and ionic conductivity for, 313

I

Inclusionary composites, experimental reconstruction of
SEM-FIB images, reconstruction using, 44–47
small-angle neutron scattering, 36–44
Infiltrated SOFC microstructures, hull space for, 318–332
microstructure generation and characterization, 319–322
contact surface density of particles, 321
direct search and multiobjective GA method, optimum microstructures in, 330*t*
gas transport factor, 321–322
gas transport factor, microstructure property closure, 327, 327*f*
microstructure hull, 322–324, 323*f*
optimization scenarios, 330*t*
optimization scheme, 324–327, 325*f*, 325*t*, 326*f*
surface density of particles, 329, 329*f*

TPB density variation, 328–329, 328*f*
Interface region, 221
Interfacial debonding, 50–61, 51*t*
CCNT's coil diameter, 57, 58*f*
CCNT's helical angle, 55–56, 57*f*
CCNT's tube diameter, 58, 59*f*
CNT, orientation of, 55
cohesive interfaces, damage in, 58–61, 59*f*, 60*f*
cohesively and perfectly bonded nanocomposites, volume fraction effect and filler type on, 54–55, 56*f*
homogenization of nanocomposites with, 50–61, 51*t*
nanofillers and matrix, modeling interface between, 52–54, 52*f*, 53*t*
Interfacial debonding-induced damage model, 31
Internal bone microstructure, 200–201
Interphase zone, finite element homogenization of, 47–50
Ionic conductivity, 308
Isotropic microstructures
random Voronoi tessellation, homogenization and elastic percolation threshold, 242–244, 242*f*, 243*f*, 244*f*, 245*f*
of Voronoi-based microstructures, mechanical characterization of, 223–249
composite structures, homogenization and elastic percolation of, 224–232, 225*f*, 226*t*
random Voronoi tessellation, homogenization and elastic percolation threshold, 232–249
with Voronoi tessellation, 223–249
Isotropic models, in maximum, minimum, and average TPBL, 102*t*
Isotropic reconstruction, three-phase microstructures, 114–117
Isotropic transversely viscohyper elastic constitutive model, 147

K

Karhunen–Loève (K-L) expansion method, 8–9
Knudsen diffusion, 322
Knudsen number (Kn), 322

L

Laguerre-Voronoi tessellation, 222
Lazy zone for bone strain, 257
Linear elastic properties, 132–133, 345–346
Liver, 146
Liver tissue, characterization of, 181–194
  application, 192–194, 194f
  artificial neural networks, 189, 189f, 192, 193f
  homogenization with combined FE-optimization method, 184–186, 185f, 191–192, 191f, 192f
  homogenized model, 188–189
  homogenized samples, generating database of, 189–190
  materials and method, 183–184, 183f
  vessels and surrounding tissue, heterogeneous model, 186–187

M

Machine learning (ML) algorithms, 299
Macroscopic bone stimulus, 257
Magnesium, 253
  corrosion of, 273f, 274t
Magnetic resonance imaging (MRI), 181–182
Mass density function, 222
Mathematical material optimization scheme, 302
MC techniques. *See* Monte Carlo (MC) techniques
MD. *See* Molecular dynamics
Mean square errors (MSEs), 104–105
Mechanical damage, 145–146
Mesh sensitivity analysis, TPMS structures and novel hybridization method, 397, 398f
Method of moving asymptotes (MMA), 305
Metropolis formula, 11
MFs. *See* Minkowski functionals
Mg-Zn-Ca alloy, 253
Microcomposites, numerical characterization of, 29–31
  applications of, 29–30
  composites realization with inclusion types
    composites loaded with ellipsoidal inclusions, 32–34, 33f

    composites loaded with helical inclusions, 34–36, 35f, 36t
  inclusionary composites, experimental reconstruction of
    SEM-FIB images, reconstruction using, 44–47
    small-angle neutron scattering, 36–44
  thermomechanical properties, numerical homogenization of, 47–84
    coupling thermomechanical properties, finite element homogenization of, 72–84
    finite element homogenization of polymer nanocomposites with interphase zone, 47–50
    homogenization of nanocomposites with interfacial debonding using FEMs, 50–61, 51t
    homogenization of nanocomposites with nonlinear matrix properties using finite element methods, 61–72
Microgravity, bone quality determination, 276–277, 277f
Micromechanics-based approach, 223
Microstructural properties, three-phase microstructures, 116f, 117f, 118–122
  active three-phase boundary length, 118, 119t, 120f
  tortuosity, 118–122
Microstructure
  finite element modeling of, 224–225, 227f
  optimal design of, 23
Microstructure hull and design, 299–306
  infiltrated SOFC microstructures, hull space for, 318–332
    contact surface density of particles, 321
    direct search and multi objective GA method, optimum microstructures in, 330t
    gas transport factor, 321–322
    gas transport factor, microstructure property closure, 327, 327f
    microstructure generation and characterization, 319–322
    microstructure hull, 322–324, 323f
    optimization scenarios, 330t
    optimization scheme, 324–327, 325f, 325t, 326f

surface density of particles, 329, 329f
TPB density variation, 328—329, 328f
periodic heterogeneous materials, practical
  approach for materials design,
  363—410
  density-based topology optimization,
    3D-PUC design using, 363—379,
    365f, 367f, 368f
  topology optimization and innovative
    building block, 3D-PUC design
    using, 379—390
  TPMS structures and novel
    hybridization method to control
    anisotropy, directional elastic
    modulus of, 391—410
random heterogeneous materials, practical
  approach for materials design
  combining two anisotropic
    microstructures, 340—341, 340f,
    341f, 341t
  combining two isotropic
    microstructures, 338—340, 338f, 339f
  combining two microstructures with
    unequal volume fractions, 341—343,
    342f, 343f
  error analysis, 343—344
  interpolation of, 332—344
  microstructure construction, 335—336,
    337f
  N-point correlation functions, 333
  numerical homogenization, 345—348
  two-phase microstructure, TPCFs for,
    334f
for SOFC microstructure, 306—318
  diffusion factor, microstructure property
    closure comparing ionic conductivity
    with, 313, 314f
  gas diffusion, 309
  high diffusion factor, microstructure
    properties with, 315, 316f
  hypothetical microstructures, gas
    diffusion factor and ionic
    conductivity for, 313
  ionic and electric conductivity, 308
  methodology, 310—311, 310f, 310t
  microstructure generation and
    characterization, 307—309
  multiobjective GA search, optimum
    microstructure based on, 318t

optimal microstructure, characteristics
  of, 317t
optimization scheme, 311—313, 312f,
  312t
TPB density, 307—308
TPBL, microstructure property closure
  comparing ionic conductivity with,
  313, 314f
solid oxide fuel cell, optimal
  microstructures for, 300
Microstructure reconstruction, 7—8, 10
MIECs. *See* Mixed ionic electric conductor
  materials
Minimally invasive surgery (MIS), 146,
  181—182
Minkowski functionals (MFs), 301—302
MIS. *See* Minimally invasive surgery
Mixed ionic electric conductor materials
  (MIECs), 320—321
Mixture random field (MRF) model,
  8—9
MMA. *See* Method of moving asymptotes
Modeling corrosion, 260—267
  cellular automata modeling, 262
  cellular automata model of corrosion, 262,
    263f
  diffusion, 261, 265—267, 266f, 267f
  spatially joint reactions, 261, 264—265
  spatially separated electrochemical
    reactions, 260—261, 263—264, 264f
Modeling of degradation, 257
Modeling interface, nanofillers and matrix,
  52—54, 52f, 53t
  finite element model, 53—54, 53f, 54t, 55f
  material, 53
Molecular dynamics (MD), 30—31
Molecular mechanics (MM) simulation,
  29—30
Monte Carlo (MC) techniques, 17, 83—84,
  96—97, 122—124, 137, 254
  multiphase random heterogeneous
    materials using, 97—107
Mooney—Rivlin model, 147
Moore neighborhood, 262, 262f
Mori—Tanaka (MT) models, 12, 15—17
MRF model. *See* Mixture random field
  (MRF) model
MT models. *See* Mori—Tanaka (MT)
  models

Multiphase random heterogeneous materials, using Monte Carlo approach, 97–107

Multipoint statistics (MPS)-based methods, 8–9

Multiscale cellular structures, 23

Multiscale homogenization, 146

Multiscale modeling, 30–31

N

Nacre-like composites, 222

Nanocomposites, 29–31
  applications of, 29–30
  composites realization with inclusion types
    composites loaded with ellipsoidal inclusions, 32–34, 33*f*
    composites loaded with helical inclusions, 34–36, 35*f*, 36*t*
  elastic modulus of, 51*t*
  inclusionary composites, experimental reconstruction of
    SEM-FIB images, reconstruction using, 44–47
    small-angle neutron scattering, 36–44
  structural characterization, SANS materials and preparation of, 39, 40*f*
  thermomechanical properties, numerical homogenization of, 47–84
  coupling thermomechanical properties, finite element homogenization of, 72–84
  finite element homogenization of polymer nanocomposites with interphase zone, 47–50
  homogenization of nanocomposites with interfacial debonding using FEMs, 50–61, 51*t*
  homogenization of nanocomposites with nonlinear matrix properties using finite element methods, 61–72

Nanofillers, 36*t*

Nanotube dispersion, 36–37

Nanotube-polymer nanocomposites, 36

Negative Poisson's ratio maximization, 374

Neo-Hookean coefficients, 169

Neumann neighborhood relation, 99

Neuron cells, 157

Neuron volume fraction (NVF), 151

Non-Gaussian random field, 8–9

Nonlinear matrix properties
  homogenization of nanocomposites with, 61–72
    numerical homogenization, 66–72, 67*f*, 68*f*, 69*f*, 70*f*, 71*f*, 72*f*, 73*t*
    SMP, constitutive equations for, 61–63, 61*f*
    three-dimensional modeling and numerical considerations, 64–66, 64*f*, 65*t*

Nonlinear viscohyperelastic model, 147

Non uniform rational B-splines (NURBS), 8–9

N-point correlation functions, 4–5, 96, 302
  and conditional probability, 109

N-point statistics, 255

Numerical characterization of tissues, 145–146
  bone microstructure, statistical reconstruction and mechanical characterization of, 194–210
  cerebral cortex tissue, finite element investigation of effective mechanical behavior of, 146–181
    brain tissue, viscoelastic homogenization of, 149–164
    heterogeneous brain tissue under quasistatic loading, homogenization of, 164–181
  liver tissue, characterization of, 181–194

Numerical reconstruction of heterogeneous materials, 7–11, 11*f*, 12*f*

NVF. *See* Neuron volume fraction

O

Oak Ridge National Laboratory (ORNL), 39

OCA-type optimizer, 370–371

Ogden constitutive model, 147

Optimal microstructures, for solid oxide fuel cell, 300

Optimization algorithm, bone density loss, 279–281, 280*f*

Orthotropic symmetry, PUCs, 376–379

Osteoblasts, 256–257

Osteoclasts, 256–257

Osteocytes, 256–257

Osteoporosis, bone density loss, 274–290
  bone microstructure data, 281

Index 431

bone microstructure, 3D reconstruction of, 275–276, 276f
microgravity, determination of bone quality in, 276–277, 277f
model objectives, 275
optimization algorithm, 279–281, 280f
PSO optimization, 282–284, 283f, 283t, 284f, 285f
rate of degradation, available bone surface on, 284–288, 286t, 287f
rate of degradation, surface contact area on, 281–282, 282f
sample scale, effective elastic modulus at, 288–290, 289f
theoretical model, 277–279, 278f

P

Parametric level set method (PLSM), 305
Particle swarm optimization (PSO) function, 170–171
Particle swarm optimization optimization, bone density loss, in osteoporosis, 282–284, 283f, 283t, 284f, 285f
Percolation, of Ni phase, 124
Percolation theory, applications of, 222–223
Percolation threshold, 269–270
modified approximation for, 222–223
Periodic boundary conditions (PBCs), 305–306
Periodic unit cell (PUC)
of cubic symmetry, 372–375
axial stiffness maximization, 374–375, 376t
bulk modulus maximization, 372, 373t
negative Poisson's ratio maximization, 374
shear modulus maximization, 372–373, 375t
orthotropic symmetry, 376–379
PGD methods. See Proper generalized decomposition (PGD) methods
Phase recovery algorithm, 112–114, 114f, 199, 348–349
Pitch-based short carbon fibers, 74
PLA. See Polylactic acid
PLA/Mg composites, 257–274
cellular automata, 271–274, 271f, 273f, 274f

corrosion test experimental setup, 267–270, 268f, 268t, 269t
modeling corrosion, 260–267
cellular automata modeling, 262
cellular automata model of corrosion, 262, 263f
diffusion, 261, 265–267, 266f, 267f
spatially joint reactions, 261, 264–265
spatially separated electrochemical reactions, 260–261, 263–264, 264f
reconstruction, 270, 270f, 271f
sample preparation, 258, 258f
statistical reconstruction of samples, 259–260
PLLA/magnesium composite, degradation of, 253–255
PLSM. See Parametric level set method
PNCs. See Polymer nanocomposites
Poisson's ratio, 48t, 83–84, 299
Polylactic acid (PLA), 253
Polymer nanocomposites, finite element homogenization of, 47–50
Polymer nanocomposites (PNCs), 31
Polyphenylacetylene (PPA), 30–31
Probability distribution functions (PDFs), 206
Prolate inclusions, isotropic random distribution of, 34f
Proper generalized decomposition (PGD) methods, 181–182
Property hull, 300, 322, 326f, 332t
PUC. See Periodic unit cell
Python, 53–54

R

Radial basis function (RBF), 104–105
Random heterogeneous materials
anisotropic representative volume element, 353–354
combining two anisotropic microstructures, 340–341, 340f, 341f, 341t
combining two isotropic microstructures, 338–340, 338f, 339f
combining two microstructures with unequal volume fractions, 341–343, 342f, 343f
design algorithm, 349–350
direct reconstruction techniques, 95

Random heterogeneous materials (*Continued*)
effective thermal conductivity, 356*f*
error analysis, 343–344
FFT approach, 131–137
linear elastic properties, 132–133
thermal properties, 134–137, 136*f*, 136*t*, 137*f*
finite element method, 122–131
elastic properties, 127–129, 127*f*, 129*f*
microstructural aspects, 125–127, 126*f*
thermal conductivity and thermal expansion, 129–131, 129*f*, 130*f*, 131*f*
interpolation of, 332–344
microstructure, accurate reconstruction of, 95–96
microstructure construction, 335–336, 337*f*
microstructure representation, 349
Monte Carlo approach, multiphase random heterogeneous materials using, 97–107
SOFC anode, realization of, 101–107, 102*f*, 102*t*, 103*f*, 103*t*, 104*f*, 106*t*, 107*f*
multiphase reconstruction of heterogeneous materials, 108–122
three-phase microstructures, approximation of TPCFs, 109
n-point correlation functions, 333
numerical homogenization, 345–348
numerical homogenization, 345–348
linear elastic properties, 345–346
thermal properties, 346–348
objective function, optimized multifunctional microstructure based on, 364*f*
reconstruction, phase recovery algorithm, 348–349
stochastic optimization technique, 96
structural problem, optimized microstructures obtained for, 362*f*
thermomechanical properties, 122–131
three reconstructed RVEs, elastic moduli of, 357*t*
two-phase microstructure, TPCFs for, 334*f*
verification study, 350–353, 351*f*, 353*t*
Random microstructure, protocol for generation of, 225–228, 227*f*, 228*t*

Random Voronoi tessellation, homogenization and elastic percolation threshold, 232–249
anisotropic microstructures, 244–249, 246*f*, 247*f*, 248*f*
isotropic microstructures, 242–244, 242*f*, 243*f*, 244*f*, 245*f*
tessellation method and finite element modeling, 235–241, 237*f*, 239*f*, 240*t*, 241*f*
RBF. *See* Radial basis function
Reconstruction algorithm, 348
Reconstruction, PLA/Mg composites, 270, 270*f*, 271*f*
Representative elementary volume (RVE), 260
Representative volume element (RVE), 5–7, 7*f*, 97, 147–148, 232–233, 302
Representative volume element, brain tissue, 149–151
Rhombus model, 231, 233*f*
Rhombus particles, 223–224
RVE. *See* Representative volume element

S

SANS. *See* Small-angle neutron scattering
SBFs. *See* Simulated body fluids
Scanning electron microscopy (SEM), 29–30
Scattering length density (SLD), 37
Second-order identity tensor, 18
SED. *See* Strain energy density
Selective laser sintering (SLS), 305–306
SEM. *See* Scanning electron microscopy
SEM-FIB images
reconstruction using, 44–47
3D reconstruction using VCAT software, 45–47, 46*f*
nanocomposite, serial sectioning of, 44–45, 45*f*, 46*f*
Semi-inverse MC reconstruction, of two-phase heterogeneous materials, 255
Semi quantitative mathematical model, 254
Sequential linear programming (SLP), 305
SERCs. *See* Short fiber-reinforced composites
Serial sectioning, nanocomposite using FIB-SEM, 44–45, 45*f*, 46*f*
Series-parallel model, 222–223

Index

433

Shear modulus maximization, 372–373, 375t

Sheet molding compounds (SMC), 8–9

Short carbon fibers (SCFs), 72–74

Short fiber-reinforced composites (SERCs), 8–9

Silica nanoparticles, 29

SIMP-based numerical homogenization, 367–369

SIMP method, 305

SIMP-OCA methodology, 304

Simulated annealing (SA) algorithm, 8–9

Simulated body fluids (SBFs), 254

Simulation, heterogeneous brain tissue under quasistatic loading, 167–169

Simulia ABAQUS, 124

Single-wall carbon nanotubes (SWCNTs), 30–31

SLD. *See* Scattering length density

Sliding separation modes, traction-separation response in, 52*f*

SLP. *See* Sequential linear programming

SLS. *See* Selective laser sintering

Small-angle neutron scattering (SANS), 36–44, 40*f*

mathematical theory of, 37–39

reconstruction of nanostructures, 39–44, 40*f*, 41*f*, 42*f*, 43*f*

structural characterization, 39

and USANS, 39

SMP modeling, constitutive equations for, 61–63, 61*f*

Solid oxide fuel cell (SOFC), 96–97

microstructure hull and design for, 306–318

diffusion factor, microstructure property closure comparing ionic conductivity with, 313, 314*f*

gas diffusion, 309

high diffusion factor, microstructure properties with, 315, 316*f*

hypothetical microstructures, gas diffusion factor and ionic conductivity for, 313

ionic and electric conductivity, 308

methodology, 310–311, 310*f*, 310*t*

microstructure generation and characterization, 307–309

multiobjective GA search, optimum microstructure based on, 318*t*

optimal microstructure, characteristics of, 317*t*

optimization scheme, 311–313, 312*f*, 312*t*

TPB density, 307–308

TPBL, microstructure property closure comparing ionic conductivity with, 313, 314*f*

of Monte Carlo approach, multiphase random heterogeneous materials using, 101–107, 102*f*, 102*t*, 103*f*, 103*t*, 104*f*, 106*t*, 107*f*

optimal microstructures for, 300

Space exploration, 255

Spaceflights, 256

Spatially joint reactions, 261, 264–265

Spatially separated electrochemical (SSE), 260–261, 263–264, 264*f*

Statistical descriptor, 1–5

correlation functions, 1

N-point correlation functions, 4–5

two-point cluster functions, 5

two-point correlation functions, 2–4, 2*f*

Stochastic optimization technique, 96

Strain energy density (SED), 203–204

Stretching-dominated cellular truss structures, 23

Strong contrast method, 17–19

for elastic modulus, 18–19

for thermal conductivity, 18

Surface contact area, on rate of degradation, 281–282, 282*f*

SWCNTs. *See* Single-wall carbon nanotubes

T

Taguchi method, 102–104, 103*t*

TEC. *See* Thermal expansion coefficient

Tessellation method, random Voronoi tessellation, homogenization and elastic percolation threshold, 235–241, 237*f*, 239*f*, 240*t*, 241*f*

Theoretical model, 277–279, 278*f*

Thermal conductivity (TC), 125

random heterogeneous materials, 129–131, 129*f*, 130*f*, 131*f*

strong contrast method for, 18

Thermal expansion coefficient (TEC), 124

Thermal expansion, random heterogeneous materials, 129–131, 129f, 130f, 131f
Thermal properties, FFT approach, 134–137, 136f, 136t, 137f
Thermomechanical properties, numerical homogenization of, 47–84
  coupling thermomechanical properties, finite element homogenization of, 72–84
  finite element homogenization of polymer nanocomposites with interphase zone, 47–50
  homogenization of nanocomposites with interfacial debonding using FEMs, 50–61, 51t
  homogenization of nanocomposites with nonlinear matrix properties using finite element methods, 61–72
Three-phase boundary (TPB), 97
Three-phase boundary length (TPBL), 97, 101
Three-phase microstructures, approximation of TPCFs, 109
  anisotropic reconstruction, 115f, 116–117
  approximation formulation, 110–112, 111f
  isotropic reconstruction, 114–117
  microstructural properties, 116f, 117f, 118–122
  n-point correlation functions and conditional probability, 109
  number of independent sets, 110
  phase recovery algorithm, modification of, 112–114, 114f
TiO_2 nanoparticles, 29
Topology optimization and innovative building block, 3D-PUC design using, 379–390
  designing representative volume elements, 383–385, 383f, 383t, 385f
  experimental verification, 390, 391f, 392f
  new structure, designing, 388–389, 390f, 390t
  optimization procedure, 385–388, 386t, 388f, 388t, 389f, 389t
  proposed building block, 379–382, 379f, 381f, 382f
Tortuosity, 118–122
TPB. See Three-phase boundary

TPBL. See Three-phase boundary length
TPCCF. See Two-point correlation cluster function
TPCFs. See Two-point cluster functions (TPCFs)
Triply Periodic Minimal Surface (TPMS)
  based structures, 393, 394f, 395t
  TPMS structures and novel hybridization method to control anisotropy, 391–410, 402f
    cell tessellation and mesh sensitivity analysis, 397, 398f
    finite element model, 396–397, 396f
    hybridization, 403–410, 405t, 406f
    hybrid structures, 393–396
    TPMS-based structures, 393, 394f, 395t
Traction-separation response, in sliding separation modes, 52f
Traumatic brain injury (TBI), 145–146
Triple phase boundary (TPB), 300, 307–308
Triple phase boundary length (TPBL), 300
Truncated-Gaussian method, 96–97
Tubular vascularization, 186–187
Two-phase random heterogeneous RVE, 136f
Two-point cluster functions (TPCFs), 2–5, 2f, 95–96, 237–238, 259, 300–301
  three-phase microstructures, approximation of, 109
    active three-phase boundary length, 118, 119t, 120f
    anisotropic reconstruction, 115f, 116–117
    approximation formulation, 110–112, 111f
    isotropic reconstruction, 114–117
    microstructural properties, 116f, 117f, 118–122
    n-point correlation functions and conditional probability, 109
    number of independent sets, 110
    phase recovery algorithm, modification of, 112–114, 114f
    tortuosity, 118–122
Two-point correlation cluster function (TPCCF), 6f, 10

Index

U

Ultra-small-angle neutron scattering (USANS), 36–37, 39, 40*f*
Universal anisotropy index (AU), 407–408
Universal inequalities, 5–7, 7*f*
USANS. *See* Ultra-small-angle neutron scattering
User-defined material subroutine (UMAT), 61

V

VCAD software, 124
VCAT software, 45–47, 46*f*, 200
Viscohyper elastic model, development, 174–181, 175*t*, 176*f*, 177*f*, 178*f*, 179*f*, 180*f*, 181*f*
Von-Mises stress, 160, 160*f*, 165*f*, 166*f*
von Neumann neighborhood, 262, 262*f*
Voronoi-based algorithm, 8–9
Voronoi-based microstructures, mechanical characterization of, 221–223
isotropic and anisotropic microstructures with Voronoi tessellation, computational elucidation of, 223–249
composite structures, homogenization and elastic percolation of, 224–232, 225*f*, 226*t*
random Voronoi tessellation, homogenization and elastic percolation threshold, 232–249
Voronoi microstructures, properties of, 222–223
Voronoi tessellation methods, 222
Voxel, 105
VUMAT ABAQUS, 148–149

X

X-ray computed tomography, 7–8

Y

Young's modulus, 48*t*, 49*f*, 50*f*, 160*t*, 288, 290, 306

Printed in the United States
by Baker & Taylor Publisher Services